T0263084

Occupancy Estimation and Modeling

Occupancy Estimation and Modeling

Inferring Patterns and Dynamics of Species Occurrence

Darryl I. MacKenzie
James D. Nichols
J. Andrew Royle
Kenneth H. Pollock
Larissa L. Bailey
James E. Hines

ELSEVIER

AMSTERDAM • BOSTON • HEIDELBERG
LONDON • NEW YORK • OXFORD
PARIS • SAN DIEGO • SAN FRANCISCO
SINGAPORE • SYDNEY • TOKYO
Academic Press is an imprint of Elsevier

Acquisitions Editor: Nancy Maragioglio
Project Manager: Jeff Freeland
Editorial Assistant: Kelly Sonnack
Marketing Managers: Linda Beattie, Philip Pritchard
Cover Design: Cate Barr
Composition: SNP Best-set Typesetter Ltd., Hong Kong
Cover Printer: Phoenix Color Corp.
Interior Printer: The Maple-Vail Book Manufacturing Group

Academic Press is an imprint of Elsevier
30 Corporate Drive, Suite 400, Burlington, MA 01803, USA
525 B Street, Suite 1900, San Diego, California 92101-4495, USA
84 Theobald's Road, London WC1X 8RR, UK

This book is printed on acid-free paper. ∞

Library of Congress Cataloging-in-Publication Data
Application submitted

British Library Cataloguing in Publication Data
A catalogue record for this book is available from the British Library

ISBN-13: 978-0-12-088766-8
ISBN-10: 0-12-088766-5

For all information on all Elsevier Academic Press publications
visit our Web site at www.books.elsevier.com

Printed and bound in the United Kingdom

Transferred to Digital Print 2011

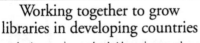

DEDICATIONS

To my parents, Ian and Linda MacKenzie; my wife Kerry; and sons, Josh, Connor and Ollie.

Darryl I. MacKenzie

To Lois, Jonathan and Christy

James D. Nichols

To my parents, Suzanne (deceased) and Jeffrey W. Royle; my wife Susan Zwicker; and daughter Abigail Suzanne

J. Andrew Royle

To my partner Barbara Brunhuber

Kenneth H. Pollock

To my parents, Bud and Velda Bailey, and my sister Tiffany.

Larissa L. Bailey

To Cindy, Ryan, Laura and Daniel

James E. Hines

"Without data, all you are is just another person with an opinion."
—*Unknown*

"An approximate answer to the right question is worth a great deal more than a precise answer to the wrong question."
—*John Tukey*

"If a tree falls in the forest, but there is no one there to see it, does it make any noise?"
—*Zen koan*

"If a bird sings in the forest, but the investigator fails to detect it, is the forest occupied?"
—*Evan Cooch*

Table of Contents

Preface

The presence or absence of a species from a collection of sampling units is a basic concept that, we now realize, is widely used in wildlife and ecological studies. We were more familiar with problems of estimating abundance, birth rates, survival probabilities, and related demographic parameters, primarily from capture-recapture data, when we began to devote substantial time to the problem in the late 1990s, and did not appreciate the generality of the concept of "occupancy" (i.e., locations where the species is present). At that time, some of us were involved with the newly formed U.S. Geological Survey's Amphibian Research and Monitoring Initiative (ARMI), advising on general study design and the types of data that should be collected as part of broad-scale monitoring programs. It quickly became apparent that it was logistically impossible to estimate changes in absolute abundance across large areas over time, so one suggestion was to simply measure the presence or absence of the species at a number of ponds or wetlands, which was soon known within ARMI as the "proportion of area occupied." With our background in capture-recapture methods, in which estimation of capture probability is the key concept for making robust inference about abundance, survival, and other parameters, questions of whether it would be reasonable to always detect the target species within an occupied sampling unit were quick to surface. The response from the biologists was a resounding "No!" Our experience made us realize that one could therefore only make reliable inferences about occupancy if the collected data and analytic methods used also accounted for the imperfect detection of the species. To us, the use of repeated "presence/absence" surveys at each sampling unit would obviously supply the type of information required to estimate detection probabilities, but at the time of our initial collaborations with ARMI, we did not have a clear idea on the best approach to efficiently analyze the resulting data, particularly if unequal numbers of surveys were conducted at the different sampling units. Our thoughts on the topic crystallized in 2000–2001, sometimes aided by a fine pint or two of Guinness, which resulted in our first paper on the general topic in 2002 (MacKenzie *et al.* 2002).

Since then, we have come to realize that the general concept of occupancy is used much more widely than simply in monitoring as a surrogate for abundance. Despite widespread acknowledgment that in many applications a species may be detected only imperfectly, we were surprised at the lack of development of appropriate analytic techniques that explicitly accounted for detection probabilities, finding only a handful of previously published methods, with all of them being developed in a monitoring context. This led us to extend our original method in an attempt to develop practical approaches for analyzing occupancy data in different situations (e.g., MacKenzie *et al.* 2003, 2004b; Royle and Nichols 2003; Dorazio and Royle 2005; MacKenzie and Royle 2005). We recognized a substantial gap in the literature that was relevant to a wide range of applications wherever reliable estimates of occupancy or changes in occupancy were required about species that are imperfectly detected. The simultaneous development of similar methodologies by independent groups (e.g., Tyre *et al.* 2003; Stauffer *et al.* 2004), and research into the effect of imperfectly detecting the species in different contexts [e.g., habitat modeling (Tyre *et al.* 2003; Gu and Swihart 2004) and metapopulation incidence functions (Moilanen 2002)], suggested that perhaps the time was right for a more comprehensive treatment of the issues associated with estimating occupancy and related parameters in wildlife and ecological studies. This book is our attempt at doing so.

We debated about the appropriate timing of this initial synthesis. A great deal of work has been completed over the last five years, but there is also much left to be done. It might be argued that an initial synthesis should await further development of methods and ideas. Our rationale for proceeding at this relatively early developmental stage is based largely on the history of the development and use of inference methods in population ecology. This field has been characterized by publication of important initial papers on topics that were followed frequently by long periods of stasis characterized by little further methodological development and little use of methods by practitioners. For some topics, a synthetic treatment was eventually published that brought together previous work and provided a common conceptual framework. Invariably, such syntheses have served as catalysts for both rapid assimilation of methods into practice and rapid development of extensions and new methods. For example, band recovery models for estimating survival rates of marked animals were initially developed by Haldane (1955), Chapman and Robson (1960), Seber (1970, 1971), Robson and Youngs (1971), and Brownie and Robson (1976), yet these methods saw little use and slow methodological development until the publication of the synthetic monograph by Brownie et al. (1978, 1985). This monograph then led to rapid assimilation of these methods by practitioners (e.g., see reviews by Johnson et al. 1992; Krementz et al. 1997) and to a variety of methodological extensions and developments (e.g., see Williams et al. 2002). Synthetic books and monographs have played similar catalytic roles for the general methodologies of closed capture-recapture models (Otis et al. 1978; White et al. 1982), distance sampling (Burnham et al. 1980; Buckland et al. 1993, 2001), and open capture-recapture models (Burnham et al. 1987; Pollock et al. 1990; Lebreton et al. 1992). Our hope is that the present book provides the kind of synthesis that will promote use of these methods and rapid methodological development in the area of occupancy estimation and modeling.

In this book you will find a synthesis of the current literature on estimating occupancy-type metrics while explicitly accounting for detection probability. We declare that one cannot make reliable inferences about such metrics when issues related to the detectability of the species are effectively ignored, and we are also of the view that detection probabilities are likely to change over time, location, and species, and hence should be estimated at every opportunity. In Chapters 1–3 we provide readers with important introductory material. Chapter 1 contains our views of modeling and the conduct of management and/or science, and of the role models play in making inference about populations. In Chapter 2, we discuss historical uses of occupancy and related concepts, noting that a number of important areas in ecology are based on species occurrence and occupancy. Chapter 3 then provides background on many of the general statistical methods used throughout this book. We strongly advise readers not to skip these introductory chapters, as they provide the necessary background upon which we build in the latter chapters. In Chapters 4–6 we consider the common problem of estimating occupancy for a single species at a single point in time, in terms of both appropriate analytic techniques (Chapters 4 and 5) and important issues of study design (Chapter 6). Next, in Chapter 7, we consider estimating the processes of change in occupancy over time for a single species, before moving on to the use of occupancy-type metrics for making inferences about multiple species in Chapters 8 and 9. Finally, we discuss topics of current and future research in Chapter 10.

Our intended audience for this book is that at the "coal face" of wildlife and ecological research, who may not necessarily have a strong background in statistics. We have thus taken great pains to use as little statistical jargon as possible, and provide detailed descriptions of how the models are developed, applied, and interpreted throughout, together with practical examples to illustrate the methods. This is, however, a technical topic, and in places some readers may struggle to understand the techniques exactly. However, we would hope that readers will at least come away with an understanding of the general concepts and what the modeling is attempting to achieve. We believe this book should be useful to biologists internationally working in government agencies, private research institutes, and universities. It may also be a useful supplemental text for graduate courses on sampling wildlife populations or wildlife biometry. Further, it would be an excellent choice for the basis of a short course on this specific topic.

We recognize that user-friendly software is necessary if the methods discussed in this book are to be widely adopted. As such, we have developed Windows-based software specifically for their application: Program PRESENCE. Currently, PRESENCE 2.0 is freely available for download via this book's companion website (http://books.elsevier.com/companions/0120887665) and is distributed with full documentation. Some techniques have also been incorporated into Program MARK (http://www.cnr.colostate.edu/~gwhite/mark/mark.htm), which was originally developed for the application of capture-recapture models to data collected on marked individuals.

Finally, we wish to stress that we do not view this book as the final word on occupancy estimation and modeling, but as one of the first. We hope that some readers will not be fully satisfied with the methods presented in this book and will seek to extend them to suit their own applications. We only ask that, when doing so, they are mindful of the general principles we develop here, in that estimation of detection probability is a necessary component for making reliable inferences about species occurrence. Such extensions will only add to the available toolbox of appropriate analytic methods. We hope that readers will find the methods detailed in the following pages as thought-provoking and useful as we have found them to be during their development over the last five years.

Acknowledgments

As with the writing of any book, there is always a long list of groups and individuals whose support and contributions are invaluable to the authors, without whom the book would be much the poorer. We would like to thank the U.S. Geological Survey's Inventory and Monitoring Program and Amphibian Research and Monitoring Initiative and the Royal Society of New Zealand's Marsden Fund for the funding that supported much of the initial development of the methods in this book; Paul Dresler, Sam Droege, Paul Geissler, Sue Haseltine, Dan James, Rick Kearney, Melinda Knutson, and Catherine Langtimm for various forms of administrative support and encouragement; Paul Geissler for his pioneering work on occupancy estimation; Alan Franklin, Erin Johnson Hyde, Ullas Karanth, Bryan Manly, John Sauer, Ted Simons, Brad Stith, Nicole Sutton, and their respective field crews for supplying data and figures used throughout this book; Regina Lanning for her help in correctly formatting 30+ manuscript pages of literature citations; Arthur Guinness and Ninkasi for providing the stimulus for many interesting discussions on this topic and many others; everybody who has asked us those "simple" or "stupid" questions that have stimulated our thoughts on this topic; and the team at

Elsevier, particularly Nancy Maragiolio and Kelly Sonnack, for their encouragement, advice, and patience.

DM would like to thank: David Fletcher, for introducing me to wildlife and ecological statistics; Bryan Manly, for setting my feet firmly on this path; James Speight, for a continuing source of inspiration; and my Ph.D. supervisor Richard Barker, for all your support and encouragement, and for introducing me to the Patuxent "mafia." I would never have written this book if you had not suggested that I take a one-year "Pre-Doc" at Patuxent in 2000–2001. And last, but not least, to Kerry, Josh, Connor, and Ollie; thanks for your continuing support throughout your husband's and father's many "absences" (both real, and present but undetected) from family life in recent times.

JDN thanks Alan Franklin, Barry Noon, and Erran Seaman for first stimulating my thinking on occupancy estimation with respect to spotted owl surveys in Olympic National Park. Thanks to Russ Hall, Catherine Langtimm, and Franklin Percival, and to Ullas Karanth for resurrecting my interest in this topic in the contexts of amphibian monitoring and tiger surveys, respectively. Thanks to Judd Howell and Graham Smith for administrative support, and to my family for their general support.

JAR thanks Robin Jung (NE ARMI) and Linda Weir (NAAMP) for involving me in various amphibian monitoring problems, and use of the anuran monitoring data in Chapter 5. Thanks to John R. Sauer who kindly provided the avian point count data used in Chapter 5. Thanks to my wife Sue for support and especially converting Latex documents to MS Word!

KHP would like to thank Dr. Barbara Brunhuber for her substantial editorial assistance.

LB thanks Velda, Bud, and Tiffany for their encouragement and unwavering support. Their insightful comments always focus me on the important things in life. Bill Link patiently explored and explained some of the impacts of unmodeled heterogeneity. Thanks to my Ph.D. and M.S. advisors, Ted Simons and Ken Pollock—your support, advice, and faith in me over the years has been so important to me. Finally, I would like to thank the researchers involved with the U.S. Geological Survey's Amphibian Research and Monitoring Initiative, specifically Cathy Langtimm, an unsung hero who invited me to join in one of ARMI's first annual meetings.

JEH would like to thank all of the folks who provided sample data and feedback on early (and future) versions of PRESENCE, and Bill Link and Gary White for help with statistical and programming problems.

CHAPTER 1

Introduction

Ecology is frequently defined as the study of the distribution and abundance of plants and animals (e.g., Andrewartha and Birch 1954; Krebs 1972). Consequently, the practice of counting animals in order to draw inferences about their numbers and distribution has a long tradition in animal ecology and management. In his classic book, *Animal Ecology*, Charles Elton (1927:173) wrote: "The study of numbers is a very new subject, and perfect methods of recording the numbers and changes in the numbers of animals have yet to be evolved." Elton then devoted 6 pages to the topic of animal "census" methods. In his equally influential classic, *Game Management*, Aldo Leopold (1933:139) listed "Census" as the first of four steps required to initiate game management on any piece of land. He then devoted a 30-page chapter to "game census" and another 25 pages to "measurement and diagnosis of productivity," a chapter that focused on assessing vital rates and population change. Methods for counting animals have indeed evolved over the last 70 years, and animal ecologists and managers now have an impressive methodological toolbox for estimating parameters associated with animal abundance and with the vital rates that produce changes in abundance (e.g., Seber 1973, 1982; Williams *et al.* 2002).

1

Today, biologists interested in understanding and managing animal populations and communities include some individuals who make full use of the methods available for drawing inferences about variation in animal numbers over time and space, and many others who do not appear to recognize the importance of appropriate inferential procedures. Because of those scientists and managers who do not take advantage of available estimation methods, the fields of animal population and community ecology, wildlife management, and conservation biology include numerous examples of substantial field efforts that do not produce reliable conclusions. These disciplines suffer not only from the failure of animal ecologists and managers to utilize the range of available methods for drawing inferences about animal abundance and associated vital rates but also from the lack of rigorous methods for estimating other quantities that may be biologically relevant. For example, other variables that could be used to quantify the current status of a community or population (we refer to these as *state variables*) include species richness (number of species) and occupancy (proportion of an area occupied by a species or fraction of landscape units where the species is present). Scant attention has been devoted to estimation of these latter state variables, with the result that there is a great need for methodological development.

In this book, we emphasize the need for estimation methods that permit inference about occupancy based on so-called presence-absence data and report results of our initial efforts to develop a set of such methods. We begin this chapter by providing brief operational definitions for some important terms, then move on to an outline of general principles for sampling animal populations, focusing on the why, what, and how of such sampling. This outline is followed by a more detailed look at the critical step of using field data to discriminate among competing hypotheses about system response to environmental variation and management actions. We note different field designs that are used to generate system dynamics for such discrimination and comment on the different strengths of inference resulting from these designs. The chapter concludes with a more detailed statement of book objectives and contents.

1.1. OPERATIONAL DEFINITIONS

The methods presented in this book should be useful to biologists involved in either science or management of biological populations. Both endeavors use the following three constructs: *hypothesis, theory,* and *model.* These terms are not always used consistently in the literature, and therefore we provide our own operational definitions for use in this book (also see Nichols 2001). We view a *hypothesis* simply as a plausible explanation (i.e., a "story") about how the world, or part of it, works. For example, we would deem density-dependent recruitment for mid-continent, North American mallards (*Anas platyrhynchos*) as a hypothesis, with density-independent recruitment as an alternative, competing hypothesis (Johnson *et al.* 1997). Once a hypothesis has withstood repeated efforts to falsify it, to the extent that we have some faith in predictions deduced from it, the hypothesis may become a *theory* (e.g., Einstein's theory of relativity). A theory can still be disproved in the future given new data or the expansion of the part of the world to which the theory is thought to be applicable (e.g., Newtonian physics).

Very generally, we view a *model* as an abstraction of a real-world system, which can be used to describe observed system behavior and predict how the system may respond to changes or perturbations. Within this broad definition we recognize many different kinds of models (Nichols 2001), three of which are especially useful within the context of this book. A *conceptual model* is a set of ideas about how the system of interest works, and may include one or more hypotheses or theories about the system. A *verbal model* is created by translating these ideas into words. Finally, a *mathematical model* results from translating a conceptual or verbal model into a set of mathematical equations, using defined parameters to symbolize the key processes of the system. In this book we derive mathematical expressions from our conceptual ideas about the processes that occur when collecting occupancy field data, placing particular attention on using the collected data to estimate the parameters of these models.

By following the logical progression above, note that a mathematical model is ultimately a representation of one or more hypotheses or theories about the system. Therefore, competing hypotheses can be formulated into competing mathematical models. Applying each model to the same set of available data, it may be possible to formally determine which model (and therefore which hypothesis) has a greater degree of support given the data at hand. Essentially this is an exercise in model selection. We advocate and use such an approach throughout this book.

1.2. SAMPLING ANIMAL POPULATIONS AND COMMUNITIES: GENERAL PRINCIPLES

It is our belief that many existing programs for sampling animal populations and communities are not as useful as they might be because investigators have not devoted adequate thought to fundamental questions associated with establishment of such programs. These failures have greatly reduced the value of efforts ranging from individual scientific investigations to large-scale monitoring programs. These latter programs are especially troubling, because they can require nontrivial fractions of the total funding and effort available for the conduct of science and management of animal populations and communities. Here we present some opinions about the sort of thinking that should precede and underlie good animal sampling programs. These opinions are structured around three basic questions to be addressed during the design of an animal sampling program (see Yoccoz et al. 2001): Why? What? and How?

WHY?

Efforts to sample animal populations are generally associated with one of two main classes of endeavor, science or conservation and management (or possibly both). Science can be viewed as a process used to discriminate among competing hypotheses about system behavior, that is, discriminating among different ideas about how the world, or a part of it, works (e.g., whether recruitment to a population is density dependent). This process typically involves mathematical models. For example, a mathematical model that could be used to represent the number of recruits to a population (r) that assumes no density dependence would be $r = N_F b$, where N_F is the number of breeding females in the population and b is the average number of female births per adult female and is viewed as a constant (with respect to current breeding female population size). A different model that conceptualizes the effect of density-dependent recruitment would be $r = N_F b(N_F)$, where $b(N_F)$ specifies a functional relationship, such that number of recruits per female is a function of total female abundance. The primary use of models is to project the consequences of hypotheses, that is, to deduce predictions about system behavior (e.g., Nichols 2001). In the case of our example, the model is used to predict the number of recruits at different levels of population density.

The key step in science, then, involves the confrontation of these model-based predictions with the relevant components of the real-world system (Hilborn and Mangel 1997; Williams et al. 2002). Faith and confidence increase for those models (and hence those underlying hypotheses) whose predictions match observed system behavior well and decrease for models that do a poor job of predicting. However, for most practical situations involving animal populations and communities, true system behavior cannot be directly observed, but must be estimated from data collected from sampling programs. Thus, sampling programs constitute a key component of scientific research.

In the conduct of management and conservation, estimates of state variables for animal populations and communities serve three distinct roles (Kendall 2001). First, estimates of system state are needed in order to make state-dependent management decisions (e.g., Kendall 2001; Williams et al. 2002). For example, the decision of which management action to take frequently depends upon the current population size. Second, system state is frequently contained in the objective functions (precise, usually mathematical, statements of management objectives) for managing animal populations and communities. Evaluation of the objective function is an important part of management, addressing the question "To what extent are management objectives being met?" Finally, good management requires either a single model thought to be predictive of system response to management actions or a set of models with associated weights reflecting relative degrees of faith in their validity. The process of developing faith in a single model or weights for members of a model set involves the confrontation of model predictions with estimates of true system response. This confrontation is the scientific component of informed management and requires animal sampling programs that provide reliable estimates of state variables and associated vital rates.

Despite the importance of being explicit about why a program for sampling animal populations or communities is needed, we believe that many studies suffer from a failure to clearly articulate specific study objectives. This is especially evident in many large-scale monitoring programs (Yoccoz et al. 2001). For example, the following objectives statements from a report on ecological monitoring programs in the United States (LaRoe et al. 1995:3, 4) are fairly typical: "The goal of inventory and monitoring is to determine the status and trends of selected species or ecosystems"; "Inventory and monitoring programs can provide measures of status and trends to determine levels of ecological success or stress." The second statement implies an interest in management and conservation, but without specification of available management actions and hypotheses about system response to those actions, the statement provides little basis for monitoring program design. Thus, we advocate clear specification of monitoring program objectives.

Objective specification is facilitated by the recognition that monitoring of animal populations and communities is not a stand-alone activity of great inherent utility, but is more usefully viewed as a component of the processes of science and/or management. This recognition leads naturally to detailed consideration of exactly how the monitoring program results are to be used in the conduct of science or management or both. Such considerations lead directly to decisions about monitoring program design, whereas vague objectives that fail to specify use of program data and estimates provide little guidance for program design and can lead to endless debate about design issues.

WHAT?

The selection of what state variable(s) and associated vital rates to estimate will depend largely on the answer to the initial question of "Why?" The selection of state variables for scientific programs will depend on the nature of the competing hypotheses and specifically on the quantities most likely to lead to discrimination among the hypotheses (i.e., for what quantities are predictions of competing hypotheses most different?). The selection of state variables for management programs will depend on the most relevant characterization of system state, on management objectives, and on the ability to discriminate among competing hypotheses about system response to management actions. Practicality must also be considered in both cases as, most likely, logistical resources will be limited.

When dealing with single species, the most commonly used state variable is abundance or population size. Estimation of abundance frequently requires substantial effort, but it is a natural choice for state variables in studies of population dynamics and management of single-species populations. Some studies of animal abundance focus directly on changes in abundance, frequently expressed as the ratio of abundances in two sampling periods (e.g., two successive years) and termed the finite rate of population increase or population growth rate, λ. In scientific studies, mechanistic hypotheses frequently concern the vital rates responsible for changes in abundance, rates of birth (reproductive recruitment), death, and movement in and out of the population. In management programs, effects of management actions on animal abundance must also occur through effects on one or more of these vital rates. Thus, many animal sampling programs involve efforts to estimate abundance and rates of birth, death, and movement for animals inhabiting some area(s) of interest.

We believe that another useful state variable in single-species population studies is occupancy, defined as the proportion of area, patches, or sample units that is occupied (i.e., species presence). Sampling programs designed to estimate occupancy tend to require less effort than programs designed to estimate abundance (e.g., Tyre et al. 2001; MacKenzie et al. 2002; Manley et al. 2004). In the case of rare species, it is sometimes practically impossible to estimate abundance, whereas estimation of occupancy is still possible (MacKenzie et al. 2004a, 2005). Thus, for reasons that include expense and necessity, occupancy is sometimes viewed as a surrogate for abundance. However, there are also a number of kinds of questions for which occupancy would be the state variable of choice regardless of the effort involved in sampling. For example, metapopulation dynamics (e.g., Hanski and Gilpin 1997; Hanski 1999) are frequently described by patch occupancy models. So-called incidence functions (e.g., Diamond 1975a; Hanski 1994a) relate patch occupancy to patch characteristics such as size, distance to mainland or some source of immigrants, habitat, etc. Occupancy is the natural state variable for use in studies of distribution and range (e.g., Brown 1995; Scott et al. 2002) and should also be useful in the study of animal invasions and even disease dynamics. Patch occupancy dynamics may be described using the rate of change in occupancy over time, and the vital rates responsible for such change are patch-level probabilities of extinction and colonization. Historical, current, and proposed uses of patch occupancy as a state variable for science and management will be discussed in more detail in Chapter 2.

When scientific or conservation attention shifts to the community level of organization, many possible state variables exist. The basic multivariate state variable of community ecology is the species abundance distribution, specifying the number of individuals in each species in the community. Many derived state variables are obtained by attributing different values or weights to individuals of different species (Yoccoz *et al.* 2001). Several common diversity indices are computed by providing a weight of 1 to every individual of each species (e.g., Pielou 1975; Patil and Taillie 1979), but it is also possible to give additional weight to individuals of species thought to be of special importance (e.g., endemic species or species of economic value) (Yoccoz *et al.* 2001). A state variable that is used commonly in community studies is simply species richness, the number of species within the taxonomic group of interest that is present in the community at any point in time or space. This state variable is used in scientific investigations (e.g., Boulinier *et al.* 1998b, 2001; Cam *et al.* 2002) and programs for management and conservation (e.g., Scott *et al.* 1993; Keddy and Drummond 1996; Wiens *et al.* 1996). The vital rates responsible for changes in species richness over time are rates of local species extinction and colonization.

In this book we focus largely on the state variable of occupancy, but note how these methods can also be applied where species richness-type metrics may be of interest (Chapter 9).

HOW?

Proper estimation of state variables and inferences about their variation over time and space require attention to two critical aspects of sampling animal populations: spatial variation and detectability (Fig. 1.1) (Lancia *et al.* 1994; Thompson *et al.* 1998; Williams *et al.* 2002). Spatial variation in animal abundance is important because in large studies and most monitoring programs investigators cannot directly survey the entire area of interest. Instead, investigators must select a sample of locations to which survey methods are applied, and this selection must be done in such a way as to accomplish two things. First, selection of study locations should be based on study objectives. In the case of scientific objectives, study locations should be selected to provide the best opportunity to discriminate among the competing hypotheses of interest (see Section 1.3 for further discussion). For example, in the case of an obser-

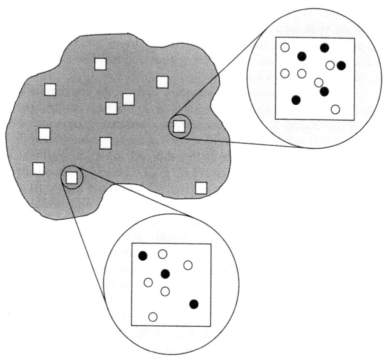

FIGURE 1.1 Illustration of the two critical aspects of sampling animal populations, spatial variation and detectability. The shaded region indicates the area or population of interest, with the small squares representing the locations selected for sampling. Within each sampling location, animals will be detected (filled circles) or undetected (hollow circles) during a survey or count.

vational study involving hypotheses about habitat variables, selected study locations might be extremes with respect to the variable(s) of interest or else might be locations at which changes in the variable(s) are anticipated. In the case of a management program, study locations should of course include the areas to which management actions are applied. Second, within larger areas selected based on study objectives, sample locations should be selected in a manner that permits inferences about the locations that are not surveyed and hence about the entire area(s) of interest. Approaches to sampling that accomplish this inferential goal include simple random sampling, unequal probability sampling, stratified random sampling, systematic sampling, cluster sampling, double sampling, and various kinds of adaptive sampling (e.g., Cochran 1977; Thompson 1992; Thompson and Seber 1996).

Detectability refers to the reality that, even in locations that are surveyed by investigators, it is very common for animals and even entire species to be missed and go undetected. Most animal survey methods yield some sort of count statistic. For example, when abundance is the quantity of interest, the count statistic might be the number of animals caught, seen, heard, or harvested. Let N_{it} be the true number of animals associated with an area or sample unit of interest, i, at time t, and denote as C_{it} the associated count statistic. This statistic can be viewed as a random variable whose expectation (basically the average value of the count if we could somehow conduct the count under the exact same conditions many times; see Chapter 3) is the product of the quantity of interest, abundance at the surveyed location, and the detection probability associated with the count statistic:

$$E(C_{it}) = N_{it}p_{it}, \tag{1.1}$$

where p_{it} is the detection probability (probability that a member of N_{it} appears in the count statistic, C_{it}). Estimation of N_{it} thus requires estimation of p_{it}:

$$\hat{N}_{it} = C_{it}/\hat{p}_{it}, \tag{1.2}$$

where the "hats" in this expression denote estimators (see Chapter 3). Expression (1.2) is very general and widely applicable. In fact, virtually all of the abundance estimation methods summarized and reviewed by Seber (1973, 1982), Lancia *et al.* (1994, 2005), Thompson *et al.* (1998), Williams *et al.* (2002), and Borchers *et al.* (2003) involve different approaches to the estimation of detection probability followed by (or integrated with) application of expression (1.2).

Frequently, interest will not be in abundance itself but in relative abundance, the ratio of abundances at two locations ($\lambda_{ijt} = N_{it}/N_{jt}$, where i and j denote locations and t still denotes time), or in rate of population change, the ratio of abundances in the same location at two times ($\lambda_{it} = N_{it+1}/N_{it}$). Sometimes count statistics are treated as indices, and their ratio is used to estimate the true ratio of abundances. For example, consider the estimator $\hat{\lambda}_{it} = C_{it+1}/C_{it}$. The expectation of this estimator can be approximated using expression (1.1) as:

$$E(\hat{\lambda}_{it}) \approx \frac{N_{it+1}p_{it+1}}{N_{it}p_{it}} = \lambda_{it}\left(\frac{p_{it+1}}{p_{it}}\right). \tag{1.3}$$

As can be seen from (1.3), the ratio of counts estimates the product of the quantity of interest, λ_{it}, and the ratio of detection probabilities. If the detection probabilities are very similar for the two sample times, then the estimator will not be badly biased, but when detection probabilities differ, then the index-based estimator will be biased. If detection probability itself is viewed as a random variable, then we still require $E(p_{it}) = E(p_{it+1})$ in order for a ratio of counts to be a reasonable estimator.

Proponents of the use of count statistics as indices for estimating relative abundance typically recommend standardization of survey methods as one means of trying to insure similar detection probabilities. Standardization involves factors that are under the control of the investigator (e.g., effort, trap type, bait, season and time of day of survey). While standardization of survey methods is usually a good idea, we believe that this approach is unlikely to produce equal detection probabilities, because there are always likely to be unidentified and uncontrollable factors that influence detection probabilities (Conroy and Nichols 1996). Sometimes it is possible to identify uncontrollable factors that could influence detection probability and incorporate them as covariates into analyses of count statistics. This approach is reasonable when dealing with factors that could only affect detection probability and not animal abundance itself. For example, differences in detection probabilities among observers are often incorporated into analyses of avian point count data (Link and Sauer 1997, 2002). However, it would not be wise to use a similar approach with habitat data, as habitat would be expected to influence not only detection probability but also animal abundance itself. Thus, "controlling" for habitat effects by incorporating them into analyses as covariates would not be appropriate. Of course, factors that we do not identify but still affect detection probability cannot be treated as covariates either.

Another common claim supporting the use of indices is that they are relatively assumption free, unlike the methods used to actually estimate abundance (e.g., Seber 1982; Williams et al. 2002). However, there are a large number of implicit assumptions to be made if the index is to be related to animal abundance. In fact, interpretation of an index as some indicator of true population size typically requires all the assumptions used to estimate abundance plus the assumption that a constant fraction of the population is counted each survey. Some uses of indices require the assumption that all animals are counted during each survey. As these assumptions are unlikely to be true, we believe that indices have a very limited use in good monitoring programs. We conclude that estimation of both absolute and relative abundance requires information about detection probability (also see Lancia et al. 1994; MacKenzie and Kendall 2002; Williams et al. 2002).

The importance of obtaining information about detection probability extends to other state variables as well. Investigations of species richness usually involve counts of the number of different species. Under some designs the counts are conducted at multiple locations within some large area to which inference is to apply, whereas other designs use counts conducted at multiple times (e.g., days) on a single area of interest (e.g., Nichols and Conroy 1996; Williams et al. 2002). In both designs, it is recognized that some species may

go undetected, and the replication (geographic or temporal) is used to estimate a species level detection probability, the probability that at least one individual of a species will be detected given that the species inhabits the area of
interest. Efforts to estimate species richness from samples of animal communities are not new (Fisher et al. 1943; Preston 1948; Burnham and Overton
1979). Nevertheless, community ecologists have tended to ignore the issue of
detection probabilities less than 1, and only recently has adequate attention
been devoted to this estimation problem (e.g., Chao and Lee 1992; Bunge and
Fitzpatrick 1993; Colwell and Coddington 1994; Walther et al. 1995; Chao
et al. 1996; Nichols and Conroy 1996; Boulinier et al. 1998a; Cam et al. 2000;
Williams et al. 2002; Dorazio and Royle 2005).

Detection probability is also very relevant to the estimation of occupancy.
Define occupancy, ψ, as the probability that a randomly selected site or sampling unit in an area of interest is occupied by a species (i.e., the site contains
at least one individual of the species). If x and s represent the number of occupied and total sites, respectively, then we can estimate occupancy as $\hat{\psi} = x/s$.
However, x is not typically known. Instead, we will have a count of sites where
the species has been detected, but this count will likely be smaller than x,
because species will not always be detected in occupied sites (i.e., due to "false
absences"). Thus, we must develop methods (e.g., based on multiple surveys
of sites) to estimate detection probability and thus to estimate x. For example,
we can use an analog of expression (1.2), where the count is the number of
sites at which the species is detected, and the detection probability is the probability that the species is detected during sampling of an occupied site. Occupancy can then be estimated as:

$$\hat{\psi} = \frac{\hat{x}}{s}. \qquad (1.4)$$

We have actually developed more direct ways to estimate occupancy (e.g.,
MacKenzie et al. 2002; Royle and Nichols 2003; Chapters 4, 5), but the basic
rationale underlying these approaches is the same as outlined here.

Inferences about occupancy may be misleading when detection probability
is not incorporated into the methods of data analysis. Not only will naïve
approaches underestimate occupancy (as above), but indices intended to
reflect relative occupancy also could be biased (MacKenzie 2006) and the effect
of casual factors or variables may be underestimated (Tyre et al. 2003) or
misidentified, particularly if detection probability covaries with the factors or
variables thought to affect occupancy (Gu and Swihart 2004; MacKenzie
2006). Inferences about the dynamic processes that drive changes in occupancy may also be inaccurate (Moilenan 2002; MacKenzie et al. 2003). Indeed,
an important theme of this book is that robust inference about occupancy and

related dynamics can only be made by explicitly accounting for detection probability.

1.3. INFERENCE ABOUT DYNAMICS AND CAUSATION

Chapter 2 will focus on the "what" of animal sampling programs and discuss the use of occupancy as a state variable. Much of the remainder of the book will then focus on the "how" question of sampling animal populations. That is, given interest in occupancy, how do we estimate this state variable and the vital rates responsible for its change in reasonable ways? Although we believe that this emphasis is justified by the absence of previous work and good guidance on drawing inferences about occupancy, we regret the need to abandon issues about "why" we sample animal populations. In our introductory discussion about why we might want to sample animal populations and communities, we emphasized that sampling programs are usefully viewed as components of the larger processes of science or management. In this section, we briefly discuss the manner in which results of animal sampling programs are used to draw the inferences needed for science or management. This discussion touches aspects of design that extend well beyond efforts to obtain reasonable estimates of state variables of interest.

The key step in the scientific process involves a comparison of estimates of state variables with model-based predictions associated with competing hypotheses. Such comparisons also constitute an important management use of estimates from animal sampling programs, as the ability to predict consequences of different management actions is critically important to informed management. Scientific programs include interest in responses of animal populations and communities to a variety of factors (e.g., changes in predators, competitors, weather, habitat, disease, toxins/pesticides). Management programs focus not only on responses to management actions but also on other factors that might improve predictive abilities. We would like to discriminate among competing hypotheses about the relevance of different causal factors to system dynamics with the ultimate goal of being able to predict the magnitude of the state variable(s) at time $t + 1$, given the magnitude of the state variable at time t and knowledge of the causal factors operating between times t and $t + 1$ (Williams 1997; Williams et al. 2002).

GENERATION OF SYSTEM DYNAMICS

The scientific process usually includes some means of generating system dynamics so that estimated changes in state variables can be compared with

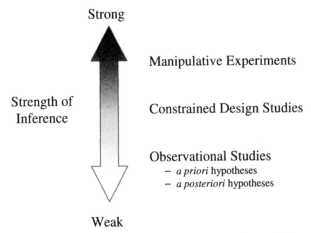

Strong

Manipulative Experiments

Strength of
Inference

Constrained Design Studies

Observational Studies
– *a priori* hypotheses
– *a posteriori* hypotheses

Weak

FIGURE 1.2 Strength of inference of different sampling designs that could be used to generate system dynamics.

the predictions of competing models. Multiple approaches are used to generate system dynamics in population and community ecology, and we classify these approaches broadly as true manipulative experiments, constrained designs or quasi-experiments, and observational studies (Romesburg 1981; Skalski and Robson 1992; Manly 1992; Williams *et al.* 2002). These approaches merit brief discussion here, as they provide different strengths of inference (Fig. 1.2).

Inferences are strongest when system dynamics are generated via the conduct of true manipulative experiments (see Fisher 1947; Hurlbert 1984; Skalski and Robson 1992; Manly 1995). Such experiments are characterized by replication, randomization in the assignment of different treatments (application of different hypothesized causal factors) to experimental units, and the use of a control or standard treatment group. In the context of population and community ecology, experimental units may be populations or communities occurring naturally or created as part of the experimental design. *Replication* refers to the application of treatments to multiple experimental units as a means of estimating the experimental error or error variance. The error variance reflects the variance among experimental units to be expected in the absence of treatment differences (i.e., the variance associated with all factors except the different treatments). *Randomization* refers to random assignment of treatments to experimental units. Randomization protects against systematic differences among experimental units receiving different treatments and represents an effort to insure that any systematic post-treatment differences among experimental units treated differently can be attributed to the treatments themselves. One treatment type is typically designated as a control and is used to provide a baseline against which other treatments can be compared. The use of a control group is especially useful in attributing causation to different treatments and permitting estimation of treatment effects on response variables. Manipulative experiments thus seek to reduce potential sources of ambiguity to the extent possible, yielding strong inferences about causation.

True manipulative experiments are frequently difficult to perform on free-ranging animal populations and communities due to cost and practical field constraints. In many instances, we may be able to manipulate systems but may be required to do so using study designs that lack replication, randomization, or both of these features (see Green 1979; Skalski and Robson 1992; Williams et al. 2002). Inferences resulting from such constrained, or quasi-experimental, designs will typically not be as strong as those based on manipulative experimentation (see examples in Nichols and Johnson 1989).

Finally, the investigator may be unable to manipulate the system at all and may be forced to rely on natural variation to generate system dynamics. For example, large-scale animal monitoring programs may provide time series of estimated state variables, and retrospective analyses can be used to try to distinguish among competing hypotheses about system dynamics (Nichols 1991a). Two general approaches to observational studies are used, and they are distinguished by the existence of *a priori* hypotheses. The observational studies that tend to be most useful to science are those for which conditional *a priori* hypotheses are specified and used to guide monitoring program design (Nichols 2001; Williams et al. 2002). The hypotheses are conditional in the sense that changes in purported causal factors are not known *a priori*, as they are when the investigator imposes a manipulation. Instead, the different hypotheses predict different relationships between suspected causal factors and system state variables, and specific predictions then emerge as changes in the causal factors occur naturally and are observed. The initial specification of the hypotheses facilitates monitoring program design, as efforts can be devoted to monitoring changes in hypothesized causal factors as well.

The other approach to observational studies involves the development of *a posteriori* hypotheses to explain observed system dynamics. Monitoring programs may yield annual estimates of quantities such as population size over relatively long time periods (e.g., 20 years), and it is commonly thought that such trajectories lead directly to an understanding of underlying population dynamics. It is a common practice to use such data with correlation and regression analyses to investigate possible relationships between population size and various environmental and management variables. The problem with this approach is that it is unlikely to yield "reliable knowledge" (Romesburg 1981), because there will typically be multiple *a posteriori* hypotheses that provide reasonable explanations for any observed time series (Nichols 1991a). Indeed, we tend to agree with Pirsig's (1974:107) assertion that The number of rational hypotheses that can explain any given phenomenon is infinite.

The potential for being misled by retrospective analysis of data exists for all kinds of observations (Platt 1964; Romesburg 1981) but is probably especially large for time series of estimates of population size and related variables. One reason for this is that population size is not observed but is estimated, often with large sampling variances and sometimes with bias. Temporal variation in point estimates of population size is thus not equivalent to temporal variation in the underlying population (Link and Nichols 1994). Another difficulty in drawing inferences from retrospective analyses of population trajectory data involves the stochastic nature of population processes. Death, for example, is typically viewed as a simple stochastic process. If a population has 100 animals at time t and if each of these animals has a probability of 0.2 of dying during the interval $(t, t + 1)$, then we do not expect exactly 80 animals to be alive at time $t + 1$. Instead, the number of survivors will be a binomial random variable with expected value 80, but with likely realized values of 78, 83, 75, etc. Reproductive processes and movement are also stochastic in nature, leading to the view of a population trajectory as a single realization of a (likely complicated) stochastic process. There is little reason for us to expect to be able to infer much about the nature of an unknown stochastic process based on a single realization of that process (Nichols 1991a). This is analogous to being handed a loaded coin, being permitted to flip it once, and then being asked to specify the probability of obtaining heads.

Another difficulty associated with inferences from retrospective studies of population monitoring data involves using correlation analysis to draw inferences about the functional relationship between variables represented by time series. A clear example of such problems involves the existence of trends and monotonicity in many environmental covariates that potentially influence animal populations. Metrics of human-related environmental variables such as habitat fragmentation, habitat degradation, and pollutant levels will frequently tend to show an increasing trend over time. Correlation analyses involving two variables, each of which shows a time trend, will tend to indicate association, although this may have nothing to do with any functional relationship between the variables. In fact, the problem of conducting association analyses of two time series extends well beyond the case of monotonic trends, and such analyses frequently lead to inappropriate inferences (Yule 1926; Barker and Sauer 1992).

These various considerations lead us to conclude that development of *a posteriori* hypotheses based on retrospective analyses of monitoring data is an approach that necessarily results in weak inferences. Certainly we do not claim that such retrospective analyses are without value, as they can sometimes provide useful insights and ideas about system behavior. Instead, our recommendation is that such analyses be viewed primarily as an approach to hypothesis generation rather than as an inferential assessment of the hypothesis as an explanation for system dynamics. We thus recommend that observational studies be guided by *a priori* hypotheses, with exploratory retrospective analyses possibly used as a means of hypothesis generation.

As noted above, distinguishing among competing hypotheses about system response to management is an important component of an informed decision process. The term *adaptive management* (e.g., Holling 1978; Walters 1986; Hilborn and Walters 1992; Williams *et al.* 2002) typically applies to management that is state dependent and that incorporates learning about system response to management actions. It is this learning component that distinguishes adaptive management from other decision processes (Kendall 2001; Williams *et al.* 2002). Estimates of system state are used not only for the purpose of making state-dependent decisions but also as a means of confronting the predictions of competing models about system response for the purpose of discriminating among their associated hypotheses. Based on objectives, potential actions, an estimate of system state, and models (with associated probabilities reflecting relative degrees of faith), managers make the decision to take a particular action at time t. This action drives the system to a new state at time $t + 1$, and this state is identified via a monitoring program. Probabilities associated with degrees of faith in the various system models are then updated based on the distance between estimated system state and the predictions of the competing models (Kendall 2001; Nichols 2001; Williams *et al.* 2002). Although this approach to multimodel inference is used in the current applications of adaptive management with which we are most familiar (Nichols *et al.* 1995; Johnson *et al.* 1997; Williams *et al.* 2002), hypothesis-testing approaches are also possible and are also based on the distance between estimated system state and model-based predictions.

In the context of the previous discussion of approaches for generating system dynamics, the learning component of adaptive management will virtually always be manipulative, in that management actions will be imposed and system response then observed. However, attainment of management objectives is of primary importance in adaptive management, and learning is valued only to the extent that it is useful in better meeting objectives. Thus, in most applications with which we are familiar, the learning components of adaptive management exhibit the features of constrained designs. However, if management is of a spatially extended system and if different actions are to be taken on different spatial units of the system, then a manipulative experimental approach might be taken as well.

In summary, the conduct of science requires some means of generating system dynamics for comparison with predictions of competing hypotheses.

True manipulative experiments represent a study design that permits strong inferences about causation. Constrained or quasi-experimental designs involve manipulations, but the absence of either randomization or replication, or both features, does not permit the strength of inference of a true experiment. Finally, observational studies based on retrospective analyses of monitoring data involve no manipulation as part of study design and rely on natural variation in purported causal factors. These analyses tend to yield weaker inferences than manipulative studies. Within observational studies using retrospective analyses, those that test predictions of *a priori* hypotheses tend to yield stronger inferences than analyses used to generate *a posteriori* hypotheses. Adaptive management represents an informed decision process incorporating explicit efforts to learn about system responses to management actions. Because learning is not the sole objective of adaptive management, management manipulations typically follow some form of constrained design.

STATICS AND PROCESS VS. PATTERN

Inferences about causation emerge most naturally from studies of system dynamics. Scientists and managers estimate the state variable at time t, apply or observe purported causal factors operating between times t and $t + 1$, and then estimate the state variable again at time $t + 1$. However, because of the difficulties in applying manipulations to animal populations and communities and in properly estimating relevant state variables over time, animal ecologists have also tried to draw inferences about dynamics based on observations of spatial pattern at a single time, t. Brown (1995:10) describes "macroecology" as a research program in ecology with "emphasis on statistical pattern analysis rather than experimental manipulation." Inferences based on such efforts have been applied to each of the state variables described above—abundance, species richness, and occupancy.

Ecologists frequently use spatial variation in abundance of animals to draw inferences about habitat "quality," based on the commonsense idea that if animals are found in higher density in one habitat than others, then that habitat is likely of high quality. For such a statement to have meaning, "quality" must be defined. In their influential work on habitat selection, Fretwell and Lucas (1969; Fretwell 1972) defined habitat quality in terms of the fitness of organisms in that habitat. The two fundamental fitness components, survival probability and reproductive rate, are also primary determinants of population dynamics, so this definition is relevant to population ecologists and managers as well. Observations of spatial variation in animal density associated with habitat variation do not yield reliable inferences about individual fitness or dynamics of populations inhabiting such areas (e.g., van

Horne 1983; Pulliam 1988). Instead, such inferences require studies of system dynamics, in this case habitat-specific demography (e.g., Franklin *et al.* 2000), preferably in conjunction with habitat manipulations.

The relationship between species richness and area is one of the oldest and most-cited static relationships in ecology (e.g., Arrhenius 1921; Preston 1948). Hypotheses about the dynamic processes responsible for this relationship include habitat selection coupled with habitat heterogeneity (e.g., Williams 1964) and increased probabilities of local extinction in small areas (e.g., MacArthur and Wilson 1967). However, these two hypotheses yield similar species-area relationships, providing no basis for distinguishing between these or other mechanistic explanations (Connor and McCoy 1979).

Occupancy appears to be used more frequently in static analyses than either of the other discussed state variables, abundance and species richness. Static analyses of occupancy data in animal ecology can be illustrated with two common applications, single-species incidence functions and multiple-species co-occurrence patterns. Incidence functions involve efforts to model dichotomous spatial occupancy pattern (presence or absence) as a function of characteristics of the sampled locations or patches. Diamond (1975a) first described incidence functions in his studies of distributional ecology of birds inhabiting islands in the area of New Guinea. He grouped islands by such characteristics as land area and total avian species richness and then plotted the proportion of islands in each category (e.g., area, richness) that was occupied by a particular species. Diamond noted that some species tended to occur only on large, species-rich islands, whereas others were found only on remote, species-poor islands. Diamond (1975a:353) viewed the incidence function as a " 'fingerprint' of the distributional strategy of a species" and used these functions to draw inferences about such processes as dispersal, habitat selection, and competition (see below and Chapter 2). These inferences have been challenged based on the consistency of observed patterns with other processes (e.g., Connor and Simberloff 1979).

Hanski (1992) adapted the incidence function for use in describing and modeling metapopulation dynamics. He noted that in an equilibrium system of many patches of similar size, the fraction of occupied patches at any point in time can be written as an explicit function of patch probabilities of extinction and colonization. He then postulated functional forms for the relationships between extinction probability and patch area and between colonization probability and patch isolation. If metapopulation dynamics can be described as a stationary Markov process, then parameters of the extinction and colonization relationships can be estimated using occupancy data from a single point in time (e.g., Hanski 1992, 1994a,b, 1998, 1999). However, the difficulties of inferring process from pattern have been noted. For example, based on analyses of year-to-year changes in occupancy of pikas (*Ochotona princeps*),

Clinchy *et al.* (2002:351) recommended that "simple patch occupancy surveys should not be considered as substitutes for detailed experimental tests of hypothesized population processes."

Use of occupancy data from multiple species to draw inferences about species interactions also has a long history in ecology. Some of the first statistical analyses adapted by ecologists were used to test the null hypothesis of independence of species occurrence using occupancy data for two species (Forbes 1907; Dice 1945; Cole 1949). Non-independent occupancy patterns of multiple species on islands have been interpreted as evidence of competition (e.g., MacArthur 1972; Diamond 1975a). For example, the "assembly rules" of Diamond (1975a) include specification of species combinations that cannot exist for reasons of interspecific competition and are based on empirical observations of species distributions on different islands. However, such inferences about process based on observed patterns have been sharply criticized. Critics argued that rejection of predictions of neutral models developed from distributional null hypotheses should precede any attempt to develop more complicated explanatory hypotheses for static species distribution patterns (e.g., Connor and Simberloff 1979, 1986; Simberloff and Connor 1981). Neutral models themselves were then criticized by proponents of the original competitive hypotheses (Diamond and Gilpin 1982; Gilpin and Diamond 1984), neutral model proponents responded (Connor and Simberloff 1984; 1986), and the entire issue of inference based on species distribution patterns was hotly debated (Strong *et al.* 1984). Such debate is not surprising, as strong disagreement is a natural consequence of weak inference, which brings us back to Pirsig's (1974) assertion about the ability to develop large numbers of plausible hypotheses to explain any given pattern.

Each of the three quantities listed as state variables of potential interest in population ecology and management (abundance, occupancy, and species richness) has been investigated with respect to its distribution over space at one point in time. Identification of spatial patterns has then led to inferences about the dynamic processes that produced these patterns. However, these inferences are always very weak, as many alternative hypotheses can be invoked to explain most ecological patterns (Fig. 1.3). Our conclusions about drawing inferences about process based on snapshots of spatial pattern are simple and straightforward. First, inferences about system dynamics should be based on estimates and observations of those dynamics, and of the vital rates that produce them, whenever possible. Second, when ecologists do try to draw inferences about dynamics based on observations of static pattern, we believe that such inferences are much more likely to be useful if the specification of model-based predictions from competing or single hypotheses precedes the investigation of pattern (e.g., see Karanth *et al.* 2004). Brown (1995:18) stated, "Macroecology seeks to discover, describe, and explain the patterns of varia-

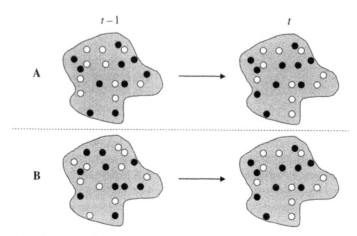

FIGURE 1.3   Illustration of how a pattern observed at time t may result from very different processes. Darkened circles represent occupied patches, and white circles represent unoccupied patches. The level of turnover between times $t-1$ and t is much greater in scenario B.

tion." We recognize that such efforts can be useful, but we recommend that they be viewed as mechanisms for hypothesis generation rather than for inference and testing.

1.4. DISCUSSION

We began this chapter by asserting that many animal sampling programs, including many large-scale monitoring programs, have deficiencies resulting from failure to adequately consider three basic questions: Why do we want to sample animal populations and communities? What quantities do we want to estimate? How should we estimate the quantities of interest? In answer to the "why" question, we suggested that animal sampling and monitoring programs should not be viewed as stand-alone activities but as components of the larger processes of science or management. This recognition forces consideration of exactly how resulting data are to be used in these processes, and this consideration leads to program designs that maximize utility of data. The answer to the "what" question will depend heavily on the answer to the "why" question, and we noted that abundance, occupancy, and species richness are reasonable state variables for a variety of objectives. Of these potential state variables, occupancy has received the least methodological attention. Indeed, our objective in this book is to provide a set of inference methods useful for investigating this state variable.

The answer to the "how" question depends on the answers to the previous two questions, but also requires attention to two basic issues. When the entire area of interest cannot be surveyed, space must be sampled in a manner that is maximally useful to study objectives and that permits inference about the entire areas of interest. Because this problem of spatial sampling characterizes a wide variety of applications in statistical inference, it has been addressed well elsewhere. Spatial sampling will be touched on throughout the other chapters but will not be emphasized in this book. The second issue involves imperfect detection, the likelihood that surveys of animal populations and communities will not result in complete counts of all individuals or species present in surveyed locations. We present a general conceptual framework that relates the various count statistics obtained in studies of animal populations and communities to the true state variables of interest. Until very recently, uses of occupancy as a state variable in animal ecology have simply not dealt with the issue that failure to find evidence of a species at a location does not necessarily mean that the species does not occupy the area. The suite of models, methods, and estimators that we develop in this book is basically designed to remedy this situation and permit inferences about occupancy that deal adequately with detection probabilities less than 1.

Because most of this book focuses on parameter estimation, we returned to the "why" question and the manner in which estimates of state variables are to be used in the conduct of science and management. We briefly addressed the general question of drawing inferences about system dynamics and causal factors responsible for these dynamics. Approaches to the generation of system dynamics for the purpose of conducting science include true manipulative experimentation, constrained design manipulative studies, and observational studies using retrospective analyses. Strength of inference is greatest for manipulative experiments and weakest for retrospective analyses of time series data from observational studies. Within the category of observational studies, those used to provide confrontations with predictions of *a priori* hypotheses are much more likely to be useful than those used solely to develop *a posteriori* hypotheses.

Finally, we noted that investigators sometimes try to draw inferences about system dynamics based on static looks at spatial patterns of state variables at single points in time. Such efforts to draw inferences about process based on observation of pattern have been used with all three state variables, abundance, occupancy, and species richness. However, such efforts suffer from the ability to develop many process-based hypotheses to explain the generation of any particular pattern. Previous uses of occupancy in animal ecology have relied heavily on inferences based on statics and pattern, and we note the shortcomings of this approach. In particular, we do not view the primary purpose of this book to be provision of methods for obtaining better estimates of static occupancy patterns for use in drawing inferences about dynamic processes. Instead, we also provide methods for drawing inferences about occupancy dynamics based on data covering multiple time periods.

In Chapter 2, we consider both historical and proposed uses of occupancy as a state variable in studies of animal populations and communities. With each use, we emphasize the need to deal adequately with detection probabilities. Chapter 3 provides an elementary overview of the statistical concepts used throughout the book. Chapters 4–6 then deal with single-species occupancy studies in which multiple locations are surveyed during a single time period or "season." The parameter of interest is the probability of occupancy of a site, given that occupancy cannot always be detected. Chapter 4 presents a basic model and estimators, and includes discussion of issues such as missing data, covariate modeling, goodness-of-fit tests, and consequences of violations of model assumptions. Chapter 5 focuses on the common assumption violation of heterogeneous detection probabilities. We present mixture models that allow for variation in detection probabilities that cannot be attributed to measured covariates. Animal abundance at a site is identified as one important source of heterogeneity in detection probability. The relationship between abundance and detection probability provides a basis for estimating abundance from occupancy survey data and for estimating occupancy itself in a manner that deals with this heterogeneity. Chapter 6 deals with the important topic of study design for single-season occupancy studies for a single species.

Chapter 7 then focuses on occupancy studies conducted over multiple years or seasons for the purpose of drawing inference about occupancy dynamics for a single species. Rate of change in occupancy over time is identified as a parameter of interest, and the vital rates responsible for such change, local probabilities of extinction and colonization, are also incorporated into estimation models. Estimation, covariate modeling, assumption violations, and study design are all considered. Chapter 8 shifts emphasis to multiple species and begins with inference procedures for two species in a single year or season. Methods permit inference about dependence in probabilities of occupancy given detection probabilities that are less than 1 and that may themselves exhibit dependence on presence or detection of the other species. These methods are then extended to multiple seasons, where the emphasis shifts to possible dependence of extinction and colonization probabilities of one species on the presence of the other species.

Chapter 9 includes some suggestions about potential uses of occupancy modeling in community-level studies. One approach exploits the analogy between the different species in a local species pool and the "locations" of typical occupancy studies in order to directly estimate the fraction of the pool that is present. If multiple time periods are available, then local extinction probability and turnover can be estimated directly using this basic approach as well. The other potential use in community-level investigations involves the synthetic treatment of Dorazio and Royle (2005) in which species richness is estimated, as is the equitability component of many diversity metrics (based on relative occupancy). This work also provides a conceptual framework for considering species-area relationships. The concluding Chapter 10 contains several ideas for future work, as well as discussion linking spatial occupancy and abundance in a common framework that facilitates consideration of the relationship between range size and abundance.

CHAPTER 2

Occupancy in Ecological Investigations

This book deals primarily with the occupancy of a sampling unit by one or more species of interest. The size and nature of the sampling unit to which the term applies may be defined either naturally or arbitrarily. For example, if interest is focused on pond-dwelling amphibians, then the pond is a likely unit of interest. If a terrestrial mammal within a national park is being studied, then the units may be defined as 1,000-hectare blocks within the park. Studies of animal range over a continent might utilize the degree block as a sampling unit for occupancy. In cases where sampling units are arbitrarily, rather than naturally, defined, the size of the sampling unit should depend on the nature of the question(s) being asked. Investigations of occupancy nearly always involve interest in a number of potential sites, so the quantity of interest is sometimes the number of sites that are occupied within a larger set of interest. More frequently, interest will focus either on the proportion of sites that are occupied or on the underlying probability that a site within a group of sites is occupied (the ψ parameter of Chapter 1).

Note that there is an important distinction between "proportion of area occupied" and "probability of occupancy." The *probability* can be considered as an *a priori* expectation that a particular site will be occupied by the species as determined by some underlying process, while the *proportion* relates to a realization of that process. For example, consider a simple coin-tossing experiment. The *probability* of a "heads" is a characteristic of the coin, while the *proportion* of "heads" is determined by conducting the experiment with multiple tosses of the coin. As the probability is generally unknown, the observed proportion can be used as an estimate of the underlying probability, and often these terms are used interchangeably (we do so ourselves in this book). In many situations the distinction is not important, and, strictly speaking, most of the models we develop in this book estimate the probability of occupancy, which can be interpreted as the proportion of area occupied without penalty. However, there are other situations in which the distinction can be very important (i.e., when a large fraction of the population of interest is sampled), and in such situations we make suggestions as to how the modeling can be used to make inference directly about the proportion of area occupied.

The basic sampling protocol commonly used for occupancy estimation simply involves visiting sites and spending time within each one looking either for individuals of the species of interest or for evidence that the species is present. This kind of sampling is sometimes referred to as a "presence-absence" survey. There are no real restrictions on sampling approaches, which may include visual observations of animals, captures of animals in traps or mist nets, observations of animal tracks, detections of animal vocalizations, and even detections based on remote methods such as camera traps (Fig. 2.1) and sound recording devices such as "frog-loggers" (an automated device to record frog calls at preset intervals during a night; Peterson and Dorcas 1994). The result of such a survey is a list of surveyed sites that are "occupied" (the species detected) and "unoccupied" (the species not detected), respectively. In most historical work, counts of occupied sites are used to compute the proportion of occupied sites among all sites visited, and this proportion constitutes the estimate of occupancy.

The problem with these count-based inferences is that occupied sites may be visited, and yet no animals (or evidence of animals) may be detected. Thus, sites classified as "unoccupied" based on survey efforts may in fact be occupied. There appears to be wide recognition of this problem, with such misclassified sites referred to as false absences (Dunham and Rieman 1999), false zeros (Moilanen 2002), false negatives (Tyre *et al.* 2003), pseudo-absences (Engler *et al.* 2004) and artifactual absences (Anderson 2003). This problem arises from the issue of detectability, articulated in Chapter 1 with respect to estimation of both animal abundance and occupancy. This book contains a suite of methods for estimating occupancy and related parameters in a manner that deals explicitly with detectability and the likelihood that some sites at

FIGURE 2.1 Setting up camera traps for large mammals at a sampling station in southern India (Eleanor Briggs).

which no evidence of occupancy was found were in fact occupied by the target species.

This chapter provides an overview of different areas of investigation in ecology for which reasonable estimates of occupancy are required; hence, areas of ecology in which the techniques outlined in this book may be useful. The review provides a clear indication of the importance of occupancy to a number of important areas of inquiry in ecology. For each class of investigation, we consider the kinds of questions being asked. We then consider possible consequences of failing to deal with detectability. That is, what sorts of inferential problems could be caused by misclassifying occupied sites as unoccupied? Finally, for each class of investigation, we indicate the chapter(s) of this book that we believe may be useful in future work.

2.1. GEOGRAPHIC RANGE

Ecology is frequently described as the study of the distribution and abundance of organisms (Elton 1927; Andrewartha and Birch 1954; Krebs 2001; also see Chapter 1). Geographic range for a single species can be viewed as the primary element describing the distributional component of ecology (Brown *et al.* 1996) and has been termed "the basic unit of biogeography" (MacArthur 1972). Species ranges have long been fundamental units of analysis used to elucidate interesting ecological patterns and to address associated questions. Ranges have been the subjects of renewed interest under the macroecological research program (Brown and Maurer 1989; Maurer 1994; Brown 1995, 1999; Rosenzweig 1995; Gaston and Blackburn 1999).

Despite the fundamental importance of "range" to questions about animal and plant distributions, explicit definitions of this term are rare, even among the scientific papers for which range is a topic of investigation (Gaston 1991). Gaston (1991) presented two ways of defining geographic species range, but both definitions are based on records of individual locations. We will modify Gaston's (1991) two "definitions" by focusing on the true geographic distribution of individuals of a species rather than on sample-based counts. It seems more appropriate to define a true quantity of interest, and then to focus on methods of estimating this quantity, rather than to mix sampling and estimation problems with definitions. The two definitions of range are thus based on knowledge of the spatial locations of all individuals of a species at some time or interval of interest.

One approach, the *extent of occurrence*, involves drawing an imaginary line that encloses an area containing all of the individuals of the species. Stated in terms of occupancy, this area should include all of the sites occupied by the species. The enclosed area should be minimal in some sense, but is very much dependent on scale and on how jagged the boundary is permitted to be. In practice, boundary lines are typically "fitted by eye" (Gaston 1991), but we can imagine several objective ways of drawing such boundaries. The other way to define geographic range is termed *area of occupancy*. If a grid is superimposed on the area containing all individuals of a species, then area of occupancy is simply defined as the set of grid cells that contain at least one individual (Fig. 2.2). Area of occupancy differs from extent of occurrence in that the latter type of range may include cells that contain no individuals of the species (Gaston 1991). Both definitions of geographic range are based on occupancy, and virtually all investigations of range are based on presence-absence (more properly, detection-nondetection) data of some sort.

A number of macroecological studies have used interspecific comparisons of the sizes of species ranges. Some studies report simple patterns in the distribution of range sizes. For example, most species appear to exhibit small ranges, whereas a relatively small number of species exhibit very large ranges (Willis 1922; Rapoport 1982; Brown 1995; Gaston 1994, 1998). Many investigations have noted or searched for interspecific correlates of range size. One of the oldest such correlates was pointed out by Darwin (1859), who noted

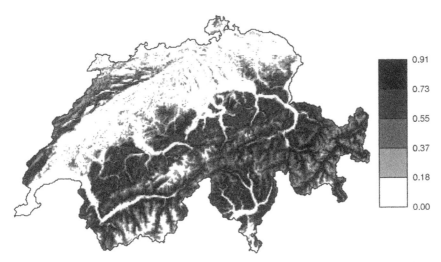

0.91
0.73
0.55
0.37
0.18
0.00

FIGURE 2.2 Estimated probability of occurrence of the Willow Tit (*Poecile montanus*) in Switzerland (Royle *et al.* 2005).

that genera containing many species tended to contain "dominant species,— those which range widely over the world, are the most diffused in their own country, and are the most numerous in individuals." This relationship has been widely investigated, leading to the general inference that within related species groups, range size appears to be related to average local abundance, such that species with large ranges tend to be abundant throughout those ranges, whereas species with small ranges also tend to be less abundant (Williams 1964; McNaughton and Wolf 1970; Buzas *et al.* 1982; Hanski 1982; Hanski *et al.* 1993; Bock and Ricklefs 1983; Bock 1984; Brown 1984, 1995; Gotelli and Simberloff 1987; Gaston and Blackburn 1996; Gaston 1996; Gaston *et al.* 1997a,b; Holt *et al.* 2002). Evidence has also been reported of a positive relationship between temporal variability of abundance and range size (Glazier 1986). Several analyses have provided evidence of a positive relationship between range size and species characteristics such as animal body mass (Brown and Maurer 1987; Brown 1995; Brown *et al.* 1996; Gaston and Blackburn 1996), dispersal capabilities of marine mollusks (Hansen 1980; Brown *et al.* 1996), and germination patterns (specifically germination niche breadth) of weedy plant species (Brandle *et al.* 2003). Geographic range size has been shown to decrease with decreasing latitude (Rapoport 1982; Stevens 1989;

Brown 1995; Brown *et al.* 1996; Rohde 1999) and decreasing elevation (Stevens 1992; Brown 1995; Brown *et al.* 1996).

Investigations of the determinants of animal distributions naturally focus on range boundaries and on the biotic and abiotic factors responsible for their locations (e.g., Bodenheimer 1938; Connell 1961; MacArthur 1972; Caughley *et al.* 1988; Root 1988ab; Gaston 1990; Repasky 1991; Brown *et al.* 1996). Macroecological research has also focused on the shapes of geographic ranges, with attention directed at such descriptors as perimeter/area ratios, fractal dimension, and north-south versus east-west orientation (Rapoport 1982; Brown 1995; Brown *et al.* 1996). The number, size, and location of holes, fragments, and other discontinuities are also features of interest with respect to range shape. In particular, the number of fragments and discontinuities tends to increase near the range periphery (Rapoport 1982; Brown 1995; Brown *et al.* 1996). Within a range, abundance typically decreases from center to periphery (Whittaker 1956; Brown 1984; Gaston 1994; Brown *et al.* 1996; Enquist *et al.* 1995), and local extinction and turnover are typically higher at the edges of ranges than in range interiors (Enquist *et al.* 1995; Curnutt *et al.* 1996; Mehlman 1997; Doherty *et al.* 2003b).

Although many of the preceding relationships and associated references reflect a view of range as a static entity, there is substantial evidence of changes in range boundaries over both geological and ecological timescales (e.g., Udvardy 1969; MacArthur 1972; Hengveld 1990; Gaston 1994; Enquist *et al.* 1995; Brown *et al.* 1996; Ceballos and Ehrlich 2002). Recent reductions in range are frequently associated with human activities and are thus of special conservation interest (Brown *et al.* 1996; Ceballos and Ehrlich 2002). A comparison of "historic" (primarily 19th century) and recent range data for terrestrial mammals indicated collective reductions in range size exceeding 70% for Africa, Australia, Europe, and Southeast Asia (Ceballos and Ehrlich 2002). Habitat fragmentation can be viewed as a problem of range reduction and has become an important conservation issue (e.g., Harris 1984; Lynch and Whigham 1984; Robbins *et al.* 1989).

Humans are also increasingly responsible for range extensions. In some cases, the spread of species to areas outside the original ranges has been inadvertent. For example, murid rodents (e.g., mice, *Mus*, and rats, *Rattus*) have been spread throughout the world via sailing vessels, and various pest insects have been introduced to new areas through imports of agricultural products. Some accidental introductions (e.g., brown treesnake, *Boiga irregularis*, on Guam; Fritts and Rodda 1998) have produced dramatic extinctions and range contractions of native species. Species introductions to new areas are sometimes intentional, such as when the Acclimatization Societies of New Zealand introduced various animal species in order to provide additional hunting and fishing opportunities (e.g., Thomson 1922; Williams 1981). Regardless of original motivation, introductions of species into new areas by humans have led to many so-called invasions, and invasive species now present an important challenge to conservation (Mooney and Drake 1986; DiCastri *et al.* 1990; Williamson 1996). Occupancy data collected over large spatial scales for multiple years are commonly used to estimate rates of spread of exotic species (e.g., Havel *et al.* 2002; Wikle 2003). Reintroductions and translocations are considered as a conservation tool as well and usually reflect attempts to restore species to previous ranges (Griffith *et al.* 1989; Chivers 1991).

Disease statics and dynamics can also be viewed as problems of range. Vaccinations recommended for humans traveling to various parts of the world are based on range maps of disease occurrence. Disease dynamics are frequently of great interest to epidemiologists, especially in the case of fast-spreading diseases such as West Nile virus (Marra *et al.* 2004). Epidemiological models for disease dynamics have been developed to predict the spread of disease organisms across host organisms and, more generally, across space (e.g., Bailey 1975; Anderson and May 1991; Elliott *et al.* 2001).

Regardless of whether range is defined as *extent of occurrence* or *area of occupancy* (Gaston 1991), the failure to detect species that are actually present in a patch or sample unit will result in biased estimates of range size and location. Range size and extent will tend to be underestimated when species detection probabilities are less than 1, with the magnitude of bias a function of sampling intensity (McArdle 1990; Anderson 2003). Perhaps a more serious problem involves the possibility of spurious relationships between range attributes and other quantities of interest that are induced by imperfect detection. For example, consider the interspecific relationship between range size and average abundance. Brown (1984:264) put forward the hypothesis that the correlations of range size and abundance that he observed could be "simply the result of statistical sampling processes." He then rejected this hypothesis, but readers may not find his reasoning to be convincing. Buzas *et al.* (1982:149) clearly recognized the difficulties in exploring the relationship between abundance and range size for fossil data: "Consequently, with rarely occurring species we are in a no-win situation. Because of the difficulties in detecting the presence of such a species, we cannot know whether or not its distribution is restricted to the locality where it is found or whether it is widespread, but undetected. The dilemma for the naturalist is obvious."

The problem arises because the probability of detecting a species in a location is often a direct function of the number of individuals of the species inhabiting the location. Formal expressions for this relationship are presented by Royle and Nichols (2003) and in Chapter 5, but the relationship is very intuitive. The consequence of this relationship is that occupancy will go undetected more frequently for species at low abundance than for abundant species, inducing the very relationship about which so much has been written. Of course this relationship between abundance and detection probability for occupancy does not mean that all inferences about the abundance-occupancy relationship are incorrect. It simply means that the evidence used to support this relationship is not very useful for this purpose. In addition, the relationship between abundance and detection probability will result in biased estimates of parameters specifying the relationship between occupancy and abundance. Such bias will likely have consequences for efforts to estimate abundance from raw occupancy data (Kunin 1998; Kunin et al. 2000; He and Gaston 2000, 2003; Warren et al. 2003; Tosh et al. 2004).

Other inferences about range may be influenced by detection probability as well. For example, the inference about increased numbers of fragments and discontinuities near range boundaries could easily be induced by changes in detection probabilities caused by decreased abundance near range boundaries. Local extinctions and, in some situations, turnover tend to be overestimated using raw detection data when detection probabilities are less than 1 (e.g., Nichols et al. 1998a). Thus, the inferences about increased extinction probabilities and turnover near range boundaries could be induced by decreased detection probabilities produced by low abundances near boundaries (Doherty et al. 2003b; Alpizar-Jara et al. 2004). In fact, virtually any gradient in abundance can produce an apparent gradient in occupancy based on raw detection data. Of course, factors other than abundance may influence detection probability as well. For example, individual animals may simply be more visible or detectable in one habitat than another, providing the potential for misleading inference about the true relationship between occupancy and habitat. Similarly, sampling effort may be higher in some areas than others (e.g., based on proximity to museums) for logistic or other reasons, again leading to the possibility of differences in occupancy in different parts of a range that do not reflect true differences in range.

The methods provided in this book can be used to address the relationships and topics reviewed above in a manner that is not nearly so vulnerable to problems produced by detection probabilities as use of raw detection-nondetection data. Methods for estimating occupancy, and thus range size, for single species are considered in Chapters 4, 5, and 6. Chapter 5 contains specific methods for dealing with the relationship between abundance and detection probability when estimating occupancy, and Chapter 10 includes discussion of the true relationship between abundance and occupancy. The methods in Chapters 4–6 permit not only the estimation of occupancy but also the estimation of parameters describing relationships between occupancy and other quantities of interest. Because detection probabilities are explicitly incorporated into the models, the procedures are not vulnerable to spurious influence by variation in detection probabilities. The methods in Chapter 7 extend the analyses to multiple years or seasons, permitting inference about changes in occupancy and range over time and about the probabilities of local extinction and colonization responsible for these changes.

2.2. HABITAT RELATIONSHIPS AND RESOURCE SELECTION

Relationships dealing with range size and configuration are necessarily very general in nature. Another approach to the investigation of animal distribution patterns involves the concept of the ecological niche (e.g., Hutchinson 1957) and the idea of each species having a unique set of requirements that must be provided by the habitat(s) in order for the species to persist there. Ecologists thus seek to identify the key habitat variables to which a species responds and to develop habitat models that can be used to predict abundance, or at least occupancy, of a particular species as a function of habitat characteristics (e.g., see Verner et al. 1986; Scott et al. 2002). Much of this work involves assessing species presence or absence on sample units and then asking whether presence can be modeled as a function of habitat characteristics measured on these units—in other words, what habitat variables best discriminate between locations that are and are not occupied by the species (e.g., Hirzel et al. 2002). This approach has been used to model habitat relationships of many vertebrates (e.g., amphibians: Johnson et al. 2002; mammals: Carroll et al. 1999; Reunanen et al. 2002; birds: Robbins et al. 1989; Klute et al. 2002; Tobalske 2002; Gibson et al. 2004; fish: Dunham and Reiman 1999; Dunham et al. 2002; reptiles: Fischer et al. 2004), invertebrates (e.g., Hanski et al. 1996; Fleishman et al. 2001; Wahlberg et al. 2002), and plants (e.g., Fertig and Reiners 2002; Edwards et al. 2004).

Many of these habitat studies based on occupancy have been directed at conservation and management and have focused on habitat conservation and habitat change resulting from such factors as human land use and climate change. For example, a number of occupancy studies have provided evidence of the importance of area of woodland and forest habitat to occupancy of certain bird species (e.g., Moore and Hooper 1975; Forman et al. 1976; Lynch and Whigham 1984; Robbins et al. 1989). These and related studies have led to recommendations about both habitat conservation and the design of nature reserves (e.g., Diamond 1975a, 1976; Wilson and Willis 1975; Robbins et al. 1989; Cabeza et al. 2004). Some studies have focused on anthropogenic determinants of occupancy such as pollutants (e.g., acid rain, Hames et al. 2002) and even human density (Chown et al. 2003).

Resource selection functions comprise a large number of specific methods that can be used in different applications (e.g., see Manly et al. 2002), with one application being to assess how a species uses resource (or sampling) units within an area of interest, rather than addressing the resource use of individual animals (i.e., in the terminology of Manly *et al.* 2002, study design I rather than designs II or III). In the resource selection context, the intent typically is to identify the relative level of use of different types of resource units. It is appropriate to consider some resource selection problems as an exercise in studying the relationship between occupancy and habitat, as the basic field protocols and analytic methods are usually very similar to those used above (MacKenzie 2006).

The studies cited above are representative of the work investigating habitat relationships with "presence-absence" data. Virtually all of this work includes false absences in which sites are occupied by the species of interest, but the species goes undetected. Approaches such as logistic regression are expected to yield biased results when applied to species presence-absence data in which the species is not truly absent from all sites at which it goes undetected (Hirzel *et al.* 2002; Tyre *et al.* 2003; Gu and Swihart 2004). Gu and Swihart (2004:195) studied this issue via simulation and found that "logistic regression models of wildlife-habitat relationships were sensitive to even low levels of nondetection in occupancy data." Bias in estimates of parameters specifying habitat relationships are expected to be greatest when detection probabilities are themselves related to the habitat variables of interest (Gu and Swihart 2004; Fig. 2.3). The bias results as logistic regression simply model the relationship between habitat and where the species is *found* (a combination of occupancy and detectability), not where the species *is* (occupancy). Such bias was demonstrated in an example presented by MacKenzie (2006; see also Chapter 4, Section 4.4) in a resource selection context. Data on the use of sites by pronghorn antelope (*Antilocapra americana*) collected during two winters were analyzed using two different methods. Assuming that resource/habitat use did not change between winters, simple logistic regression (where detection in at least one winter indicated a site was "used," and nondetection was equated to "unused") revealed that distance from a water source appeared to be the most important habitat variable, but when using a method that accounted for detection probability, slope and sagebrush density were the variables considered most important for habitat use. Distance from a water source was, however, found to be the most important variable in terms of detection probability, which would explain why simple logistic regression identified it as an important variable. In addition, approaches that ignore the issue of imperfect detections should yield variance estimates that are too small, as they do not incorporate all components of uncertainty (e.g., detection probabilities <1).

The models discussed in Chapters 4–6 of this book permit the same kind of logistic regression modeling, but this modeling is accomplished in a manner that accounts for the fact that species are not always detected when present.

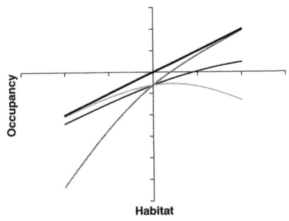

FIGURE 2.3 Illustration of the effect of imperfect detection on the apparent relationship between "occupancy" and a habitat variable. The thick black line represents the true relationship (on the logistic scale); the thin black line indicates the apparent relationship when detection probability is constant; the thin dark-grey line indicates the apparent relationship when detection probability positively covaries with habitat; and the thin light-grey line indicates the apparent relationship when detection probability negatively covaries with habitat.

Estimates of parameters describing habitat relationships should exhibit little bias, and variance estimates should appropriately incorporate detection uncertainty. Moving beyond static habitat-occupancy relationships, the models of Chapter 7 permit the modeling of both changes in occupancy and vital rates responsible for those changes as functions of either static habitat variables or changes in habitat variables. Chapter 8 permits models of habitat relationships that also account for influences of other species (e.g., competitive or predation interactions) on both occupancy and detection probability.

2.3. METAPOPULATION DYNAMICS

Subdivided populations connected by some degree of movement were first considered by Wright (1931, 1951), who focused on the genetic structure produced by such systems. Genetic structure in such a system will be partially determined by the nature of migration among subpopulations, with different structures expected to result from different patterns of migration (e.g., Wright's "island model" as opposed to the "stepping-stone" and more general "isolation by distance" models of Kimura and Weiss 1964 and Crow and Kimura 1970). Ecologists Andrewartha and Birch (1954) focused on local extinctions and colonizations as determinants of population dynamics in such systems.

Levins (1969, 1970) formalized the concept of a metapopulation, a system of subpopulations, any of which can go extinct and later become recolonized. The metapopulation concept is now thought to provide a useful description of a large number of natural populations, and the literature on metapopulations has become substantial (e.g., see McCullough 1996; Hanski and Gilpin 1997; Hanski 1999; Hanski and Gaggiotti 2004).

Metapopulation structures involve systems of patches or sites that are sometimes occupied by the species of interest and sometimes not, depending on the dynamic processes of extinction and colonization. Indeed, the number or proportion of sites that are occupied is the state variable of interest in the original models of Levins (1969, 1970) and in various subsequent models of metapopulations (e.g., Lande 1987; Hanksi 1991, 1997, 1999). Thus, patch occupancy is of primary importance in investigations of metapopulations, as are local rates of extinction and colonization, the vital rates responsible for changes in occupancy. Until the last two to three years, occupancy estimation was viewed as a simple problem and was based on a binomial estimator of patches at which the species was detected, divided by total patches surveyed. The literature of metapopulation ecology has been more focused on estimation of local rates of extinction and colonization and on functional relationships between these vital rates and patch characteristics. Two primary approaches have been used to draw inferences about extinction and colonization, and they are distinguished by the temporal scale of the occupancy data used in estimation. One approach is based on static detection/nondetection data for multiple sites collected from a single season, whereas the other is based on detection/nondetection data from multiple sites collected in multiple seasons.

INFERENCE BASED ON SINGLE-SEASON DATA

The single-season approach to estimation of rates of extinction and colonization precedes interest in metapopulations and has its origins in community ecology. Diamond (1975a) introduced the concept of "incidence functions," in which the probability of occurrence of bird species on islands was modeled as a function of factors (e.g., island area, distance between island and mainland) hypothesized to be relevant to species-level vital rates and, hence, to species distributions. Hanski (1991, 1992) used incidence functions to describe single-species metapopulations, relating patch occupancy to patch characteristics hypothesized to influence probabilities of patch extinction and colonization. If the metapopulation system is in dynamic equilibrium, if the primary patch characteristics affecting extinction and colonization have been identified, and if the functional forms of these relationships have been adequately specified, then data on the fraction of patches occupied at a single point in time can be used to estimate the parameters defining these relationships, as well as the extinction and colonization probabilities themselves (Hanski 1991, 1992, 1994a, 1999). Hanski (1992:660) has noted that direct study of the extinction and colonization processes is "expensive and time-consuming," in contrast to incidence functions "constructed from 'snap-shot' information collected during one sampling period." However, the decreased expense associated with single assessments of pattern comes at a cost, as attempts to infer process from pattern can easily lead to incorrect inferences (see Chapter 1; also Clinchy et al. 2002).

Early work on incidence function estimation of extinction and colonization assumed that colonization probability was a constant for all patches but that extinction probability decreased as a function of patch area (e.g., Hanski 1991, 1992). Substantial interest was focused on the coefficient specifying the rate of decrease in extinction probability with increasing patch area. Later work then expressed colonization rate for specific patches as a function of distances to, and occupancy of, other patches (Hanski 1994a&b, 1998, 1999). The "rescue effect" of Brown and Kodric-Brown (1977) was considered by including the possibility of recolonization in the extinction function (Hanski 1994a, 1999).

In addition to Hanski's (1991, 1992, 1994a&b, 1999; Hanski et al. 1995; Wahlberg et al. 1996, 2002) direct estimation of vital rates and of parameters describing relationships between these rates and patch characteristics, several authors have drawn inferences about colonization and extinction processes indirectly by investigating the relationship between occupancy and patch characteristics. This work is thus very similar to that described above for habitat relationships, except that the patch characteristics of interest are those hypothesized to be important determinants of colonization (e.g., patch isolation) and extinction (e.g., patch area) within a metapopulation context. For example, one prediction is that, other characteristics being roughly equal, occupancy should increase as a function of patch area because of the reduced extinction probabilities of large patches. Support for this prediction comes from several studies of metapopulation systems (e.g., Lomolino et al. 1989; Peltonen and Hanski 1991; Hanski 1992, 1998; Hanski et al. 1995, 1996; Thomas and Hanski 1997; Smith and Gilpin 1997; Dunham and Rieman 1999; Wahlberg et al. 2002; Bradford et al. 2003). Occupancy has been modeled as a function of various measures reflecting patch isolation (Hanski's [1994a, 1999] connectivity measures, distance to nearest occupied patch, etc.), with the prediction that greater isolation should reduce colonization, leading to lower occupancy. Most results have provided evidence of a negative relationship between occupancy and isolation, although such evidence is not always found (Lomolino et al. 1989; Peltonen and Hanski 1991; Hanski et al. 1995; Whitcomb et al. 1996; Smith and Gilpin 1997; Thomas and Hanski 1997;

Hanski 1998; Dunham and Rieman 1999; Clinchy *et al.* 2002; Wahlberg *et al.* 2002).

Studies of metapopulation dynamics based on single-season analyses have also included information on habitat availability. Lande (1987, 1988) extended the Levins' (1969, 1970) metapopulation models by designating some fraction of patches that were suitable for the species. Remaining patches were unsuitable and could not be occupied. Lande (1987) was able to compute a threshold proportion of suitable patches (termed the "extinction threshold"), such that systems with a smaller proportion of suitable patches were doomed to extinction. This type of joint occupancy and habitat modeling has received substantial attention (e.g., Ovaskainen *et al.* 2002; Merila and Kotze 2003) and has been applied, for example, to Northern spotted owl (*Strix occidentalis caurina*) populations in fragmented habitat (Lande 1988; Noon and McKelvey 1997). Hanski and Ovaskainen (2000) extended the concept and defined "metapopulation capacity," a metric reflecting the relative capacities of different landscapes to support viable metapopulations.

This thinking has been extended to model habitat destruction as a specified fraction of suitable patches that is suddenly destroyed such that they become permanently unsuitable. Such models lead to inferences about "minimum viable metapopulations," "minimum amount of suitable habitat," and metapopulation capacity (Hanski *et al.* 1996; Hanski and Ovaskainen 2000), as well as to related inferences about reserve designs that maximize time to metapopulation extinction (Ovaskainen 2002) or minimize the probability of extinction over a finite time frame (Moilanen and Cabeza 2002; Cabeza *et al.* 2004). Some work on reserve design has used probability of occurrence as a surrogate for local persistence (Araujo and Williams 2000; Williams and Araujo 2000; Araujo *et al.* 2002). A modeling effort by Ellner and Fussmann (2003) emphasized the potential importance of within-patch succession to metapopulation persistence. Such models have also been used to explore interactions among species in a community, such as competitive coexistence in two-species systems (Nee and May 1994; Moilanen and Hanski 1995) and persistence in predator-prey systems (Nee 1994). All of these modeling efforts have not necessarily involved parameter estimation, but they are included in this section on single-season data because they employ the same models that are used for estimation based on incidence functions.

The described use of incidence function data to estimate rates of extinction and colonization assumes that incidence (probability that a patch is occupied) is known or estimated without bias. Failure to detect a species when present will result in negatively biased estimates of proportional occupancy and thus in biased estimates of rates of both extinction and colonization. Moilanen (2002:524) studied three problems with data used in conjunction with incidence functions to estimate vital rates from detection-nondetection data and concluded that "all effects seen so far pale in comparison with biases caused by false zeros in the data set." He reported the possibility of substantial positive bias in estimated rates of both extinction and colonization (Moilanen 2002). Functional relationships such as those between extinction probability and patch size and between colonization probability and patch isolation will be estimated with bias as well (Moilanen 2002).

Biased estimates of proportional occupancy also influence inferences based on joint consideration of occupancy and proportion of habitat that is suitable. In addition to the problems caused by use of biased estimates of occupancy, the proportion of suitable patches may not be known in all cases, creating additional problems. For example, some insects are restricted to feeding on particular host plant species (e.g., Hanski et al. 1995; Wahlberg et al. 1996, 2002), as are various types of parasites to feeding on particular species of animals (Deredec and Courchamp 2003). We would not expect detection probabilities for habitat patches to be 1 in these situations. Such metrics as extinction threshold, metapopulation capacity, minimum viable metapopulation, and minimum amount of suitable habitat will all be estimated with bias in the presence of nondetections of both occupancy and site status (suitable and not suitable).

The models discussed in Chapters 4–6 of this book permit unbiased estimation of occupancy in a manner that accounts for detection probabilities less than 1. These estimates could be incorporated into the incidence function modeling of Hanski (1991, 1992, 1994a, 1998, 1999) as a means of obtaining improved estimates of rates of colonization and extinction and of the parameters governing the relationships between these rate parameters and patch characteristics such as size and isolation. We prefer to avoid assumptions about stationarity and known relationships with patch characteristics and to instead estimate rates of local extinction and colonization from detection history data that cover multiple seasons or time periods (Chapter 7). Using such an approach, we can view stationarity and relationships between rate parameters and patch characteristics as hypotheses to be tested, rather than as assumptions required for estimation.

We can envision situations in which proportion of suitable habitat should be treated as an occupancy estimation problem. In such cases, detection-nondetection sampling might be used for estimating both occupancy of the species of interest and presence and suitability of habitat. For example, in the case of host-specific parasitic animals or insects feeding on particular host plant species, we might use the methods of Chapters 4–6 to estimate occupancy for both the focal and host species, with host species occupancy providing an estimate of the proportion of suitable patches. The methods of Chapter 8 permit simultaneous inference about occupancy for both focal and host species.

INFERENCE BASED ON MULTIPLE-SEASON DATA

Detection-nondetection data for a set of patches over seasons or years can be used to directly estimate local extinction and colonization probabilities without requiring assumptions of system equilibrium or stationarity. In most previous work with such data, extinction rate between two seasons or years, t and $t + 1$, has been estimated by conditioning on a set of patches at which the focal species is detected at time t, and then computing the fraction of these at which the species is not detected at time $t + 1$. Similarly, colonization rate has been estimated by conditioning on the set of patches at which the focal species is not detected at time t, and then computing the fraction of these at which the species is detected at time $t + 1$. In addition to direct estimation of these rate parameters, logistic regression and related approaches have been used to relate "observed" extinctions and colonizations to covariates such as patch area and degree and isolation. These approaches to inference (summarized by Morris and Doak 2002) have been used in a variety of study situations including for shrews (*Sorex* spp.) on islands in Finnish lakes (Peltonen and Hanski 1991); pikas (*Ochotona princeps*) inhabiting ore dumps created by miners in California (Smith and Gilpin 1997; Clinchy *et al.* 2002); various plant species in European and North American study areas (Kéry 2004 and references therein); European nuthatches (*Sitta europaea*) in woodlots in the Netherlands (Verboom *et al.* 1991); passerine bird species in 10 km × 10 km grid cells in Great Britain (Araujo *et al.* 2002); scarlet tanagers (*Piranga olivacea*) in North American forest fragments (Hames *et al.* 2001); various breeding bird species in woodlots of southern Finland (Haila *et al.* 1993); plant species found on serpentine seeps in California (Harrison *et al.* 2000); several amphibian species inhabiting ponds in southwestern Ontario (Hecnar and M'Closkey 1996), southeastern Michigan (Skelly *et al.* 1999), and a montane park in central Spain (Martinez-Solano *et al.* 2003); pool frogs (*Rana lessonae*) in permanent ponds along the Baltic coast of Sweden (Sjogren-Gulve and Ray 1996); frogs and toads at survey sites throughout Wisconsin (Trenham *et al.* 2003); and Glanville fritillary butterflies (*Melitaea cinxia*) in dry meadow patches of southwestern Finland (Hanski 1997; Thomas and Hanski 1997).

Some workers have extended the conditional binomial modeling of detection-nondetection data to develop methods for estimating local extinction and colonization parameters using a sequence of such data over time, over species, or over multiple locations. Rosenzweig and Clark (1994) obtained maximum likelihood estimators for Markov process models governed by extinction and colonization rate parameters that corresponded to the period separating successive samples. Clark and Rosenzweig (1994) extended the estimation to the more difficult problem in which detection-nondetection sampling occurs at irregular intervals that do not always correspond to the time period for which the rate parameters are defined. Erwin *et al.* (1998) extended this modeling to include estimation for ultrastructural models of rate parameters as functions of site-specific covariates. Ter Braak and Etienne (2003) applied a Bayesian approach to parameter estimation under Markov modeling of detection-nondetection data. Other approaches to estimation under this kind of patch occupancy modeling include Verboom *et al.* (1991) and Ferraz *et al.* (2003).

Estimates obtained using any of the above approaches generally will be biased by failure to detect the species at all occupied patches. Extinction rate estimates based on conditional binomial models should always be biased high, as some patches will appear to represent extinctions when this is not the case (see Kéry 2004). Parameters expressing the relationship between extinction rate and covariates such as population size can also be biased when detection probability is not accounted for (Kéry 2004). Binomial estimation of colonization is conditional on patches at which the species is not detected, and the species will really be present in some of these patches. Similarly, the species may be present, yet undetected, in the second sampling period as well. The result is that estimates of colonization probability will tend to be biased. Estimation based on the full Markov models will similarly tend to yield biased estimates of the rate parameters.

The methods of Chapter 7 were developed to permit estimation under the types of Markov models considered by others (e.g., Clark and Rosenzweig 1994; Erwin *et al.* 1998), with the addition of permitting detection probabilities to be less than 1. The discussions of proportion of suitable habitat in the section on single-season estimation bring up an interesting possibility for multiple-season models. We can consider modeling habitat dynamics as well (Chapter 10), either with or without the assumption that suitable habitat can always be detected perfectly. Markov models of habitat suitability would permit estimation of transition probabilities associated with habitat changes. The species occupancy dynamics could then include constraints reflecting habitat changes. For example, if an occupied patch at time t made the transition to an unsuitable patch at time $t + 1$, then the species extinction probability is known to be 1 and would be so constrained for that time interval. If unsuitable habitat can be occupied, but with lower probability than suitable habitat, then the multiseason models for two species (Chapter 8) could be modified or used directly to explore the dependency of the focal species on habitat suitability. We believe that joint modeling of habitat dynamics and patch occupancy dynamics will be an interesting and useful extension of the work presented in this book.

2.4. LARGE-SCALE MONITORING

Large-scale monitoring programs have become popular in recent years. In some cases, monitoring is tied closely to management decisions (e.g., Nichols

1991a; Nichols *et al.* 1995), whereas in other cases, the reasons underlying monitoring efforts are not as clear (e.g., see comments by Krebs 1991; Yoccoz *et al.* 2001). Monitoring efforts that are not so well defined are frequently viewed as surveillance tools and, as such, are focused on detection of change in animal populations. Often, density or abundance is the state variable for which estimates of change are sought. However, estimation of density and abundance often requires substantial effort (e.g., Lancia *et al.* 1994; Pollock *et al.* 2002), leading some to view occupancy as a surrogate for abundance. In territorial species, when sample units are selected to be the approximate sizes of territories, estimates of the number of occupied sites are virtually equivalent to estimates of numbers of territorial animals or pairs (e.g., spotted owls *Strix occidentalis*, Azuma *et al.* 1990; MacKenzie *et al.* 2003). Indeed, previous workers have used presence of an individual animal in a territory as a "mark" and applied capture-recapture models to detection-nondetection data from repeat samples as a means of estimating population size (e.g., Hewitt 1967 for red-winged blackbirds, *Agelaius phoeniceus*; Thompson and Gidden 1972 for American alligators, *Alligator mississippiensis*).

Even for species that are not territorial, occupancy is sometimes viewed as a surrogate for abundance in monitoring programs (e.g., Bart and Klosiewski 1989). Occupancy is clearly related to abundance, as it focuses on one tail of the distribution of abundance of animals across space or patches (i.e., the portion of the distribution associated with the probability a patch contains one or more individuals: $Pr(N_i > 0)$, where N_i is abundance for patch i). Others view occupancy not as a surrogate for abundance but as an appropriate state variable for large-scale monitoring (Hall and Langtimm 2001; Manley *et al.* 2004). In other cases, occupancies of multiple species have been summed, and species richness has been used as a state variable (e.g., Martinez-Solano *et al.* 2003; Weber *et al.* 2004).

Detection-nondetection surveys have been recommended or used for various large-scale monitoring programs. Such surveys are currently used throughout Switzerland to monitor multiple vertebrate and invertebrate taxa (Weber *et al.* 2004). They have been used and recommended for amphibians in various regions (Hall and Langtimm 2001; Martinez-Solano *et al.* 2003; Trenham *et al.* 2003; Bailey *et al.* 2004). Occupancy surveys of potential territory sites are used to monitor spotted owls (e.g., Azuma *et al.* 1990; MacKenzie *et al.* 2003) and marbled murrelets (*Brachyramphus marmoratus*) (Stauffer *et al.* 2002) in northwestern North America. Camera-trapping and track plates have been used by the U.S. Forest Service to monitor occupancy of fishers (*Martes pennanti*) and martins (*Martes americanus*) in the Klamath region of the western United States (Zielinski and Stauffer 1996; Carroll *et al.* 1999). The New Zealand Department of Conservation has designed a pilot program for monitoring occupancy of an endangered insect, the Mahoenui giant weta

(*Deinacrida mahoenui*), on the North Island of New Zealand (MacKenzie *et al.* 2004a, 2005). Zonneveld *et al.* (2003) discuss the design of occupancy surveys for insect species such as the endangered Quino checkerspot butterfly (*Euphydryas editha quino*). Occupancy surveys based on animal sign (e.g., tracks, scat) have been recommended for large-scale monitoring of tigers (*Panthera tigris*; Figure 2.4) and their prey species in India (Nichols and Karanth 2002). The count-based North American Breeding Bird Survey (Robbins *et al.* 1986) is sometimes analyzed from an occupancy perspective that focuses on whether or not a species is detected at a stop or route (Bart and Klosiewski 1989).

Biased estimates of occupancy resulting from failure to detect animals that are present have the potential to lead to biased estimates of change in occupancy over time. In particular, temporal variation in detection probability will be confounded with any true temporal variation in studied populations. The problems associated with variable and unknown detection probabilities in occupancy monitoring programs generally have been recognized (e.g., Azuma *et al.* 1990; Zielinski and Stauffer 1996; Hall and Langtimm 2001; Nichols and Karanth 2002; Stauffer *et al.* 2002; Zooneveld *et al.* 2003; Bailey *et al.* 2004;

FIGURE 2.4 Camera-trap photograph of a tiger (*Panthera tigris*) from southern India (K. Ullas Karanth, Wildlife Conservation Society).

Kawanishi and Sunquist 2004; MacKenzie *et al.* 2003, 2004a, 2005). As with monitoring other state variables (Yoccoz *et al.* 2001; Pollock *et al.* 2002), the solution to this problem is simply to incorporate detection probabilities as parameters in models used to estimate change in occupancy over time. The models of Chapter 7 were developed for this purpose and can be parameterized to directly estimate rate of change in occupancy over time.

2.5. MULTISPECIES OCCUPANCY DATA

Most of the occupancy applications discussed thus far have focused on questions related to single species ecology. They can also be applied to studies of multiple species in which each species is considered independently. However, often occupancy data are also used to address questions about possible interactions or other relationships among multiple species. As was the case for metapopulation studies, investigations of multiple species can be classified into two types of studies: those based on static patterns of occupancy and those based on occupancy dynamics.

INFERENCE BASED ON STATIC OCCUPANCY PATTERNS

Investigations of static occupancy patterns for multiple species may be based on data from a single season or on data accumulated over years as species range maps. In both cases, analyses of patterns are used to draw inferences about the underlying dynamics that produced them. Use of occupancy data from multiple species to draw inferences about species interactions also has a long history in ecology (Chapter 1). Consider a two-species system, and assume that survey efforts to detect occupancy have been conducted on a set of sample units. The data obtained from such a survey are typically viewed as a contingency table, with both species having been detected on some units, neither species on other units, and still other units with detections of only species A and others with only species B. The null hypothesis of independence of species occurrence is then tested with a 2×2 contingency table χ^2 test, and interaction indices have been developed to reflect the strength of the alternative hypotheses that species co-occur more or less frequently than expected (see Forbes 1907; Dice 1945; Cole 1949; Pielou 1977; Hayek 1994). Similar thinking has been applied to the analysis of more complex multispecies occupancy data (Table 2.1). In such multispecies systems, "null models" are typically developed to deduce occupancy patterns expected under a null hypothesis of independence, or no interspecific interactions (Connor and Simberloff 1979, 1984, 1986; Simberloff and Connor 1981; Diamond 1982;

TABLE 2.1 Species Incidence Matrix Specifying the Occurrence (1) and Nonoccurrence (0) of M Different Species at Each of s Specific Sites

Site	Species						
	1	2	3	4	5	...	M
1	1	0	0	1	0	...	1
2	1	1	0	0	1	...	0
3	0	1	1	1	0	...	0
4	1	0	0	0	1	...	1
5	1	1	0	1	0	...	0
...	...	...	...	...	...	...	...
s	1	0	0	1	1	...	0

Diamond and Gilpin 1982; Gilpin and Diamond 1982, 1984; Stone and Roberts 1990, 1992; Kelt et al. 1995; Manly 1995; Gotelli and Graves 1996; Gotelli 2000; Gotelli and McCabe 2002).

A potential problem with attempts to draw inferences about interspecific interactions from species incidence (presence-absence) matrices involves other factors (e.g., habitat preferences and physiological tolerances) that are likely to result in nonrandom patterns of species co-occurrence, yet have nothing to do with interspecific interactions. This class of problem is inherent in all attempts to draw inferences about process based on pattern and has been recognized in previous efforts to analyze species incidence matrices (e.g., Connor and Simberloff 1984; Gilpin and Diamond 1984). One approach to dealing with such factors is to identify them a priori and incorporate them into analyses. For example, regression models have been developed to predict detections of one species as a function of both habitat variables and detections of other species (Schoener 1974; Crowell and Pimm 1976). Other approaches have incorporated geographic and habitat characteristics directly into null models (Kelt et al. 1995; Gotelli et al. 1997; Peres-Neto et al. 2001).

Static occupancy data collected for multiple species have also been used to draw inferences about relationships among multispecies extinction and colonization processes that do not involve interspecific interactions. For example, occupancy data for insular locations can be tested for the existence of a "nested subset" structure, such that smaller biotas contain a nonrandom subset of larger ones (Patterson and Atmar 1986; Patterson 1987, 1990; Bolger et al. 1991; Wright and Reeves 1992; Andren 1994; Cook and Quinn 1995, 1998; Lomolino 1996; Wright et al. 1998; Fleishman and Murphy 1999). Common ecological explanations for the existence of such patterns involve differences among species in susceptibility to extinction and/or ability to colonize vacant habitat. The subset of species found in the locations of lowest richness (e.g., the island at the distal end of an archipelago, or the island most distant from the mainland) may be those species with the lowest extinction probabilities or the highest rates of colonization. As in the case of testing for evidence of interspecific interactions using species incidence data, substantial effort has been devoted to statistical methods providing inferences about existence of a nested pattern in a set of insular occupancy data for multiple species (Patterson and Atmar 1986; Wright and Reeves 1992; Cook and Quinn 1998; Wright et al. 1998; Cam et al. 2000b). However, only Cam et al. (2000b) used an approach for inference about nestedness that explicitly dealt with species detection probabilities. This approach is based on estimating the fraction of species in one sample unit that is present at another, despite not detecting all species at either site (Nichols et al. 1998b; Cam et al. 2000b).

We believe that detection probabilities can have an important influence on results of observed species co-occurrence patterns based on occupancy data (MacKenzie et al. 2004b; Chapter 8). Species frequently have different detection probabilities. In addition, we have encountered several field situations where detection probability of species A is actually thought to depend on presence, or even detection, of species B. For example, in occupancy surveys of northern spotted owls in the Pacific northwestern United States, biologists have hypothesized that presence of larger barred owls (Strix varia) in or near northern spotted owl (Strix occidentalis caurina) territories may cause northern spotted owls to vocalize less frequently. With respect to past investigations using multispecies occupancy data, we find it surprising that such a large amount of attention has been devoted to the statistical issues associated with inferences about patterns indicative of interactions, while virtually no attention has been devoted to the well-known problem that species are not always detected when present.

It is clear that nested subset analyses that do not account for detection probability can lead to poor inferences. For example, the detection (or nondetection) of a species on an island may be as much a function of detection probability (high or low) as of the ecological processes of local extinction and colonization. The statistical approaches developed to investigate nestedness rely heavily on the appearance of zeros or "gaps" in species incidence matrices. Such gaps can simply represent failures to detect species that are present, a possibility that has been recognized (Grayson and Livingston 1993; Kodric-Brown and Brown 1993) but not dealt with (except by Cam et al. 2000b). Single-season occupancy data can thus be used to estimate occupancy for each species separately at each location using the methods of Chapters 4–6. These occupancy probabilities can then be used to draw inferences about the possibility of a nested structure in a manner not confounded by variation in detection probabilities (Chapter 9). It is not clear whether an approach based on occupancy estimates for each species will necessarily perform better than the approach of Cam *et al.* (2000) based on occupancy data aggregated over species. However, it is clear that a single-species occupancy approach will permit more flexibility in modeling.

The models of Chapter 8 were developed specifically for multispecies occupancy surveys, in which interest is on species-specific occupancies and/or evidence of spatial segregation or aggregation among species. The models permit inferences about these issues in the presence of detection probabilities that are less than 1 and that may vary by species and depend on presence and even detection of other species. Chapter 8 also considers multispecies occupancy data collected over seasons or years. For such data, attention shifts from single-season co-occurrence patterns to whether the vital rates of occupancy dynamics, local extinction and colonization, of one species depend on the presence of another. Detection probabilities are also incorporated into this modeling.

INFERENCE BASED ON OCCUPANCY DYNAMICS

Attempts to deduce process from observation of pattern are inherently difficult (Chapter 1 and above). The use of multispecies occupancy data from several points in time (e.g., years) may be more useful for drawing inferences about processes associated with interspecific interactions, especially when such studies include an event hypothesized to alter community organization. For example, Sanders *et al.* (2003) collected annual occupancy data for members of a native ant community on sample plots in northern California for a period of seven years. Some of the sampled areas were invaded by a nonnative species, the Argentine ant (*Linepithema humile*), during the course of the study. On areas not invaded by the Argentine ant, occupancy data for native ant species showed evidence of species segregation consistent with a hypothesis of competitive interactions. However, on areas that were invaded, native ants showed evidence of segregation before invasion, yet appeared to co-occur randomly or even with aggregation following invasion (Sanders *et al.* 2003). This study thus provided relatively strong inferences about effects of the invasive species on community organization. Although relatively rare, experimental manipulations have been used to directly investigate the roles of interspecific interactions and habitat in species incidence patterns using occupancy data (Syms and Jones 2000).

Multispecies occupancy data collected over multiple seasons or years can also be used to draw inferences about rates of extinction and colonization in the absence of any hypothesis about interspecific interactions. For example, a number of the studies cited in Section 2.3 deal with inferences about rates of extinction and colonization for one or more species at one or more locations over a sequence of seasons or years. In addition to the use of inference methods that ignore the issue of detection probabilities, some workers have developed estimators for rates of extinction and turnover that do incorporate detection probabilities and that are based on aggregations of species (Nichols *et al.* 1998a; Williams *et al.* 2002). For these aggregate methods, each species within a group of interest is viewed as a replicate (similar to an individual in a single-population study of survival) and assumed to exhibit a common rate of extinction or turnover. Such approaches have been used with multiple-season occupancy data for multiple species to draw inferences about effects of fragmentation on avian extinction and turnover (Boulinier *et al.* 1998b, 2001), the influence of sexual dichromatism on local rates of species extinction of birds (Doherty *et al.* 2003a), and the influence of location of species within their geographic range (edge vs. interior) on local rates of extinction and turnover (Doherty *et al.* 2003b; Karanth *et al.* 2005). This general approach has also been used with fossil data to estimate rates of global extinction and origination for various taxa (Nichols and Pollock 1983; Conroy and Nichols 1984; Nichols *et al.* 1986; Connolly and Miller 2001ab, 2002).

When occupancy data are collected for multiple species over multiple seasons or years, failure to deal with detection probabilities leads to biased estimates of rates of both extinction and colonization (see Section 2.3). The multi-season models of Chapter 7 can be used to obtain estimates of these rate parameters for each species in a community. Inferences about rates of extinction or colonization for groups of species can then be made using an approach that deals better with interspecific variation within species groups (Chapter 9) than does the aggregated approach of Nichols *et al.* (1998a). Some analyses that can be conducted using the methods of Chapter 7 are not possible with the aggregation-based approach to estimation. For example, in the work of Doherty *et al.* (2003a) on extinction probabilities of bird species that do and do not exhibit sexual dichromatism, the authors would have preferred an analysis that dealt with phylogenetic relationships among the species while addressing the question of primary interest. Phylogeny could not be dealt with using the estimation approach based on aggregation, but it can be dealt with using occupancy-based estimates of extinction rate for each species.

Other multi-species questions not amenable to an approach based on aggregating species may be addressable using multi-species occupancy data collected over time. For example, the idea of species "guilds" is used widely in ecological work. A "guild" is regarded as a group of species exhibiting similar characteristics with respect to some specified life history trait such as foraging habit or nest site selection. One way of testing the guild concept, or perhaps even defining "guild," might be to ask whether local rates of extinction and/or colonization seem to exhibit parallel temporal variation for species in a group.

For example, it would be possible to fit an additive "species + time" model for species within an *a priori* group. Good fit of such a model might provide support for the hypothesis that the species were members of a guild, whereas rejection of this model in favor of a full "species × time" interaction model would lead to rejection of the guild hypothesis and to the conclusion that the species were not responding to environmental variation in the same manner.

Detection probabilities less than 1 will produce bias in estimates of local rates of extinction and colonization based on presence-absence data. We thus recommend the multiple-season models of Chapter 8 as a means of drawing inferences about the colonization and extinction processes for multiple species. These models should be useful for the more complicated cases in which multiple species are believed to interact, as well as in the simpler case of no interspecific interactions.

2.6. DISCUSSION

The purpose of this chapter has been to review a number of ecological questions that are typically addressed using occupancy or detection-nondetection data. Based on this abbreviated review, it is clear that occupancy data are widely used. For some topics (e.g., use of occupancy in large-scale monitoring), occupancy is viewed as a surrogate for abundance. However, for other important topics, including studies of range size, species-habitat relationships, metapopulation dynamics, interspecific interactions, and interspecific variation in extinction and colonization, occupancy is the state variable of primary interest. For many questions, the dynamics of this state variable, and the vital rates that underlie these dynamics (patch-level probabilities of extinction and colonization), are the subjects of primary interest.

It is also clear from this review that the vast majority of studies in which occupancy and its associated vital rates are estimated include no mechanism for dealing with the likelihood that the species of interest goes undetected at some locations at which it is present. The review indicates that many workers are aware of this problem, but the usual approach is to either assume the problem away or to hope that its likelihood is sufficiently small to cause few problems. Some workers try to minimize the probability of missing a species by visiting sites multiple times. What is somewhat surprising is that, conditional on the assumption that nondetection is equivalent to absence, substantial statistical rigor has been applied to the analysis of occupancy data for both single-species and multi-species problems. Thus, it cannot be claimed that those who deal with occupancy data simply shy away from statistical methods. Yet, for some reason, until recent years collectors and analysts of occupancy data have not followed the approach of scientists studying animal population ecology at the level of the individual animal and incorporated detection parameters directly into their inference methods.

Consider the following statements from a study of the influences of incomplete data sets (meaning missed species presence) on species-area relationships and nested subset analyses:

Most large data sets on species distributions and community composition will be incomplete. . . . The question becomes, then, how complete the data must be to change the qualitative results and the inferences that are drawn from them. (Kodric-Brown and Brown 1993:741)

The authors then recommend "additional field studies" to "help to assess the completeness of the data and to detect and correct for certain kinds of bias" (Kodric-Brown and Brown 1993:741). Our recommendation throughout this book will be to instead incorporate the collection of additional information (typically replicate surveys of sites within a relatively short time frame) directly into the sampling protocol and to then utilize models that explicitly incorporate detection probabilities.

The approaches for estimation of occupancy from detection-nondetection sampling at a single season (Chapters 4–6) will be useful for virtually all of the questions discussed in the present chapter. Studies of range at a fixed point in time should be based on occupancy estimates, rather than on raw detection data. Similarly, studies of range dynamics should be based on the models of Chapter 7 that explicitly incorporate detection probabilities that may change over time. Efforts to model occupancy as a function of habitat covariates should use the methods in Chapters 4–6, as these methods separate effects of habitat on occupancy and detection probability. The methods of these chapters should be useful in investigating incidence functions for metapopulation modeling, whereas the models of Chapter 7 permit the direct modeling of patch occupancy dynamics. Hypotheses that are frequently incorporated as assumptions into previous estimation methods can be represented as candidate models in a model set that also includes competing ideas, and model selection methods or hypothesis testing can be used to discriminate among such ideas. The methods of Chapter 7 should be especially useful in large-scale monitoring programs, as parameters reflecting rate of change in occupancy over time can be directly estimated and modeled, as can the vital rates responsible for such change.

Methods of Chapter 8 will be especially relevant to attempts to deduce inferences about process from occupancy patterns of multiple species in the face of detection probabilities that may vary among species and even depend on the presence of certain other species. Many of our ideas on how this could be done are given in Chapter 9. In addition, the methods of Chapter 8 should be useful in addressing various questions about occupancy dynamics for multiple species in the case not only of independent dynamics but also of dependent dynamics, in which the vital rates of one species are a function of the presence of another species. Thus, our intention with this book is to provide a beginning toolbox for addressing the ecological questions of this chapter in ways that deal adequately with the sampling reality that species are not always detected in sample units when present.

CHAPTER 3

Fundamental Principles of Statistical Inference

In this chapter we give an overview of the important statistical concepts that we will be using in this book. Our intent here is to provide an introduction to many basic ideas in terms that those less familiar (or comfortable) with statistics can understand. We have also included a short appendix to provide details on some topics that are secondary to the main thrust of this chapter but that some readers may not be familiar with (e.g., sigma and pi notation for sums and products, basic matrix algebra, and differentiation and integration). In the following pages there are a number of equations that may appear daunting at first glance; however, we provide ample explanation so that readers may interpret what the equations represent. After all, equations are simply a mathematical shorthand for conveying sets of ideas or concepts. Once one learns to read and interpret the mathematical language, understanding of the information being conveyed follows naturally (just like learning any other language, such as French, Japanese, or Spanish).

We strongly encourage readers to take the time to fully understand the concepts discussed in this chapter. It is our experience that those with a sound grasp of the underlying statistical methods and assumptions make the best use of modeling procedures like those included in this book. They also tend to be the people who are less likely to misuse the methods or attempt to apply them incorrectly. Sound use of statistical inference is at the heart of successful analyses of all wildlife studies. We are usually basing our inferences on a small sample from the population and obtaining only uncertain estimates of the parameters of interest. Statistical inference can be viewed as a rigorous method of studying and drawing conclusions about the quantities of interest in the face of that uncertainty, which is due to sampling variability (i.e., by not being able to census the entire population).

We begin with some background and definitions of key statistical concepts before discussing the desirable properties of parameter estimators. We then consider the important role of the likelihood in parameter estimation, with particular emphasis on the use of maximum likelihood estimation. We briefly discuss hypothesis testing and in particular likelihood ratio tests and goodness of fit tests. We discuss methods of assessing the bias and precision of estimators using simulation and related computer-intensive methods. We also discuss methods of including the effects of auxiliary variables on the parameters using appropriate link functions. For ecologists with little statistical background, a good introduction to many of these standard methods is Chapter 4 of Williams et al. (2002). A much more detailed treatment suitable for ecologists with a very strong background in statistics is Casella and Berger (2002).

We discuss Bayesian methods of estimation but do not consider them in detail, as they are not widely used in this book. They are, however, very important and are becoming much more widely used in wildlife investigations. For theoretical and applied details on Bayesian methods, we recommend the book by Gelman et al. (1998). We also refer the interested reader to Brooks et al. (2000, 2002) for good, recent, applied treatments of this topic.

When applying the methods detailed in this book to real studies, we will often have to consider a whole suite of statistical models (possibly representing competing hypotheses about the system or population) and choose the "best" model(s) from among them. One approach for doing so is the use of information-theoretic methods. As part of the estimation and inferential procedure, an "honest" approach would also account for the fact that the "best" model was not selected a priori: a different set of data from the same population may well result in a different model being selected as "best." Therefore, the model selection procedure itself introduces a source of uncertainty that should be accounted for. We discuss the use of "model averaging" as one approach for doing so. Following Burnham and Anderson (2002), we will emphasize the use of the Akaike Information Criteria (AIC) for assessing the weight of evidence for members of the model set for the purposes of model choice and model averaging. Generalizations to allow for small sample sizes (AICc) and overdispersion (QAIC) will also be considered.

3.1. DEFINITIONS AND KEY CONCEPTS

Statistical inference always involves the study of *unknown parameters* from a *population* of interest, using some kind of *sampling* procedure in which *sampling units* are drawn from the population. For example, a sampling unit might be a site or plot where we would like to establish the presence or absence of a species of plant or animal, and the population would then be all of the potential sites over the range distribution of the species. One key parameter of interest would be the probability that a site is occupied by the plant. Some important definitions are now presented.

- *Population*: The complete set of sampling units that we are interested in studying.
- *Sample*: The sample is the set of sampling units that we actually study. The sample is a part of the population and is chosen based on some assumed probability mechanism. For example, if simple random sampling is used, then all sampling units in the population are equally likely to be chosen (or the probabilities of inclusion are equal for all the units).
- *Parameter*: A parameter is a characteristic of the population that we would like to know about. For our earlier example, the parameter would be the probability that a site is occupied by the species. Generally, a population parameter is never known exactly. In our example, the symbol ψ may be used to represent the parameter that we refer to as the probability of occupancy.
- *Estimator of a Parameter*: As we never know the value of a parameter exactly, we have to use an estimator of the parameter based on a sample. In our notation, $\hat{\psi}$ may be used to represent an estimator of the parameter ψ. For example, if the individuals of the species are very easy to detect, then the sample proportion of sites occupied by the species can be used to estimate the population probability of occupancy. Clearly ψ, the parameter, and $\hat{\psi}$, the estimator, are distinct, and use of the circumflex or "hat" to represent an estimator has become standard notation. In general, detection probability has to be considered and this simple estimator is unsatisfactory, but we defer this discussion to the next chapter.

RANDOM VARIABLES, PROBABILITY DISTRIBUTIONS, AND THE LIKELIHOOD FUNCTION

A random variable is the observed outcome of a stochastic process (hence the term *random* variable), collected in our sampling scheme. Our statistical inference is based upon these random variables and the underlying stochastic process that is assumed to have generated them. Estimators are functions of these random variables. To illustrate, let us consider again our example of estimating the probability of occupancy for a species. Suppose you had five sites that are a random sample of sites from the range of the species and when you visited them you found that the five sites had the following results in terms of occupancy by that species:

$$UOUOU,$$

where U indicates the site is unoccupied and O the site is occupied. If we assume that all the sites visited have perfect detection for the species, then the probability of getting this particular sequence of occupied and unoccupied sites is:

$$Pr(UOUOU) = (1 - \psi)\psi(1 - \psi)\psi(1 - \psi)$$
$$= \psi^2(1 - \psi)^3 \tag{3.1}$$

In this instance, the exact sequence is being considered as the random variable. Alternatively, rather than being concerned with the sequence of occupied and unoccupied sites, we could summarize this information by defining our random variable (x) to be the number of occupied sites. The probability of observing exactly two occupied sites out of five follows the well-known binomial distribution:

$$Pr(x = 2) = \binom{5}{2}\psi^2(1 - \psi)^3 \tag{3.2}$$

with general form

$$Pr(x|\psi) = \binom{s}{x}\psi^x(1 - \psi)^{s-x}, \tag{3.3}$$

where s is the total number of sites surveyed, ψ is the probability of site being occupied (or more generally the probability of a "success"), and

$$\binom{s}{x} = \frac{s!}{x!(s-x)!}$$

is the number of ways x "successes" could be observed from s independent samples (also known as the binomial coefficient in this situation). For example, by considering only the number of occupied sites in our sample, rather than the exact sequence of observations, the difference between Eq. (3.1) and (3.2) is the binomial coefficient. Here it equals

$$\binom{5}{2} = \frac{5!}{2!\,3!} = \frac{120}{2 \times 6} = \frac{120}{12} = 10.$$

That is, by sampling five sites, there are 10 possible sequences of Os and Us with 2 Os (e.g., OOUUU, OUOUU, OUUOU, ...). In Eq. (3.3), the term "$|\psi$" on the left-hand side simply denotes that the probability of observing x is conditional on (or depends upon) the value for $|\psi$. To simplify notation, we do not always include this term [as in Eq. (3.2)], but it is usually implied.

To summarize, we denote the discrete probability distribution of the observed random variable(s) (x; which may be a vector, as indicated by the bold font) as a function of the parameter(s) (θ) as $Pr(\mathbf{x}|\theta)$. An alternative, more general, notation for the probability distribution is $f(\mathbf{x}|\theta)$. Random variables may be discrete as in our example or continuous. Distributions may be univariate (relating to only a single random variable) or multivariate (relating to multiple random variables).

A very important discrete multivariate distribution is the multinomial distribution that is an extension of the binomial distribution to the case of more than two categories. If there are k categories, then:

$$Pr(\mathbf{x}|\theta) = \frac{n!}{x_1! x_2! \dots x_k!} \prod_{i=1}^{k} \pi_i^{x_i}.$$

This distribution can be viewed as providing a probabilistic description of possible outcomes resulting from distributing n objects into k cells or bins, with x_i indicating the number of the n entities or objects that are found in cell i, and π_i indicating the probability that any one of the n entities is found in cell i. The π_i can also be viewed as the expected proportion of the n entities that are found in cell i.

The likelihood function is a very important statistical concept used throughout this book. It is also a very simple concept. As mentioned above, the purpose of statistical inference is to draw conclusions about population parameters (θ) based upon the observed data or random variables ($\mathbf{x}$), and using the likelihood function is one approach to achieving that inference. Compared to the probability distribution for a random variable(s) above, the roles of the data and parameters are reversed in a likelihood function. Now interest lies in the values of θ conditional upon the observed data $\mathbf{x}$; we denote this as $L(\theta|\mathbf{x})$. Once the probability distribution for a random variable has been determined, calculating the likelihood function is simply a matter of changing notation; no additional mathematics is required (i.e., $L(\theta|\mathbf{x}) \equiv f(\mathbf{x}|\theta)$). Hence, constructing the likelihood function is equivalent to determining the probability of observing a particular set of data (although interpretations of the two concepts are very different). For example, in the above case where two out of five sites were occupied by our plant of interest, the likelihood function would be:

$$L(\psi|x = 2) = \binom{5}{2} \psi^2 (1-\psi)^3.$$

EXPECTED VALUES

Before we can consider statistical properties of estimators, we need to define expected values. The expected value of a random variable x can be viewed as a weighted average of the random variable, where the weights are simply the probabilities of observing x, as given by the probability distribution $f(x|\theta)$ (note here we are only concerned about the expected value of a single random variable). For a discrete distribution, the expected value can be calculated as:

$$E[x] = \sum_x xf(x|\theta) = \mu,$$

where the summation is performed over all allowable values of x. In our example, the expected number of occupied sites would be:

$$E[x] = \sum_{x=0}^{s} xf(x|\psi)$$

$$= \sum_{x=0}^{s} x \binom{s}{x} \psi^x (1-\psi)^{s-x}$$

$$= \sum_{x=1}^{s} \frac{xs(s-1)!}{x(x-1)!(s-x)!} \psi^x (1-\psi)^{s-x}$$

$$= s\psi \sum_{x=1}^{s} \frac{(s-1)!}{(x-1)!(s-x)!} \psi^{x-1} (1-\psi)^{s-x}$$

$$= s\psi \sum_{y=0}^{r} \binom{r}{y} \psi^y (1-\psi)^{r-y}, \text{ where } r = s-1 \text{ and } y = x-1$$

$$= s\psi$$

The simplification is possible because the summation in the second-to-last line $\left(\sum_{y=0}^{r} \binom{r}{y} \psi^y (1-\psi)^{r-y} \right)$ is equal to 1. This is because our new random variable $y (= x - 1)$ can take the possible values of $0, 1, \ldots, r (= s - 1)$, which is the range of values that the summation is performed over (i.e., the summation is conducted across all possible values for y), and the term being summed is the probability of observing y. Therefore the summation is simply adding together the probabilities of observing all possible values of y, which must equal 1 (i.e., y must have a value between 0 and r, inclusive). The result is that the expected number of sites occupied by the plant is the number of sites surveyed times the probability the plant occupies a site (e.g., if $s = 5$ and $\psi = 0.6$, the expected number of occupied sites would be $5 \times 0.6 = 3$).

Calculating the expected value for a continuous random variable is very similar, although now integration over all possible values of x is required rather than summation:

$$E[x] = \int_x xf(x|\theta)\,dx = \mu.$$

Note that integration can be viewed as the summation of infinitely many, infinitely small terms (see Appendix); in fact, many computer algorithms for calculating integrals use such an approach.

The expected value of a function or transformation of a continuous random variable, $g(x)$, is $E[g(x)] = \int_x g(x)f(x|\theta)\,dx$, or an analogous equation for a discrete random variable, and unless $g(x)$ is linear $E[g(x)] \neq g(E[x])$. For example, $E[3x] = 3\mu$ but $E[\log(x)] \neq \log(\mu)$. However in many practical cases, using the transformation of the expected value may be a reasonable approximation, for example, $E[\log(x)] \approx \log(\mu)$.

The variance of a random variable is:

$$Var(x) = E\left[(x - E(x))^2\right] = E\left[(x - \mu)^2\right] = \int_x (x - \mu)^2 f(x|\theta)\,dx = \sigma^2.$$

For our plant example above:

$$Var(x) = \sum_{x=0}^{s} (x - \mu)^2 f(x|\theta) = s\psi(1 - \psi).$$

The variance thus provides a metric for the variation or spread of a random variable about its average or expectation.

The covariance between two random variables x and y is:

$$Cov(x, y) = E[(x - E[x])(y - E[y])] = \int_y \int_x (x - \mu_x)(y - \mu_y)\,dx\,dy.$$

For a multinomial distribution:

$$Cov(x_i, x_j) = E[(x_i - E[x_i])(x_j - E[x_j])] = -np_i p_j.$$

$$Var(x_i) = np_i(1 - p_i).$$

$$Var(x_j) = np_j(1 - p_j).$$

The correlation between two random variables is:

$$\rho(x, y) = \frac{Cov(x, y)}{\sqrt{Var(x)Var(y)}}.$$

For a multinomial distribution:

$$\rho(x_i, y_j) = \frac{Cov(x_i, y_j)}{\sqrt{Var(x_i)Var(x_j)}} = \frac{-np_ip_j}{\sqrt{np_i(1-p_i)np_j(1-p_j)}} = \frac{-\sqrt{p_ip_j}}{\sqrt{(1-p_i)(1-p_j)}}$$

INTRODUCTION TO METHODS OF ESTIMATION

An intuitive method of obtaining estimates is to use *the method of moments*, and many simple estimators can be derived in this way. This method involves equating sample moments to the corresponding population moments. Formally, the kth population moment is defined as $E[x^k]$. Consider again our example on site occupancy of a conspicuous plant species. Remember we had five randomly sampled sites from the range of the species, and we visited them and found that two of the five sites were occupied. In this example (five sites where two were occupied), we have shown above that the expected value of x (the number of occupied sites) is $E(x) = s\psi$, which is the first population moment. The first sample moment is just the value of the random variable x (note the second sample moment would be x^2). Equating the sample and population moments gives $x = s\psi$. Rearranging to solve for ψ gives the following method of moments estimator:

$$\hat{\psi} = x/s,$$

which for our example gives $\hat{\psi} = 2/5 = 0.4$.

In some cases, the method of moments estimators do not have good properties, in that they are not fully asymptotically efficient because they may not be a function of the *minimal sufficient statistics* (Casella and Berger 2002:285). Method of moments estimators are an older estimation method that has been largely superseded by other techniques with the advent of modern computers. However, in many simple situations the method of moments estimators are equivalent to estimators derived by other means.

A second approach to estimation is based on the likelihood function. This is the approach taken largely in this book. Maximum likelihood estimators (MLEs) are the values for the parameters that maximize the likelihood function given the observed data. The properties of MLEs are derived from the view that the values for the observed random variables are one realization from an infinitely many number of possible realizations that could have been sampled from the populations. We provide details of this estimation procedure in Section 3.2.

Bayesian parameter estimation has a different philosophical basis. The probability of observing the random variable, or data, given the parameter value(s) (which, recall, is equivalent to the likelihood function), is used in combination with the *prior* probability distribution for the parameters to obtain a *pos-*

terior probability distribution for the parameters. We provide more details of this estimation approach in Section 3.3.

PROPERTIES OF POINT ESTIMATORS

The key properties of a point estimator are the *bias*; the *variance* (and *standard error*) that measure the *precision* of an estimator; and the *mean squared error* (and *root mean squared error*) that measure the overall *accuracy* of an estimator (Fig. 3.1). We will show that the accuracy of the estimator is a combination of both the bias and precision of the estimator. The formal statistical definitions are now presented.

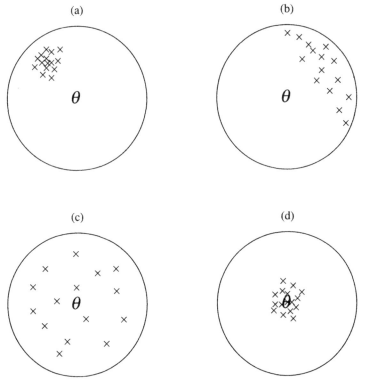

FIGURE 3.1 Illustration of estimator bias, precision, and accuracy. Parameter estimates, $\hat{\theta}$ (denoted by ×), compared to true parameter value, θ. (a) Precise and biased: inaccurate. (b) Imprecise and biased: inaccurate. (c) Imprecise and unbiased: inaccurate. (d) Precise and unbiased: accurate.

Bias

$$Bias[\hat{\theta}] = E[\hat{\theta}] - \theta.$$

Expressed in words, this means the *bias* of a parameter estimator ($\hat{\theta}$) is the difference between the expected value of the estimator and the parameter. For our example, a natural estimator for the proportion of sites occupied would be the fraction of sites occupied by the plant, that is, $\hat{\psi} = x/s$, which is a linear function of the random variable x. Using the above results, the expected value of $\hat{\psi}$ would be $E[\hat{\psi}] = E\left[\dfrac{x}{s}\right] = \dfrac{s\psi}{s} = \psi$, and therefore $Bias(\hat{\psi}) = E[\hat{\psi}] - \psi = 0$, so that we have an *unbiased* estimator. However, recall that here we have assumed the plant is always detected when present at a site, and we argue that assumption will not hold generally.

Precision (Variance and Standard Error)

$$Var(\hat{\theta}) = E\left[(\hat{\theta} - E[\hat{\theta}])^2\right],$$

and in words this means the variance is the expected squared deviation of the estimator from its expected value. This is a measure of how "close" an estimator is to its expectation. To obtain a measure that is on the original scale, we define the square root of the variance of the estimator to be the standard error of the estimator:

$$SE(\hat{\theta}) = \sqrt{Var(\hat{\theta})}$$

For our example,

$$Var(\hat{\psi}) = E\left[\left(\frac{x}{s} - \psi\right)^2\right] = \frac{E\left[(x - s\psi)^2\right]}{s^2}$$

$$= \frac{E\left[(x - E[x])^2\right]}{s^2} = \frac{Var(x)}{s^2}$$

$$= \frac{s\psi(1 - \psi)}{s^2} = \frac{\psi(1 - \psi)}{s}$$

and therefore:

$$SE(\hat{\psi}) = \sqrt{Var(\hat{\psi})} = \sqrt{\frac{\psi(1 - \psi)}{s}}.$$

The standard error decreases as 1 over the square root of the sample size (s), which is the standard relationship in many statistical problems.

Accuracy (Mean Squared Error)

$$MSE(\hat{\theta}) = E\left[(\hat{\theta} - \theta)^2\right] = Var(\hat{\theta}) + \left[Bias(\hat{\theta})\right]^2$$

and, in words this means the mean squared error denotes the expected squared deviation of the estimator from the parameter. This measure combines bias and precision in one overall measure of how "close" an estimator is to the parameter it is estimating (note the subtle differences in the interpretation of the mean squared error and the variance). The root mean squared error just converts the measure to the original scale:

$$RMSE(\hat{\theta}) = \sqrt{MSE(\hat{\theta})}$$

For our example,

$$MSE(\hat{\psi}) = Var(\hat{\psi}) + \left[Bias(\hat{\psi})\right]^2$$
$$= \frac{\psi(1 - \psi)}{s} + 0 = \frac{\psi(1 - \psi)}{s}.$$

Generally, for an unbiased estimator the mean squared error is equal to the variance of the estimator.

COMPUTER-INTENSIVE METHODS

Monte Carlo or simulation methods can be very useful approaches to evaluating the properties of estimators, and they are widely used. In particular, they can be used to evaluate bias in estimators, due either to small samples or to the failure of assumptions.

Further, simulation may be very valuable for calculating or evaluating variances (standard errors) and confidence intervals for situations in which there are questions about the validity of large sample approaches (Manly 1997). Also, in some cases, the simulations may be used to include other sources of uncertainty in variance or interval estimation that were not included in the estimation algorithm.

Another related, widely used procedure is "bootstrapping," which is a computer-intensive resampling procedure (Efron 1979; Manly 1997). Bootstrapping is often used to obtain estimates of the bias and precision of estimators. It can also be used to obtain interval estimators with better properties than those from profile likelihood or asymptotic normality approaches. We direct those interested in reading more on computer-intensive methods within a biological context to Manly (1997).

3.2. MAXIMUM LIKELIHOOD ESTIMATION METHODS

Here we provide greater detail on maximum likelihood estimation and related topics that follow directly from it.

MAXIMUM LIKELIHOOD ESTIMATORS

A very widely used method of estimation that produces estimates with good properties is the method of maximum likelihood. As discussed above, we calculate the probability distribution of the observed data as a function of the parameters $Pr(x|\theta)$ (or more generally $f(x|\theta)$), where x denotes the data and θ denotes the parameters. This probability distribution is then viewed as a function of the parameters conditional on the data, and this is called the likelihood function $L(\theta|x)$. We need to find the values of the parameters that maximize this function; that is, given the underlying model, for what values of the parameters are these data most likely? These are the maximum likelihood estimators (MLEs).

A very simple example of a likelihood function involves our example on site occupancy of a conspicuous plant species. Recall that from a sample of five sites, two were found to be occupied, and that the probability for two sites being occupied was:

$$Pr(x = 2|\psi) = \binom{5}{2}\psi^2(1-\psi)^3$$

with the likelihood (a function of ψ), given the data:

$$L(\psi|x = 2) = \binom{5}{2}\psi^2(1-\psi)^3$$

where ψ is the probability the site is occupied. It can be shown that the likelihood is maximized at the method of moments estimator given in above, $\hat{\psi} = x/s$, which for our sample gives $\hat{\psi} = 2/5 = 0.4$. In this case, the MLE is the same as the method of moments estimator, but that is not always true.

Here it is easy to find the maximum by just plotting the likelihood function (Fig. 3.2). Also, regardless of whether we consider our random variable to be the sequence of occupied and unoccupied site or the summarized number of occupied sites [i.e., the likelihood is based on Eq. (3.1) or (3.2)], note that by plotting the functions we would obtain the same MLE. The two functions are proportional, and the constant of proportionality $\left(\text{i.e., }\binom{5}{2}\right)$ does not affect

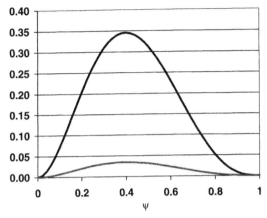

FIGURE 3.2 Plot of the likelihood functions where the random variable is considered either as the number of occupied sites (—) or as the specific sequence of occupancy and unoccupied sites (—). Note that in both cases the likelihood is maximized when $\psi = 0.4$.

the resulting MLE. However, in practice, other methods must generally be used to obtain MLEs, as the likelihood may be very complex. One approach to finding MLEs for likelihoods that do not have many parameters is to use standard results from calculus. Set the partial derivatives of the likelihood with respect to each parameter equal to zero, and then solve the resulting equations (Williams *et al.* 2002:47). Another general approach that is used for more complex likelihoods is to use a software package that may use a variety of numerical algorithms to find the maximum.

PROPERTIES OF MAXIMUM LIKELIHOOD ESTIMATORS

The asymptotic (large sample) properties of MLEs (which were derived by R. A. Fisher in the 1920s) are very powerful. To quote Williams *et al.* (2002:48),

The MLE $\hat{\theta}$ has an approximately normal distribution for large sample sizes. Furthermore, its distribution converges asymptotically to a normal distribution as sample sizes increase. Though the estimator may be biased, it is asymptotically unbiased in the sense that the expected value of $\hat{\theta}$ converges to the parameter θ as sample sizes increase. The variance of the estimator $\hat{\theta}$ is asymptotically minimum, in that $\hat{\theta}$ has the least variance of all unbiased estimators of θ when sample size is large.

VARIANCES, COVARIANCE (AND STANDARD ERROR) ESTIMATION

Exact variances and covariances can be computed for some simple estimators (in our example, $Var(\hat{\psi}) = \psi(1 - \psi)/s$). However, most often only large sample results are possible in practice. From the asymptotic likelihood theory just discussed, the approximate variance-covariance matrix of a set of MLEs can be obtained from functions of the second partial derivatives evaluated at the MLEs (Williams *et al.* 2002:735). Most computer software uses numerical algorithms based on this principle to estimate the variance-covariance matrix for all of the parameters being estimated from the likelihood.

Sometimes we will also need to obtain the MLE of a function of the original parameters. To obtain a point estimator here is straightforward, as the MLE of a function is simply the function evaluated at the MLEs of the original parameters. For example, if $h(\theta)$ is the function, and $\hat{\theta}$ is the MLE of θ, then the MLE of $h(\theta) = h(\hat{\theta})$. To obtain the estimated variance (if there is one parameter) or variance-covariance matrix of this estimated function (if, as more commonly, there is a whole set of parameters) is a little more difficult. In many cases, the "delta method" (which takes advantage of a Taylor series expansion to linearize the function) is used as an alternative method to obtain large sample (approximate) variances and covariances (Seber 1982:8; Williams *et al.* 2002:736). Here we present the key result first in one dimension and then in matrix form, as these methods will be used later in the chapter.

For one parameter we have:

$$Var[h(\hat{\theta})] \approx [h'(\hat{\theta})]^2 Var(\hat{\theta}),$$

where $h'(\hat{\theta})$ is the derivative of $h(\theta)$ with respect to θ, evaluated at its MLE, $\hat{\theta}$.

The matrix equivalent is:

$$\mathbf{Var}(h(\hat{\theta})) = [h'(\hat{\theta})][\mathbf{V}][h'(\hat{\theta})]^T,$$

where $\mathbf{V}$ is the estimated variance-covariance matrix for the vector of parameter estimates $\hat{\theta}$, and $h'(\hat{\theta})$ is the row vector of partial derivatives of $h(\theta)$ with respect to θ evaluated at the MLEs, $\hat{\theta}$. We illustrate the use of the delta method in Section 3.4.

CONFIDENCE INTERVAL ESTIMATORS

Confidence intervals are one method of expressing the uncertainty in an estimator. While it is sometimes possible to obtain exact confidence intervals, in many cases we obtain approximate or large sample confidence intervals based either on the asymptotic normal distribution of the ML estimator or on the profile likelihood method, which is based on an appropriate log-likelihood ratio having an asymptotic chi-square distribution.

The asymptotic $(1 - \alpha)\%$ confidence interval based on the normal distribution takes the form:

$$\hat{\theta} \pm z_{\alpha/2} SE(\hat{\theta}),$$

with $z_{\alpha/2}$ the appropriate critical value from the standard normal distribution. With 95% confidence, the z value is 1.96, while with 90% confidence it is 1.645.

A profile likelihood confidence interval can be based on the following quantity (Williams et al. 2002:49):

$$\varphi(\theta_0) = 2 \ln \left[\frac{L(\hat{\theta}_0, \hat{\theta})}{L(\hat{\theta}_0, \hat{\theta}^*)} \right].$$

$L(\hat{\theta}_0, \hat{\theta})$ is the likelihood evaluated at the MLEs for all the parameters, and $L(\theta_0, \hat{\theta}^*)$ is the likelihood evaluated at the MLEs of all the other model parameters. θ_0 is allowed to vary over its range (i.e., the parameter θ_0 is not estimated in $L(\theta_0, \hat{\theta}^*)$ but can take any allowable value). We find the two values of θ_0 that satisfy the equation $\varphi(\theta_0) = \chi_1^2(\alpha)$ where $\chi_1^2(\alpha)$ is the $(1 - \alpha)\%$ percentile of the chi-square distribution (Williams et al. 2002:728) with one degree of freedom. The profile likelihood method typically gives confidence intervals with somewhat better coverage than the confidence intervals based on the normal approximation.

Alternatively, confidence intervals can be calculated using computer-intensive methods such as those discussed above (e.g., the bootstrap).

3.3. BAYESIAN METHODS OF ESTIMATION

The "classical" (i.e., non-Bayesian) view of statistics supposes that parameters are fixed, and seeks to find procedures with desirable properties for estimating those parameters. Usually, probability is used as a basis for evaluating the procedures, under a scenario in which replicate realizations of the data are imagined. This view supposes that it is sufficient to draw inferences about parameters based on what might have happened (but did not), not on what actually did happen (i.e., the observed data). Classical theory in widespread use relies heavily on asymptotic (large sample) arguments to assess the operating characteristics of procedures. For example, the parameter ψ is fixed, and a 95% confidence interval for ψ will contain the true value 95% of the time (in large samples). Thus, while the interval may or may not (we cannot know) contain the true value, if we were able to repeat our study an infinite number of times and collect very large quantities of data, we would expect the stated interval to contain the truth 95% of the time. Note that the view of the parameter as fixed and the data random is manifest as a probability statement about the interval, not the parameter.

An alternative and increasingly common view of statistics seeks to provide a direct probabilistic characterization of uncertainty about parameters given the specific data at hand. This view is commonly referred to as the "Bayesian" view. Bayesian statistical methods have become increasingly popular in recent years, partially due to the advent of fast computers and efficient methods for solving Bayesian inference problems that, typically, require solving complex integration problems. Similar to classical views of statistics, the Bayesian does view the data as realizations of a random variable. However, the Bayesian also views the parameters of a model as random variables and provides a probabilistic characterization of the state of knowledge of these parameters by statement of a prior distribution. Then, with both data and parameters viewed as random variables, a conceptually simple calculation known as Bayes Rule yields the probability distribution of the parameters given the data, a quantity known as the posterior distribution. The Bayesian formulates inferences for the parameters based on this posterior distribution, conditional on the observed data, and not by entertaining notions of repeated experimentation or data collection.

THEORY

The *prior* distribution of the parameters is denoted $f(\theta)$, while the distribution of the random variables (data) given the parameters is $f(x|\theta)$ (recall that this is essentially the likelihood function $L(\theta|x)$). Therefore, by use of Bayes' Theorem, the *posterior* distribution of the parameters is:

$$f(\theta|x) = \frac{f(\theta)f(x|\theta)}{f(x)}$$
$$\propto f(x|\theta)f(\theta).$$

Bayesian inference is based on this posterior distribution. For example, the posterior mean of θ is commonly used as a point estimate, and one may construct "Bayesian confidence intervals" (also referred to as *credible intervals*) using quantiles of the posterior distribution.

When we use constant or uniform priors (i.e., priors that assume all parameter values have the same probability of occurring and that may be "improper" if the prior does not integrate to 1), notice that the posterior and the likelihood function are proportional, that is, $f(\theta|x) \propto f(x|\theta) = L(\theta|x)$. Uniform priors are widely used and are sometimes referred to as "noninformative" priors, although this term is not precise and there are other prior distributions that are often viewed as being noninformative. While there are major differences in the underlying philosophies between Bayesian and likelihood approaches, what the posterior distribution being proportional to the likelihood means for applied useage is as follows. When sufficient quality data have been collected and when constant or uniform priors have been used, then resulting inferences from the Bayesian and likelihood methods tend to be very similar.

We believe that the justification of Bayesian methods based on the above argument is incomplete and very misleading. In many situations there will be much prior information available based on other similar field studies and on strong but diffuse knowledge from "expert opinion." This knowledge can be very helpful and can be used to define strong prior distributions, leading to much less uncertainty in the posterior distribution of the parameters. Ignoring all this knowledge and using "noninformative priors" in such cases seems illogical, yet this is what maximum likelihood methods are effectively doing. For example, consider a simple coin-tossing experiment in which we wish to estimate the probability of obtaining a head, p, when 7 out of 10 tosses of the coin resulted in a head. Assuming a "noninformative" prior distribution, where all possible values for p are equally likely (i.e., p could be reasonably expected to have any value between 0 and 1), the resulting posterior distribution is given in Fig. 3.3(a). Note the most likely value for p from this posterior distribution is 0.70, with a 95% credible interval of (0.39, 0.89). However, generally it would be reasonable to assume that prior to conducting a coin-tossing experiment, p should about 0.5. Fig. 3.3(b) illustrates the effect of assuming a much more informative (but reasonable) distribution for p. Now the most likely value for p is 0.52 with a 95% credible interval of (0.43, 0.61). Assume that an informative prior distribution has clearly reduced the uncertainty in the posterior distribution, although also note that in this instance the prior and posterior distributions are very similar. This indicates that the collected data has had little impact upon our preconceived notions about p, which is partially due to the relatively small sample size. Had the coin been tossed 100 times resulting in 70 heads, then the posterior distribution in Fig. 3.3(c) would be obtained. While there is clear justification in this case for using an informative prior distribution for p, one should always be aware of the potential that resulting inferences based upon a posterior distribution may be sensitive to the choice of prior distribution, "noninformative" or otherwise.

Another key advantage of Bayesian estimation methods is that the Bayesian view of parameters as themselves arising from a distribution, rather than being fixed quantities, is especially useful when considering models with "random effects." For example, instead of assuming that all sites within some area of

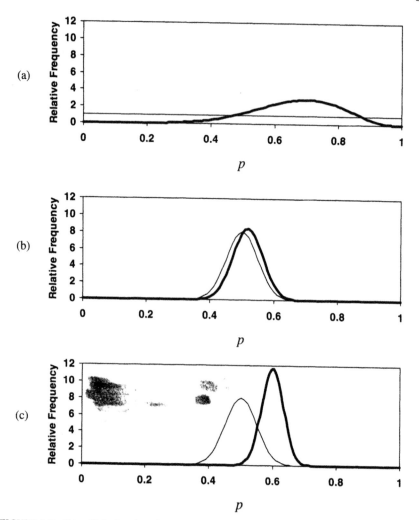

FIGURE 3.3 Prior (light lines) and posterior (dark lines) distributions for the probability of a head, p, from a coin-tossing experiment in which 7 out of 10 tosses were heads with (a) a "non-informative" prior, and (b) an informative prior. Prior and posterior distributions for p from a coin tossing-experiment where 70 out of 100 tosses were heads with (c) the same informative prior used in (b).

interest have the exact same probability of being occupied by a species of interest, we might consider random effects models, in which we view the occupancy probabilities for all sites as coming from the same underlying distribution but with different realized probabilities as governed by that distribution. In long-term studies, year-to-year variation in either occupancy or rates of extinction and colonization are likewise usefully viewed as random effects in many situations. Such thinking and modeling is possible from a frequentist or likelihood perspective as well, but it is a very natural outcome of Bayesian thinking.

COMPUTING METHODS

In recent years the power of modern computation has led to a revival of interest in Bayesian methods and an emphasis on use of the methods in a wide variety of applied problems. Modern Bayesian inference sometimes uses numerical integration methods to obtain posterior distributions if the number of parameters in the model is fairly small. More typically, simulation methods based on Markov chain Monte Carlo (MCMC) and in some cases the software WinBUGS can be used (e.g., Section 4.5.2; Link *et al.* 2002). Regardless of the investigator's views about Bayesian philosophy, the pragmatic fact is that MCMC can readily provide estimates under models that would be extremely difficult to deal with from a strictly frequentist perspective (e.g., by using maximum likelihood estimation).

3.4. MODELING AUXILIARY VARIABLES

In many examples, important parameters such as species occupancy (ψ) or nuisance parameters such as detection probability (p) may be functions of auxiliary variables or covariates (e.g., habitat type, patch size, elevation, rainfall in previous 24 hours, time of day, or distance to nearest road). Modeling these relationships can be viewed as a type of generalized linear regression technique. A *link function* is used as a function of the parameter that is more convenient for expressing the linear relationship with the covariates. Typically here, the parameters that we wish to be functions of covariates are probabilities, and the use of a special link function called the logit link has become almost standard practice (although there are other link functions that could be used). One advantage of this link function is that the estimated probabilities are bounded between 0 and 1. If we use a log link or no link at all, then this is not guaranteed.

THE LOGIT LINK FUNCTION

The logit link function is used to model the probability of "success" as a function of covariates (e.g., logistic regression). The purpose of the logit link is to take a linear combination of the covariate values (which may take any value between $\pm\infty$) and convert those values to the scale of a probability, that is, between 0 and 1. The logit link function is defined in Eq. (3.4):

$$\text{logit}(\theta_i) = \ln\left(\frac{\theta_i}{1-\theta_i}\right) = \beta_0 + \beta_1 x_{i1} + \beta_2 x_{i2} + \ldots + \beta_U x_{iU}, \tag{3.4}$$

where θ_i is the probability (or parameter) of interest for the ith sampling unit and $x_{i1}, x_{i2}, \ldots, x_{iU}$ are the values for the U covariates of interest measured at the ith sampling unit. The regression coefficients $\beta_1, \beta_2, \ldots, \beta_U$ determine the size of the effect of the respective covariates, and β_0 is the intercept term. Eq. (3.4) can be rearranged such that θ_i has the following relationship to the covariates:

$$\theta_i = \frac{\exp(\beta_0 + \beta_1 x_{i1} + \beta_2 x_{i2} + \ldots + \beta_U x_{iU})}{1 + \exp(\beta_0 + \beta_1 x_{i1} + \beta_2 x_{i2} + \ldots + \beta_U x_{iU})}. \tag{3.5}$$

The logit of θ_i is also known as the log-odds for "success." The term $\frac{\theta_i}{1-\theta_i}$ is the *odds* of success (i.e., how much greater the probability of "success" is compared to that of a "failure") and is often expressed as a ratio. For example, odds of $3:1$ suggest the probability of "success" is three times that of a failure. The probability of success can be calculated from the odds as:

$$\theta_i = \frac{odds}{1+odds},$$

so in this instance the probability of success would be, $\theta_i = 3/4 = 0.75$. In some instances, the odds may simply be stated as a single number rather than as a ratio (e.g., 3), in which case the ":1" is implied. While not commonly used in ecological applications, most people will be familiar with the concept of odds through gambling and games of chance (e.g., horse racing or roulette). Interestingly, in most gambling situations the odds are usually given in terms of the player *losing*. For example, a "long shot" in a horse race is one that is thought unlikely to win, and a bookie may offer odds of $30:1$ on such a horse. Here the odds reflects the amount of money to be paid out for every $1 bet should the horse win, but undoubtedly also reflects the bookie's belief about the probability that the horse will *not* win the race (i.e., the probability of the "long shot" *not* winning is 30/31, or in terms of winning the race, 1/31).

A common method for comparing odds is with an *odds ratio*. As the term implies, the odds ratio is simply the ratio of two sets of odds (i.e., $OR = odds_2/odds_1$), or alternatively, the value of one set of odds is multiplied by to get a second set of odds (i.e., $odds_2 = OR \times odds_1$). Hence, an odds ratio close to 1 would suggest the two odds (and therefore the two probabilities) are similar.

Now Eq. (3.4) could be rearranged in terms of the odds of success for sampling unit i as:

$$\frac{\theta_i}{1-\theta_i} = \exp(\beta_0 + \beta_1 x_{i1} + \beta_2 x_{i2} + \ldots + \beta_U x_{iU})$$
$$= \exp(\beta_0)\exp(\beta_1 x_{i1})\exp(\beta_2 x_{i2})\ldots\exp(\beta_U x_{iU}).$$

Expressed in this form, $\exp(\beta_0)$ is simply the odds of success at a sampling unit when the value for all covariates is zero, and $\exp(\beta_1 x_{iu})$ is simply an odds ratio that depends upon the covariate x_{iu}. In fact, $\exp(\beta_u)$ can be interpreted as the odds ratio for a one-unit change in the covariate x_{iu}. For instance, suppose in our example of a conspicuous plant species the probability of occupancy could be expressed as $\text{logit}(\psi_i) = -1.0 + 1.5 Rocks_i$, where $Rocks_i$ is an indicator covariate that equals 1 if site i had a predominately rocky soil type, and 0 otherwise. At a site without a rocky soil type (i.e., where $Rocks_i = 0$), the odds of occupancy is $\exp(-1.0) = 0.37$ (:1), indicating the site is less likely to be occupied than unoccupied. At sites with a rocky soil type, these odds are multiplied by a factor of $\exp(1.5) = 4.48$; hence the odds of occupancy at a rocky site will be $0.37 \times 4.48 = 1.66$ (:1), or approximately $5:3$. We suggest that when using the logit-link function, interpreting the effect of covariates on the probability of success via Eq. (3.5) can be complicated and difficult to explain. An alternative is to interpret their effect in terms of odds and odds ratios.

ESTIMATION

Estimation of the regression parameters ($\hat{\beta}_u$) and their standard errors is straightforward. These parameters are in the likelihood, so that computer software can be used to obtain the estimators and their standard errors using the asymptotic properties of MLEs, or their posterior distribution can be obtained when using a Bayesian analysis.

Using maximum likelihood theory, estimation of the probabilities at a site ($\hat{\theta}_i$) and their standard errors are rather complex. These are derived parameters that are not directly in the likelihood (i.e., their values are computed for use in the likelihood, but the likelihood is now maximized with respect to the β_u values, not θ_i). The derived parameters ($\hat{\theta}_i$) are functions of the MLEs, and we can use the theorem quoted earlier that states that the MLE of a function is the function of the MLE. To obtain standard errors we need to use the delta method in matrix form (Appendix). Recall that:

$$\text{Var}(h(\hat{\boldsymbol{\beta}})) = [h'(\hat{\boldsymbol{\beta}})][V][h'(\hat{\boldsymbol{\beta}})]^T$$

and here we have:

$$h_i(\hat{\boldsymbol{\beta}}) = \hat{\theta}_i = \frac{\exp(\hat{\beta}_0 + \hat{\beta}_1 x_{i1} + \hat{\beta}_2 x_{i2} + \ldots + \hat{\beta}_U x_{iU})}{1 + \exp(\hat{\beta}_0 + \hat{\beta}_1 x_{i1} + \hat{\beta}_2 x_{i2} + \ldots + \hat{\beta}_U x_{iU})},$$

with the partial derivative with respect to β_k being:

$$h'_{ik}(\hat{\boldsymbol{\beta}}) = x_{ik}\hat{\theta}_i(1 - \hat{\theta}_i),$$

and V, the estimated variance-covariance matrix of the $\hat{\boldsymbol{\beta}}$. For example, consider once again the equation given above, in which the probability of occupancy for our conspicuous plant species depended upon soil type. Let us now further consider that the coefficients are estimated quantities and we wish to calculate standard errors for the estimated probabilities of occurrence. In other words, we have the equation:

$$\hat{\psi}_i = \frac{\exp(-1.0 + 1.5 Rocks_i)}{1 + \exp(-1.0 + 1.5 Rocks_i)}$$

where $\hat{\beta}_0 = -1.0$ and $\hat{\beta}_1 = 1.5$. For a site without rocky soil type, the covariate $Rocks_i$ ($= x_1$) will be 0, and the "covariate" x_0 will be 1 (as β_0 represents an intercept term that applies to all sites, though it is often omitted from equations as above for notational convenience). Hence, the estimated probability for a site with non-rocky soil type (NRST) is:

$$\hat{\psi}_{NRST} = \frac{\exp(-1.0 + 1.5 \times 0)}{1 + \exp(-1.0 + 1.5 \times 0)}$$
$$= \frac{\exp(-1.0)}{1 + \exp(-1.0)}$$
$$= \frac{0.37}{1 + 0.37}$$
$$= 0.27$$

To calculate the standard error, we need to evaluate the vector of partial derivatives ($h'(\hat{\boldsymbol{\beta}})$) and the variance-covariance matrix for the $\hat{\beta}$s, V. V would generally be outputted (or at least obtainable) from the software used to fit the model to the data. Suppose here we obtained the variance-covariance matrix of:

$$V = \begin{bmatrix} 0.30 & -0.05 \\ -0.05 & 0.40 \end{bmatrix}$$

where $Var(\hat{\beta}_0) = 0.30$, $Var(\hat{\beta}_1) = 0.40$, and $Cov(\hat{\beta}_0,\hat{\beta}_1) = -0.05$. The vector of partial derivatives would be:

$$h'_{NRST}(\hat{\beta}) = [x_0\hat{\psi}_{NRST}(1-\hat{\psi}_{NRST}) \quad x_1\hat{\psi}_{NRST}(1-\hat{\psi}_{NRST})]$$
$$= [1\times0.27\times0.73 \quad 0\times0.27\times0.73]$$
$$= [0.20 \quad 0],$$

which can then be used to obtain the variance of $\hat{\psi}_{NRST}$ by application of the delta method:

$$Var(\hat{\psi}_{NRST}) = [0.20 \quad 0]\begin{bmatrix} 0.30 & -0.05 \\ -0.05 & 0.40 \end{bmatrix}\begin{bmatrix} 0.20 \\ 0 \end{bmatrix}$$
$$= 0.01$$

and hence a standard error of 0.11 (after some allowance for rounding error). For sites with rocky soil type (RST) a similar procedure can be used, where $\hat{\psi}_{RST} = 0.62$ and $h'_{RST}(\hat{\beta}) = [0.24 \quad 0.24]$, to give $Var(\hat{\psi}_{RST}) = 0.03$ and a standard error of 0.18. We leave the mechanics of doing so as an exercise for the reader.

Estimation of the derived probabilities is, however, relatively straightforward in a Bayesian analysis when using computer-intensive methods such as MCMC. Here the probabilities can be defined as additional "nodes" that the computer algorithm simply calculates and stores as part of the MCMC procedure. As such, the posterior distribution for $\hat{\theta}_i$ and $\hat{\beta}_u$ can be obtained simultaneously.

3.5. HYPOTHESIS TESTING

In this chapter, we focus mainly on estimation methods and choices between models with similar parameter structures; however, we will need to discuss likelihood ratio tests for comparing models and goodness of fit tests for assessing the "fit" of models, as these techniques will be used in later chapters. Therefore, we now present a very brief introduction to hypothesis testing.

BACKGROUND AND DEFINITIONS

In hypothesis testing, there is a null hypothesis and then an alternative hypothesis. The idea is to develop a test statistic that has known distribution under the null hypothesis and see if the observed value of the test statistic based on the data is unusual when compared against this known distribution.

Null Hypothesis (H_0): $\theta = \theta_0$
Alternative Hypothesis (H_1) $\theta \neq \theta_0$

The alternative hypothesis may be one-sided (e.g., $\theta > \theta_0$) or two-sided, but often takes the two-sided form presented above. The first step after one defines the null and alternative hypotheses is to define a test statistic that has a known distribution under the null hypothesis. The next step is to define a critical region of values for the test statistic where the probability of obtaining values in this region is equal to α (the size of the test). If the observed test statistic is within the critical region one rejects the null hypothesis and accepts the alternative.

Hypothesis testing leads to a dichotomous decision; one either rejects the null hypothesis or not. The investigator's decision (to reject or not) can be either right or wrong with respect to the truth. Therefore, statisticians provide a nomenclature for the errors that can be made, depending on the decision of whether or not to reject the null hypothesis.

- *Type 1 Error*: The probability that the null hypothesis is rejected when it is true (α).
- *Type 2 Error*: The probability that the null hypothesis is not rejected when it is false (β).
- *Power of a Test*: The probability that the null hypothesis is rejected when it is false is called the power of the test and is $(1 - \beta)$. A good test will have high power even for values of θ close to, but distinct from, θ_0.

LIKELIHOOD RATIO TESTS

To compare a model 0 with another model 1, when model 0 is nested inside model 1, we can use a likelihood ratio test. The idea of nesting simply means that the less general model 0 (typically associated with the null hypothesis) can be obtained by constraining some of the parameters of the more general model 1 (typically associated with the alternative hypothesis). The likelihood ratio test statistic is:

$$\chi^2 = -2\ln[L(\theta_0|\mathbf{x})/L(\theta_1|\mathbf{x})],$$

and if the model 0 is true and sample size large, then the distribution is chi-square with degrees of freedom equal to the number of additional parameters in model 1 versus model 0. This is related to the profile likelihood method used for computing approximate confidence intervals. Both are based on a likelihood ratio having an asymptotic chi-square distribution.

GOODNESS OF FIT TESTS

An old, common, and simple goodness of fit test to assess the fit of a model to data that is still widely used is Pearson's chi-square test. It has a test statistic of the form:

$$\chi^2 = \sum_{i=1}^{k} \left[\frac{(O_i - E_i)^2}{E_i} \right],$$

where O_i is the observed number of sample units found in class i, E_i is the number of sample units expected to appear in class i under the hypothesis of interest, and k is the number of classes being considered (e.g., these could be the cells or bins of a multinomial distribution). Under the null hypothesis, χ^2 is asymptotically distributed as central chi square and the degrees of freedom are df $= k - \delta$, where k is the number of classes or cells, and δ is the number of parameters fitted in the model. A statistic with similar properties is the *deviance* (as used in Chapter 5).

The use of goodness of fit tests is also important in accounting for overdispersion in model selection methods discussed and defined below. MacKenzie and Bailey (2004) used the Pearson statistic as a goodness of fit test and to estimate overdispersion for single-species, single-season occupancy problems. However, they used a parametric bootstrap procedure to estimate the exact distribution rather than the chi square distribution. Often the expected values in some of the k classes or cells were too small for the chi square to be a good approximation to the exact distribution.

3.6. MODEL SELECTION

In many classic statistical inference books, the focus is on inference for a particular model, which can be viewed as a hypothesis, with the parameter structure totally specified. Maximum likelihood estimation has become a benchmark for such inference, although Bayesian estimation methods are starting to become widely used. However, in many of the examples we consider in this book, we may have multiple hypotheses about our biological system, and the first question we have to consider is how to decide which model (or hypothesis) from a set of candidate models is the "best" for a given data set. Only once we have settled that question can we present good point or interval estimates of the parameters required from the selected model.

Burnham and Anderson (2002) have studied model selection and multimodel inference in detail. Here we give a very brief treatment based on their approach using the Akaike Information Criterion (AIC). We agree with them that this approach is better than older approaches such as forward or backward stepwise methods.

A key concept at the basis of their approach is *the principle of parsimony*. Statisticians view the principle of parsimony as a "bias versus precision" tradeoff. In general, bias (of parameter estimates) decreases, but variance (of the parameter estimates) increases as the dimension of the model (number of parameters) increases. The fit of any model can be improved (and hence a reduction in bias achieved) by increasing the number of parameters; however, a tradeoff or cost is increasing variance and must be considered when selecting a model(s) for inference. Parsimonious models achieve a proper tradeoff between bias and variance. All model selection methods are based to some extent on the principle of parsimony (Burnham and Anderson 2002:31). One older method of model choice used by quantitative ecologists selecting from a set of candidate capture-recapture models, for example, is to come up with a global model that fits the data adequately based on a goodness of fit test and then use sequential likelihood ratio tests to decide whether simpler nested models can be used for the data (for examples see Pollock *et al.* 1990; for a more detailed description of the approach see Williams *et al.* 2002:54). However, this method cannot handle nonnested model structures very well, and it tends to favor models with more parameters than some other approaches, such as AIC (Burnham and Anderson 2002). Burnham and Anderson (2002) emphasize that model selection is really distinct from traditional hypothesis testing and present the AIC based on the pioneering work of Akaike in the 1970s and 1980s (e.g., Akaike 1973).

THE AKAIKE INFORMATION CRITERION (AIC)

The AIC method is based on the likelihood but has a penalty added to encourage parsimony (i.e., models with as few parameters as necessary). The objective is to compare a suite or set of models (not necessarily nested models) and choose the one that minimizes AIC. The criterion can be expressed mathematically as:

$$AIC = -2\ln[L(\theta|x)] + 2\delta$$

where δ is the number of parameters estimated in the model. The absolute magnitude of AIC is not very relevant, but the differences in AIC among different models are the focus of model selection. Usually all the models are compared to the model with minimum achieved AIC by using a table of differences in AIC. Thus, for a particular model (k) the difference would be:

$$\Delta AIC_k = AIC_k - AIC_{min}.$$

Burnham and Anderson (2002) note as a rough rule of thumb that all models with AIC differences of less than 2 have a substantial level of empirical support, 4 through 7 substantially less support, and greater than 10 essentially no support. The ΔAIC_k values also form the basis of the normalized AIC model weights:

$$w_k = \frac{\exp\left(-\frac{1}{2}\Delta AIC_k\right)}{\sum_{r=1}^{R} \exp\left(-\frac{1}{2}\Delta AIC_r\right)}$$

$$r = 1, \ldots, R$$

for a suite of R models, and w_k is interpreted as "the weight of evidence in favor of model k being the actual Kullback-Leibler best model for the situation at hand" (Burnham and Anderson 2002:75). The AIC weights sum to 1 for all of the members of the model set, again emphasizing the relative nature of AIC comparisons, conditional on the model set. AIC model weights may also be interpreted (heuristically) as the probability that model k is the "best" model in the candidate set (Burnham and Anderson 2004). When multiple models in the candidate set have some common feature (e.g., different formulations of the same hypothesis, or multiple models containing the same factor or covariate), one approach for determining the overall level of support for that common feature (given the model set) is to sum the model weights for each of the respective models (Burnham and Anderson 2004). For example, suppose there are five models in the candidate set and the AIC model weights for Models 1–5 are 0.3, 0.3, 0.2, 0.1, and 0.1, respectively. All models have a comparatively similar level of support; hence, there is no clear group of models (and therefore hypotheses) that may be considered as "better" representations of the data. However, only Models 1, 2, and 3 include habitat as a covariate for occupancy, and adding these model weights could be used to assess the total level of support that occupancy is related to a site's habitat. This gives a combined model weight of 0.8 (= 0.3 + 0.3 + 0.2), which translates to substantial overall support for the habitat hypothesis.

Burnham and Anderson (2002) also suggest a small sample second-order bias adjustment to AIC:

$$AIC_C = AIC + \frac{2\delta(\delta+1)}{n-\delta-1},$$

where n is the effective sample size. Unless the sample size is very large compared to the number of parameters ($n/\delta \leq 40$), then AIC_C is generally recommended for use in practice. We note that the exact nature of sample size is not always obvious, for example, in capture-recapture and occupancy modeling. In fact, the "effective sample size" may also vary for different parameters in the model (e.g., be different for occupancy and detection probabilities). Because of this dilemma, we currently have no firm suggestions on what should be regarded as an "effective sample size" for the occupancy models we describe in this book; hence we simply use AIC. The effect of this is that more complex models are perhaps ranked higher than they should be in the examples we present.

GOODNESS OF FIT AND OVERDISPERSION

In models of overdispersed data, the mean or expectation structure of the model is adequate, but the variance structure $(\sigma^2(\theta))$ is inadequate. One approach is to think of the true variance structure as following the form $(\gamma(\theta)\sigma^2(\theta))$; however, it is complex to fit such a form. As a simpler approach, we suppose $\gamma(\theta) = c$, so that the true variance structure $c\sigma^2(\theta)$ is some constant multiplier of the theoretical variance structure $\sigma^2(\theta)$. For example, in capture-recapture models, a multinomial likelihood is often used, and while the expectation structure of the multinomial model may be adequate (i.e., point estimate of parameters may be valid), the variance structure may be inadequate. Values of c in the range of 2–4 would not be unreasonable due to positive correlations among individuals within a group or flock of animals.

A common method of estimating overdispersion is to use the observed chi-squared goodness of fit statistic for a global model (the most general model in the model set; the model with the most parameters) divided by its degrees of freedom:

$$\hat{c} = \chi^2 / df.$$

If there is no overdispersion or lack of fit, then c should equal 1, and $\hat{c}$ should be approximately 1 (because the expected value of a chi-square statistic is equal to its degrees of freedom). Many popular capture-recapture programs such as MARK (see, e.g., White and Burnham 1999; White et al. 2001) give a variety of methods (such as bootstrapping) for estimating $\hat{c}$, but we will not present that level of detail here. In Chapter 4 we discuss one way of estimating $\hat{c}$ specifically for single-species, single-season occupancy models. In the next section we show how the AIC criterion is modified by overdispersion.

QUASI-AIC

It is our experience that overdispersion is very common in modeling ecological data and that it needs to be included in our model selection criterion. The

AIC and AIC_C criterion can be modified for overdispersion ($\hat{c}$) (Burnham and Anderson 2002:70) using:

$$QAIC = \frac{-2\ln[L(\theta|x)]}{\hat{c}} + 2\delta$$

and

$$QAIC_C = QAIC + \frac{2\delta(\delta+1)}{n-\delta-1}.$$

Once QAIC or $QAIC_C$ have been used, the empirical estimates of variances and covariances can be obtained by multiplying the theoretical model-based variances and covariances by $\hat{c}$. Note that, although $\hat{c}$ is estimated based on the most general model in the set, $\hat{c}$ should be used to estimate variances and covariances for parameters of all models in the set. Burnham and Anderson (2002) point out that this approach has been used since at least the 1970s.

MODEL AVERAGING AND MODEL SELECTION UNCERTAINTY

In many ecological situations, multiple models in the candidate set may be reasonable; that is, the "best" model is not always apparent. Instead of choosing a single "best" model to draw inferences from, we may use estimates from multiple models in the candidate set, calculating "model averaged" estimates. In this case, the AIC weights of the candidate models are used to obtain a weighted average of the individual parameter estimates. For a suite of R models, the estimator is:

$$\hat{\theta}_A = \sum_{r=1}^{R} w_r \hat{\theta}_r,$$

with the subscript indicating model r. The estimate of the variance of the model-averaged estimator allowing for model uncertainty is:

$$Var(\hat{\theta}_A) = \left[\sum_{r=1}^{R} w_r \sqrt{Var(\hat{\theta}_r|g_r) + (\hat{\theta}_r - \hat{\theta}_A)^2} \right]^2.$$

$Var(\hat{\theta}_r|g_r)$ is the variance of the estimate obtained from model r. The notation indicates that this variance is conditional on the model r. The second component of the variance then reflects model uncertainty by focusing on the difference between a model-specific estimate and the weighted mean estimate based on all models. We refer the interested reader to Burnham and Anderson (2002:158) for more detail.

3.7. DISCUSSION

We now proceed to use many of these statistical techniques in the remainder of the book. The general approach will be to derive a general likelihood for each sampling design we will consider and then to use model selection methods and goodness of fit tests to choose a particular model for use on a data set. In Chapter 4 we consider the simplest practical case of sampling a series of sites for occupancy on multiple occasions in a single season where detection probability for the species at each site is less than 1.

Single-species, Single-season Occupancy Models

In the preceding chapters we have introduced many of the basic concepts that readers should keep in mind when estimating and modeling species occupancy. It is important that these concepts are well understood so that the approaches presented in the following chapters may be used correctly and to their fullest utility. By covering the underlying principles early in this book, we hope readers will be mindful of how these issues interrelate with the analytic techniques that we shall cover for estimating occupancy and related parameters. Further, without a firm understanding of the underlying statistical and design concepts, it would be very easy to misuse these techniques, potentially resulting in erroneous conclusions.

In the remainder of this book, we develop a series of models that can be used to estimate and model occupancy patterns and dynamics. Many of the models we present in the latter chapters have evolved from considering the problem of estimating the proportion of area (or patches) occupied by a single species, at a single point in time. In this chapter we outline the general sampling problem; review previous two-step ad hoc methods; and then present a new flexible, robust, and powerful model-based approach. Several practical examples are given throughout to illustrate these methods. We conclude the chapter with a discussion of the important issues raised and show how this general approach leads us to more complex topics that will be covered in later chapters.

4.1. THE SAMPLING SITUATION

We wish to estimate the proportion of an area, or proportion of suitable habitat within an area, that is inhabited by a target species. We use the term *area* in the general sense of a statistical population, that is, a collection of sampling units about which we wish to make inference. Sampling units may constitute arbitrarily defined spatial units (such as grid cells of a specified size) or discrete naturally occurring sampling units (e.g., forest remnants, ponds, or islands). From this population of S sampling units, s units are selected at which the intent is to establish the presence (occupied) or absence (unoccupied) of the target species. Throughout this book we generally regard that S is very large in comparison to s, and that we wish to make inference about the population of S units, not the sample of s units (this may not always be appropriate, and we detail a potential solution in Section 4.5). Henceforth, we shall generically refer to sampling units as *sites*. The manner in which sites are selected has important consequences for the appropriateness and accuracy of resulting inferences, although at this point we shall assume sites have been selected in such a manner that they are representative of the entire population (e.g., a random sample). We shall revisit issues related to site selection in Chapter 6.

While species presence may be confirmed by detecting the species at a site, it is usually impossible to confirm whether a species is absent. The nondetection of the species may result either from the species being genuinely absent, or from the species being present at the site but undetected during the surveying. Unless the species is so conspicuous that it will always be detected when present (a situation we believe to be very rare), or unless a very intensive level of surveying has been conducted at each location, there may be a good chance that the species was indeed present but not detected. This has long been recognized by many field biologists who have used repeated surveys to minimize the possibility of declaring a species "falsely absent" from a location.

We therefore consider a basic sampling scheme in which s sites are each surveyed K times for the target species (later we generalize to allow an unequal number of surveys). During each survey, appropriate methods are used to detect the species at the sites. Such methods include the visual, aural, or indirect confirmation of at least one individual of the species. Indirect methods may involve scent stations, tracking tunnels, or detecting other species sign, such as fresh droppings. It is assumed that the target species is never falsely detected at a site when absent (i.e., by misidentification of the species), which is likely to be a reasonable assumption in many situations (Tyre *et al.* 2003; Wintle *et al.* 2004).

The K surveys are conducted over a suitable time frame during which the s sites are all closed to changes in the state of occupancy; in other words, sites are either always occupied or unoccupied during the surveying period (this may be relaxed in some situations; see Section 4.4). We define this period of population closure as a *season*; the actual time frame encompassed by a season will vary from situation to situation. For example, during a study of breeding bird colonies, a season may last for two or three months, while for small mammal studies the closure assumption may only be reasonable for a week. In effect, the concept of a "season" enables us to consider a snapshot of the population at a given point in time, from which we can attempt to infer patterns about the level of occupancy.

The sequence of detections (1) and nondetections (0) of the target species from the K surveys of site i is recorded as a detection history (h_i). For example, suppose three surveys were conducted at a site with the species being detected in the first and third surveys, but not detected during the second. The detection history for this site could be expressed as $h_i = 101$. Similarly, the detection history for a site where the species was not detected in any of the three surveys would be $h_i = 000$.

We consider that there are two processes occurring in the general sampling situation: occupancy and detectability. *Occupancy* relates to the presence or absence of the species from sites during the sampling period (the "season"), which will be the quantity of prime interest in many situations. *Detectability* is an aspect of the sampling/surveying protocols that will generally be regarded as a nuisance parameter. However, as illustrated in Chapter 2, not accounting for the imperfect detection of the species may result in misleading conclusions about the population. In the remainder of this chapter we first consider how occupancy could be estimated if the target species was always detected when present (i.e., no false absences) or if the probability of detection was known. We then review two-step ad hoc approaches for estimation when detectability is unknown before moving to a full model-based approach for the simultaneous estimation of both occupancy and detectability.

4.2. ESTIMATION OF OCCUPANCY IF PROBABILITY OF DETECTION IS 1 OR KNOWN WITHOUT ERROR

In most practical situations, we consider it unlikely that a target species will always be detected when present at a site. This is especially true for those rare or cryptic species that are often of special interest to managers and researchers.

However, it is instructive to consider this best possible case because it provides useful insights and a "gold standard" for determining how well an estimator may be performing in a given situation. The precision of any estimator that incorporates detectability could not do better than the estimator for which the occupancy state of sites is known without error.

Let us assume that all sites have a common probability of being occupied by the species, ψ. Therefore, the number of sites that are occupied by the target species (x) from a random sample of s sites will follow a binomial distribution, with $E[x] = s\psi$ and $Var[x] = s\psi(1 - \psi)$ (Chapter 3). Using the standard results for a binomial proportion, an estimate of ψ when the species is detected perfectly will be:

$$\hat{\psi}_B = \frac{x}{s} \qquad (4.1)$$

with an associated variance of:

$$Var(\hat{\psi}_B) = \frac{\psi(1-\psi)}{s}, \qquad (4.2)$$

which can naturally be approximated by substituting in the estimated value for ψ.

Now let us assume that the target species is detected imperfectly and that the probability of detecting the species in a single survey of an occupied site is equal to p, which has a value that is known exactly (i.e., has no associated sampling error). Therefore, the probability of detecting the species at least once after K surveys of the site will be $p^* = 1 - (1 - p)^k$. This is 1 minus the probability of the species being undetected in all K surveys. The number of sites at which the target species is *detected* (s_D) from a random sample of s sites will again follow a binomial distribution with $E[s_D] = s\psi p^*$ and $Var[s_D] = s\psi p^*(1 - \psi p^*)$, that is, the probability of the species being present and detected at a site will be ψp^*. Based on this, an estimator for the proportion of sites occupied when p is known would be:

$$\hat{\psi}_p = \frac{s_D}{sp^*} \qquad (4.3)$$

with variance:

$$Var(\hat{\psi}_p) = \frac{\psi(1-\psi p^*)}{sp^*},$$
$$= \frac{\psi(1-\psi)(1-\psi p^*)}{s(1-\psi)p^*} \qquad (4.4)$$
$$= \frac{\psi(1-\psi)}{s} + \frac{\psi(1-p^*)}{sp^*}$$

Eq. (4.4) highlights that this variance consists of two components. The first component corresponds to the binomial variation associated with the true underlying value of ψ. The second component is due to the imperfect detection of the species and, in effect, is the result of having to estimate the number of sites that were occupied in the sample. As we shall see, the variance for most occupancy estimators can be written in a similar form.

Another important point, highlighted by Eq. (4.4), is that when a species is detected imperfectly, the variance of an occupancy estimator can never be smaller than the binomial variation term. This is because the second component must be greater than 0 (although it will tend to 0 as p^* tends to 1.0, that is, as it becomes almost certain that the species will be detected at least once in K surveys of an occupied site). This is also useful information from a study design perspective, to which we shall return in Chapter 6.

However, rarely will the detection probability be exactly known prior to a study being conducted, so while the above discussion does provide some interesting insights to which we shall return later, the estimators and associated variances above are likely to be of little practical value. More often than not, both occupancy and detection probabilities will be unknown and must be estimated jointly from the collected data.

4.3. TWO-STEP AD HOC APPROACHES

There have been a number of independent efforts to estimate the proportion of sites occupied by a target species in the face of an unknown level of imperfect detection. These approaches can be loosely grouped into two classes. In the first class of methods, a two-step approach is taken, in which the probability of detecting the species is estimated in the first step and is then used in a second step to estimate the occupancy parameter. We define such methods as two-step, ad hoc estimation procedures. They do provide valid estimators, but they do not provide a framework that is as flexible for comparing competing hypotheses about the system as offered by considering the full sampling process. Further, they do not lend themselves to generalization to more complex sampling situations. The second type of approach, and in our mind the superior one, involves directly modeling the sampling process in a way that enables the simultaneous estimation of both occupancy and detectability parameters in a single model-based framework. This class of model is considered in the next section and for the remainder of this book.

GEISSLER-FULLER METHOD

The earliest method for occupancy estimation of which we are aware is that proposed by Geissler and Fuller (1987). They considered a sampling situation in which site i is surveyed K_i times for the presence/absence of the species, and the species is first detected at the site in survey t_i. A conditional probability of detection can be estimated for each site as the proportion of surveys in which the species is detected, following the first detection of the species at the site, that is:

$$\hat{p}_{GF,i} = \frac{\sum_{j=t_i+1}^{K_i} h_{ij}}{K_i - t_i},$$

where h_{ij} is a binary indicator for whether the species was detected (1) or not (0) in survey j of site i. For example, if a site was surveyed 6 times, with the species being detected for the first time in the second survey and detected subsequently only once, then $K_i = 6$, $t_i = 2$, and $\hat{p}_{GF} = 1/(6-2) = 0.25$. The probability of detecting the species at least once during the K_i surveys can then be calculated as:

$$\hat{p}^*_{GF,i} = 1 - (1 - \bar{\hat{p}}_{GF})^{K_i},$$

where $\bar{\hat{p}}_{GF}$ is the simple average (across all sites) of the $\hat{p}_{GF,i}$'s. A Horvitz-Thompson based–estimate of the occupancy probability is then given:

$$\hat{\psi}_{GF} = \frac{\sum_{i=1}^{s} \dfrac{w_i}{\hat{p}^*_{GF,i}}}{s},$$

where w_i is an indicator variable equaling 1 if the species was observed at the site, and 0 otherwise.

Geissler and Fuller (1987) then suggest that the actual estimator should be the median or the mean value obtained from a large number of nonparametric bootstraps, and that the standard error for the estimators can also be approximated from a nonparametric bootstrap procedure as the sample standard deviation of the B values of ψ_{GF} obtained from the bootstrapped data sets (Manly 1997).

One of the key assumptions of the Geissler-Fuller approach is that the probability of detecting the species at a site is constant for all surveys, and especially that the detection probability does not change after the first detection of the species. This is a particularly important assumption in many practical situations. Once observers have detected the species at a site, they may be more or less likely to detect the species in subsequent surveys. One could readily imagine a situation in which once the observer has found an indication that the species occupies a site, the observer uses that information to make detection of the species easier in future visits, for example, by returning to an animal's den or nest that was discovered during a previous survey. In addition, while Geissler and Fuller (1987) define the detection probabilities $\hat{p}^*_{GF,i}$ as site specific, by using the average of these values in the calculation of $\hat{p}^*_{GF,i}$, they effectively assume that the probability of detecting the species in any given survey is equal across all sites. Another implicit assumption they make is that the probability that a site is occupied by the species is also constant across all sites. These assumptions are not uncommon, and indeed the majority of the approaches in this book make similar assumptions. However, as we shall see, the model-based approach discussed below provides a flexible framework that allows some of these factors to be incorporated directly into the method of analysis.

AZUMA-BALDWIN-NOON METHOD

Azuma et al. (1990) consider a slightly different sampling scheme to that given above. For the monitoring of spotted owls (Strix occidentalis) in Washington, Oregon, and California, they consider a situation in which s sites are monitored for the presence/absence of owls with up to K repeated surveys of the sites. However, the site is not surveyed again once the presence of a pair of spotted owls has been confirmed at the location (by a detection). In later sections we refer to this general design as a removal study design, as sites are "removed" from the pool of sites being actively surveyed once the species is first detected, and also because of the analogy with "removal studies" in mark-recapture (Otis et al. 1978).

Azuma et al. (1990) assumed a constant probability of detecting the species (given presence), p, and then modeled the number of surveys required to detect the species (Y) as a truncated geometric distribution (TGD):

$$\Pr(Y = t | 0 < Y \le K) = \frac{p(1-p)^{t-1}}{1-(1-p)^t}, \quad t = 1, 2, 3, \ldots, K$$
$$= 0, \text{otherwise.}$$

This provides a model for the time to first detection for those locations where the species was detected during the K surveys. Azuma et al. (1990) suggest a method-of-moments–based estimator for p, by equating the expected value of Y to the average number of surveys required to detect the species (conditional on the species being detected) and iteratively solving for p:

$$E(Y | 0 < Y \le K) = \frac{1}{p} - \frac{K(1-p)^K}{1-(1-p)^K}$$

An estimate of the proportion of sites occupied would then be:

$$\hat{\psi}_{ABN} = \frac{s_D}{s\hat{p}^*_{ABN}},$$

where s_D is the number of sites where the species was detected at least once and $\hat{p}^*_{ABN} = 1 - (1 - \hat{p}_{ABN})^K$ is the estimated probability of detecting the species at least once during K surveys based upon the TGD. Azuma et al. (1990) present a number of equations for estimating the variance of their occupancy estimator, although provided that the number of sites surveyed is reasonably large (i.e., $s \approx s - 1$), they should all provide similar results. Assuming that S, the total number of "sites" in the population, was very large (i.e., an infinite population), the variance for their estimator will be approximately:

$$Var(\hat{\psi}_{ABN}) = \frac{\hat{\psi}_{ABN}}{s-1}\left((1 - \hat{\psi}_{ABN}) + \frac{(1 - \hat{p}^*_{ABN})}{\hat{p}^*_{ABN}}\right).$$

Note that $Var(\hat{\psi}_{ABN})$ can be rearranged into a form similar to Eq. (4.4), the estimator variance for when p is known exactly, but with $s - 1$ rather than s in the denominator. Azuma et al. (1990) suggest this adjustment as estimated quantities for occupancy and detection probabilities have been used, though no justification for the adjustment is given. They also provide extra details on correcting for finite sampling (where s represents a substantial fraction of the total population of possible sites); estimating the number of locations to sample for a given level of precision; and assessing whether the TGD assumption is reasonable (see Azuma et al. 1990 for these details).

NICHOLS-KARANTH METHOD

Nichols and Karanth (2002) suggest a more general approach to the problem of estimating the proportion of area occupied, in the context of large-scale monitoring of tiger and their prey species (Fig. 4.1) in India. They advocate that an estimate of the number of occupied sites (x) can be obtained by using closed population mark-recapture methods (e.g., Otis et al. 1978; Williams et al. 2002). By focusing on the detection histories for sites at which the species was detected at least once, and considering them as the capture histories of individuals encountered during a mark-recapture experiment (White et al. 1982), estimating the number of occupied sites where the species was not detected is completely analogous to estimating the number of individuals in the population that were never captured. Once x has been estimated using an

FIGURE 4.1 Camera-trap photograph of a tiger prey species, a muntjac (*Muntiacus muntjak*), from southern India (K. Ullas Karanth).

appropriate technique, an occupancy estimator and its associated variance are obtained by:

$$\hat{\psi}_{NK} = \frac{\hat{x}}{s},$$

and:

$$Var(\hat{\psi}_{NK}) = \frac{\psi(1-\psi)}{s} + \frac{Var(\hat{x})}{s^2}.$$

Note the general form of the variance equation as compared to Eq. (4.4), although here uncertainty, with respect to detection probability, is taken into account directly via $Var(\hat{x})$.

4.4. MODEL-BASED APPROACH

An alternative to the ad hoc estimation methods given previously is to model the probability of the observed outcomes resulting from the stochastic sampling process. Doing so, it is possible to estimate both the occupancy and detection parameters simultaneously. This framework also provides a direct means of investigating potential relationships between the probabilities of occupancy and detection and factors such as habitat type or weather conditions at the time of surveying. That is, competing hypotheses about the system can be easily explored and compared within a model-based context, unlike in the ad hoc approaches given above. In addition, the flexibility of the approach enables unequal survey effort at different sites, providing a great deal of flexibility for realistic study designs.

BUILDING A MODEL

A model can be constructed in the manner outlined in Chapter 1: Take a set of hypotheses to create a conceptual model about the system, construct a verbal model from this set of ideas, and then translate the verbal descriptions into a set of mathematical equations that may be used to estimate the parameters of interest. The development of the model here is presented slightly differently to MacKenzie *et al.* (2002), with the various steps in the process being outlined more explicitly. The underlying principles of the model-building procedure are, however, the same.

The basis of our conceptual model is that there are two stochastic processes occurring that affect the outcome of whether the species is detected at a site. First, a site may be either occupied (with probability ψ) or unoccupied (with probability $1 - \psi$) by the species. If the site is unoccupied, then obviously the species cannot be detected there. If the location is occupied, then at each survey (j) there is some probability of detecting the species (p_j; therefore, the probability of not detecting the species in the survey is $1 - p_j$); this is the second process. This conceptual model also includes the assumption that the occupancy status of sites does not change between surveys (i.e., closure of the site to changes in occupancy state), but we will be able to relax that assumption later.

For any given detection history, using this conceptual model we can develop a verbal description of the underlying processes that gave rise to the observed data. For example, consider the detection history $h_i = 0101$ resulting from four detection/nondetection surveys for the target species at a site, i. A verbal description of this detection history is:

"the site was occupied by the species, the species was not
detected in survey 1, was detected in survey 2, was
not detected in survey 3 and was detected in survey 4."

The probability of observing this particular detection history can then be obtained by translating this verbal description into a mathematical description through a series of equations. This is simply achieved by substituting the parameters that represent the relevant stochastic processes for the appropriate phrases. Hence the above description of the process could be translated to the following probability statement:

$$\Pr(h_i = 0101) = \psi(1 - p_1)p_2(1 - p_3)p_4$$

Comparison with the above verbal description should make it clear that to create the probability statement all that has happened is that the phrase "the site was occupied by the species" has been represented by the parameter ψ; the phrase "species detected in survey j" is represented by p_j; and the phrase "species not detected in survey j" is represented by $(1 - p_j)$. Probability statements for the observed detection histories can be obtained in a similar manner for all sites at which the species was detected at least once.

For sites at which the species was never detected, the same procedure of translating a verbal description into a probability statement is used, although it is marginally more complicated as there are now two possibilities for why the species was never detected at a site. Consider the history $h_i = 0000$, for which the verbal description of the data would be:

"the site was occupied by the species and the
species was not detected in any of the 4 surveys",

or

"the site was unoccupied by the species."

The first possibility could be translated to $\psi \prod_{j=1}^{4}(1 - p_j)$, while the second option translates to $(1 - \psi)$. Because the two possibilities are indistinguishable from the data, both must be represented in the probability statement as:

$$\Pr(h_i = 0000) = \psi \prod_{j=1}^{4}(1 - p_j) + (1 - \psi).$$

The fact that both options are possible for the observed data is represented by the addition of the two terms within the probability statement. We use this technique frequently throughout this book, whenever there are multiple possibilities for the same set of observed data.

Once the probability statement for each of the s observed detection histories is formed, assuming that the detection histories for the s sites are independent, the model likelihood for the observed data can be constructed in the usual manner (see Chapter 3), that is:

$$L(\psi, p|h_1, h_2, \ldots, h_s) = \prod_{i=1}^{s} \Pr(h_i),$$

which reduces to:

$$L(\psi, p|h_1, h_2, \ldots, h_s) = \left[\psi^{s_D} \prod_{j=1}^{K} p_j^{s_j} (1-p_j)^{s_D-s_j} \right] \left[\psi \prod_{j=1}^{K} (1-p_j) + (1-\psi) \right]^{s-s_D},$$

where s_D is the number of sites where the species was detected at least once, and s_j is the number of sites where the species was detected during the jth survey.

Under the assumption that the probability of detection is constant for all occasions, the above approach is equivalent to modeling the number of detections at each site (y_i) as a binomial random variable with an inflated zero class, that is:

$$\Pr(Y = y_i) = \psi \binom{K}{y_i} p^{y_i} (1-p)^{K-y_i}, \quad y_i > 0$$

$$= \psi(1-p)^K + (1-\psi), \quad y_i = 0$$

Such an approach has been used recently by a number of authors, including Stauffer et al. (2002, 2004), Tyre et al. (2003), and Wintle et al. (2004). However, by considering the sampling process more generally as a series of surveys with specific detection probabilities, we maintain a greater degree of flexibility.

The main assumptions of this model are: (1) the occupancy state of the sites does not change during the period of surveying; (2) the probability of occupancy is equal across all sites; (3) the probability of detecting the species in a survey, given presence, is equal across all sites; (4) the detection of the species in each survey of a site is independent of detections during other surveys of the site; and (5) the detection histories observed at each location are independent.

ESTIMATION

As discussed in Chapter 3, the likelihood equation defined above could be used to estimate model parameters using either a frequentist or a Bayesian philosophy (by regarding the likelihood function as the probability of observing the data given the parameters). In this section we focus primarily on estimation using maximum likelihood techniques. This is not a reflection of our beliefs about which statistical inference philosophy is more appropriate, but is a matter of convenience. For the type of model being considered here, computer-intensive methods such as Markov chain Monte Carlo (MCMC) are required to obtain the posterior distribution of the model parameters. This makes it more difficult to discuss general results and comparisons with other methods without using extensive computer simulation studies. By contrast, the maximum likelihood estimates (MLEs) of the model parameters can be written in a relatively simple form that makes discussion of results somewhat easier. However, in practice, as computer software will usually be used as part of the data analysis, it may be feasible to consider either approach.

Below we consider obtaining maximum likelihood estimates of the model parameters for two different situations: where detection probabilities are assumed to be either constant or survey specific. We also present the formula for calculating the asymptotic variance of the MLE of occupancy in the former situation for comparison with the ad hoc occupancy estimators given above. In most circumstances we envision that readers will use computer software to obtain parameter estimates; hence, we only present relatively rudimentary estimating equations to illustrate the intuitive nature of the estimators. It should also be noted that, for some data sets, the equations given below may result in estimates of occupancy that are greater than 1. This is because the equations do not enforce the constraint that the probabilities must take values between 0 and 1, which is further encouragement for using software with which this constraint can be enforced automatically.

Constant Detection Probability Model

Assuming a constant detection probability, the model likelihood becomes:

$$L(\psi, p | \mathbf{h}_1, \mathbf{h}_2, \ldots, \mathbf{h}_s) = \left[\psi^{s_D} p^{\sum_{j=1}^{K} s_j} (1-p)^{K s_D - \sum_{j=1}^{K} s_j} \right] \left[\psi(1-p)^K + (1-\psi) \right]^{s - s_D}. \quad (4.5)$$

By taking first derivatives with respect to each parameter, and equating to zero, we obtain the following equations:

$$\hat{\psi}_{\text{MLE}} = \frac{s_D}{s \hat{p}^*_{\text{MLE}}}, \quad (4.6)$$

where $\hat{p}^*_{\text{MLE}} = 1 - (1 - \hat{p}_{\text{MLE}})^K$ is the estimated probability of detecting the species at least once during a survey (given the species is present); and

$$\tilde{p}_{MLE} = \frac{\hat{p}_{MLE}}{\hat{p}^*_{MLE}} = \frac{\sum_{j=1}^{K} s_j}{Ks_D}, \tag{4.7}$$

where $\tilde{p}_{MLE}$ is the estimated probability of detecting the species during a survey, conditional upon the species being detected at least once at a site. Eq. (4.7) can be rearranged to obtain a closed-form estimate of $\hat{p}_{MLE}$, but requires finding the roots of a $K - 1$ order polynomial (which would typically involve using numerical methods whenever $K > 2$). However, note the intuitive form of the equations as they are given. Eq. (4.6) is very similar to the estimators given in Sections 4.2 and 4.3; thus, while not derived from a likelihood perspective, the other estimators should approximate the MLE whenever detection probabilities have been estimated appropriately. Eq. (4.7) shows that a conditional estimate of detection probability ($\tilde{p}_{MLE}$; from which we may numerically derive $\hat{p}_{MLE}$) is given by the ratio of total number of species detections to the total number of surveys conducted at sites where the species was detected at least once.

Likelihood theory suggests that the asymptotic variance formula for $\hat{\psi}_{MLE}$ can be obtained by inverting a quantity known as the *information matrix* (a matrix whose (i, j)th element is the expected value of the negative second partial derivative of the log-likelihood with respect to parameters i and j). We do not go through the mechanics of this here, but note that many software packages use a numerical method based on this theory to obtain variance and standard error values for parameter estimates for a wide range of likelihood-based statistical methods (e.g., mark-recapture modeling, generalized linear models).

The variance formula for $\hat{\psi}_{MLE}$ [Eq. (4.8)] can be expressed in two forms, with either two or three components. The two-component form has the familiar features of a component due to the binomial proportion and the second component related to uncertainty in the number of sites that were actually occupied in the sample. This second component can be rearranged to express $Var(\hat{\psi}_{MLE})$ as a function of three components. Now, the second component is the uncertainty in the number of occupied sites, assuming that p is known, and the third component is the contribution to $Var(\hat{\psi}_{MLE})$ from having to estimate p from the data simultaneously. That is, in the latter form, the first two components give the variance formula above for the situation where p is known [Eq. (4.4)]:

$$\begin{aligned} Var(\hat{\psi}_{MLE}) &= \frac{\psi(1-\psi)}{s} + \frac{\psi(1-p^*)(1-p)}{s[p^*(1-p)-Kp(1-p^*)]} \\ &= \frac{\psi(1-\psi)}{s} + \frac{\psi(1-p^*)}{sp^*} + \frac{\psi(1-p^*)Kp(1-p^*)}{sp^*[p^*(1-p)-Kp(1-p^*)]} \end{aligned} \tag{4.8}$$

Clearly, for a given set of data, the variance may be approximated by substituting in the estimated values for ψ and p (and p^*).

Survey-specific Detection Probability Model

Based upon the model likelihood with survey-specific detection probabilities, using the same techniques as before, we obtain the following estimating equations:

$$\hat{\psi}_{MLE} = \frac{s_D}{s\hat{p}^*_{MLE}}.$$

where now $p^*_{MLE} = 1 - \prod_{j=1}^{K}(1 - \hat{p}_{j,MLE})$,

and $\tilde{p}_{j,MLE} = \dfrac{\hat{p}_{j,MLE}}{1 - \prod_{i=1}^{K}(1 - \hat{p}_{j,MLE})} = \dfrac{s_j}{s_D}$.

Again, $\hat{p}_{j,MLE}$ cannot be simply expressed in a closed form; hence, a numerical method should be used, but note the intuitive form of $\tilde{p}_{j,MLE}$. Here s_D is equivalent to the number of surveys conducted at time j at sites where the species was detected at least once during the K surveys. Hence $\tilde{p}_{j,MLE}$ is the fraction of surveys conducted at time j at sites where the species was eventually detected that resulted in a detection.

We do not present an equation for $Var(\hat{\psi}_{MLE})$ when detection probabilities are survey specific here, but note that similar methods to those described above (inverting the information matrix) could be used.

Probability of Occupancy Given Species Not Detected at a Site

In many instances, one quantity that will be of interest is the probability that the species was present at a site given it was never detected. From Bayes' Theorem we have:

$$\psi_{condl} = Pr(\text{species present}|\text{species not detected})$$

$$= \frac{Pr(\text{species present and not detected})}{Pr(\text{species not detected})}$$

$$= \frac{\psi\prod_{j=1}^{K}(1-p_j)}{(1-\psi)+\psi\prod_{j=1}^{K}(1-p_j)}$$

This can be simply calculated from the estimated parameters. For example, suppose a site was surveyed twice and the species was never detected. Given an estimated occupancy probability of $\hat{\psi} = 0.65$ and detection probability estimates of $\hat{p}_1 = 0.3$ and $\hat{p}_2 = 0.6$, the estimated probability of the species being present given it was never detected at the site would be:

$$\frac{\hat{\psi}\prod_{j=1}^{2}(1-\hat{p}_j)}{(1-\hat{\psi})+\hat{\psi}\prod_{j=1}^{2}(1-\hat{p}_j)} = \frac{0.65\times(1-0.3)\times(1-0.6)}{(1-0.65)+0.65\times(1-0.3)\times(1-0.6)}$$

$$= \frac{0.65\times0.7\times0.4}{0.35+0.65\times0.7\times0.4}$$

$$= \frac{0.182}{0.532}$$

$$\approx 0.34$$

That is, the unconditional estimated probability of the species being present was 0.65, but by taking into account the fact that the species was not detected in two surveys, the estimated probability of presence is now 0.34. Hence, the fact that the species was not detected at a site can be incorporated into our inferential procedure about the occupancy status of specific sites. An approximate asymptotic variance for ψ_{condl} could be obtained by use of the delta method (Section 3.2), by differentiating the expression for ψ_{condl} with respect to ψ and the p's and given the variance-covariance matrix for ψ and the p's. The derivative of ψ_{condl} with respect to ψ is:

$$\frac{\partial\psi_{condl}}{\partial\psi} = \frac{1-p^*}{(1-\psi p^*)^2}$$

where p^* is the probability of detecting the species at least once in the K surveys (as defined above), and when detection probability is constant:

$$\frac{\partial\psi_{condl}}{\partial p} = -\frac{\psi(1-\psi)K(1-p)^{K-1}}{(1-\psi p^*)^2},$$

or when detection probability is survey specific:

$$\frac{\partial\psi_{condl}}{\partial p_j} = -\frac{\psi(1-\psi)\prod_{k\neq j}(1-p_k)}{(1-\psi p^*)^2}.$$

EXAMPLE: BLUE-RIDGE TWO-LINED SALAMANDERS

The blue-ridge two-lined salamander (*Eurycea wilderae*) is one of over 30 species of salamanders that occur within the Great Smoky Mountains National Park (Dodd 2003). Like many amphibians, *E. wilderae* has a dual life strategy, with a one- to two-year aquatic larvae period, followed by a more terrestrial adult phase, with individuals often occurring far from water (Petranka 1998). In the late 1990s researchers, working in cooperation with the National Park Service, conducted a series of studies aimed at developing efficient long-term monitoring methods for a suite of salamander species in the southern Appalachians. Occupancy was one state variable explored, and more detailed occupancy analysis of this data can be found in Bailey *et al.* (2004) and MacKenzie *et al.* (2005). Here we use detection information from one year (2001) and one species (blue-ridge two-lined salamander) to illustrate simple, single-season models described above.

Salamanders were sampled using two detection methods: an area-constrained natural cover transect ($50 \times 3\,m$) and a $50\,m$ coverboard transect, consisting of five coverboard stations spaced at $10\,m$ intervals. The two transects were parallel to one another and separated by approximately $10\,m$. Natural cover objects (wood and rock) or coverboards (pine boards) were carefully turned, and encountered salamanders were identified as to species or species complexes. Together, the area sampled by these transects constituted a "site" or sample unit. Sites were located near trails, approximately $250\,m$ apart to ensure independence among sites. Thirty-nine sites ($s = 39$) were sampled once every two weeks from April to mid-June ($K = 5$ surveys), when salamanders are believed to be most active and near the surface.

We explore two simple models: one model assumes that occupancy and detection probabilities are constant across sites and surveys (denoted $\psi(\cdot)p(\cdot)$), and the second model assumes constant occupancy among sites, but detection probabilities are allowed to vary among the five surveys (denoted $\psi(\cdot)p(t)$). The data were analyzed using the software PRESENCE 2.0.

Of the 39 sites surveyed, blue-ridge two-lined salamanders were detected at 18, which leads to a naïve occupancy estimate (s_D/s) of 0.46. Compare this to the parameter estimates given in Table 4.1, where the two models have been ranked according to AIC and model averaged estimates have been calculated (see Chapter 3 for details of these procedures). Estimates from both models indicate that occupancy is underestimated by approximately 20% when detection probability is not accounted for. The model with constant detection probability has much greater support as indicated by the model weights, but there is still sufficient support for model $\psi(\cdot)p(t)$, suggesting some variation in detection probabilities among surveys. Model averaged estimates suggest the probability of detecting blue-ridge two-lined salamanders during a single survey of

TABLE 4.1 Summary of Models Fit to Blue-ridge Two-lined Salamander Example Data

Model	ΔAIC	w	$NPar$	$-2l$	$\hat{\psi}$	$SE(\hat{\psi})$	$\hat{p}_1$	$\hat{p}_2$	$\hat{p}_3$	$\hat{p}_4$	$\hat{p}_5$
$\psi(\cdot)p(\cdot)$	0.00	0.73	2	161.76	0.60	0.12	0.26	0.26	0.26	0.26	0.26
$\psi(\cdot)p(t)$	1.95	0.27	6	155.71	0.58	0.12	0.18	0.13	0.40	0.35	0.27
Model Averaged Estimate					0.59	0.12	0.24	0.22	0.30	0.28	0.26

ΔAIC is the relative difference in AIC values compared with the top-ranked model; w is the AIC model weight; $NPar$ is the number of parameters; and $-2l$ is twice the negative log-likelihood value. Estimates of occupancy ($\hat{\psi}$) and its standard error ($SE(\hat{\psi})$) are given, along with estimates of detection probabilities ($\hat{p}$). Naïve occupancy estimate is 0.46 (18 of 39 sites had one or more detections).

occupied sites is only 0.2–0.3; thus, the probability of missing the species over the course of the sampling season is over 0.15 (i.e., the probability of a false absence at an occupied site).

Further, suppose we wanted to estimate the probability of a site being occupied, given blue-ridge two-lined salamanders were not detected there in any of the five surveys ($\hat{\psi}_{condl}$). For simplicity, here we shall estimate ψ_{condl} based upon the "best" model, but note that the model averaged estimates could be used equally well, although this would require the variance-covariance matrix for the model-averaged estimates to be calculated. From the model $\psi(\cdot)p(\cdot)$, an estimate of ψ_{condl} would be:

$$\hat{\psi}_{condl} = \frac{\hat{\psi}(1-\hat{p})^K}{1-\hat{\psi}\left[1-(1-\hat{p})^K\right]} = \frac{0.59(1-0.26)^5}{1-0.59\left[1-(1-0.26)^5\right]}$$
$$= 0.25$$

The standard error for ψ_{condl} can be approximated by application of the delta method, where the variance-covariance matrix for ψ and p given by PRESENCE 2.0 is:

$$V = \begin{bmatrix} 0.0150 & -0.0038 \\ -0.0038 & 0.0033 \end{bmatrix}$$

and

$$h'(\hat{\theta}) = \left[\frac{\partial \hat{\psi}_{condl}}{\partial \hat{\psi}} \quad \frac{\partial \hat{\psi}_{condl}}{\partial \hat{p}} \right]$$
$$= [0.7720 \quad -1.2551],$$

giving

$$Var(\hat{\psi}_{condl}) = [0.7720 \quad -1.2551]\begin{bmatrix} 0.0150 & -0.0038 \\ -0.0038 & 0.0033 \end{bmatrix}\begin{bmatrix} 0.7721 \\ -1.2551 \end{bmatrix}$$

$$= 0.0215$$

or $SE(\hat{\psi}_{condl}) = 0.15$.

MISSING OBSERVATIONS

MacKenzie *et al.* (2002) developed their likelihood equation by assuming that the probability of detecting the species during survey j was the same at all sites. The logic behind this assumption is that for many species, detection probabilities will vary with environmental conditions (e.g., level of precipitation, temperature, etc.) and that these conditions will affect all sites in a similar manner at the same instant. It is, therefore, important to realize that from a biological viewpoint, the model with survey-specific detection probabilities only makes sense when sites are surveyed simultaneously (or within a short time frame).

However, in many situations this may not be the case. From a logistical perspective it is unlikely that all, or even a substantial fraction, of the sites can be surveyed at any one time. Data are often collected by small teams of surveyors who must travel from one site to another, precluding the simultaneous surveying of all sites. We use this practicality of the sampling to introduce the concept of missing observations. For example, suppose that over a five-day period, a number of sites are to be searched multiple times for the target species. Consider the data below for two of the sites, where the first site was searched on days 1, 2, 3, and 5, whereas the second site was searched on days 2, 4, and 5. For the first site, day 4 could be treated as a missing observation, and days 1 and 3 as missing observations for the second site.

Site	Day 1	Day 2	Day 3	Day 4	Day 5
1	1	0	1	—	0
2	—	0	—	1	1

By considering the general form of the likelihood above, where the stochastic processes that resulted in an observed detection history are considered to construct an associated probability statement, missing observations can be easily accommodated. In essence, if site i was not surveyed on occasion j, then the probability of detecting the species at that occasion must be zero, that is, $p_{ij} = 0$. By imposing this constraint, whenever a site is not surveyed, the jth survey occasion is effectively ignored for that site, and neither p_j nor $(1 - p_j)$ appear in the probability statement. This fairly reflects the fact that no information regarding the detection (or nondetection) of the species has been collected from that site at that time. Returning to the above example, the probability of observing the respective detection histories could be expressed as: $\Pr(h_1 = 101\text{--}0) = \psi p_1(1 - p_2)p_3(1 - p_5)$ and $\Pr(h_2 = -0\text{--}11) = \psi(1 - p_2)p_4 p_5$, where the "–" in the detection history denotes the missing observation(s).

While we have introduced the concept of missing observations from a design perspective in which logistical constraints prohibit sites being surveyed simultaneously, missing observations may arise through a myriad of other circumstances, such as a change in weather, a vehicle or equipment breakdown, or hungover field technicians (or principal investigators!). To date, we have treated the two types of missing observations (design-based and random) in a similar manner, although we note that missing observations arising from random events could also be modeled by an additional set of parameters. Such an approach would be more in line with one of the most important tenets of statistical inference, that the method of analysis should reflect the manner in which the data were collected. However, in this situation we doubt that such an extension would have a major effect on the resulting inference, provided that the probability of a missing observation does not depend upon the occupancy and detection probabilities.

The ability to accommodate missing observations has important ramifications for the design of occupancy studies based on this method of analysis. Equal sampling effort is not required across all sites. Indeed, as the purpose of the repeated surveys is to collect appropriate information allowing detection probabilities to be estimated, going to all sites an equal number of times may not be an efficient use of resources. This and other design considerations are explored more fully in Chapter 6.

One special design that we note here is the "removal design" in which a maximum of K surveys will be conducted at a site, but surveying for the species halts once the species is first detected. The surveys that are not conducted after the species has been detected at a site can be regarded as a form of missing observations. In this special case, survey-specific detection probabilities can no longer be estimated (technically, they are not identifiable), although imposition of a single constraint (e.g., $p_{K-1} = p_K$) renders the remaining detection probabilities estimable, or they could be a function of measured covariates (see below). However, in the case where detection probabilities are constant, the model likelihood can be expressed as:

$$L(\psi, p | h_1, h_2, \ldots, h_s) = \left[\psi^{sD} p^{sD} (1-p)^{\sum_{i=1}^{sD} t_i - sD} \right] \left[\psi(1-p)^K + (1-\psi) \right]^{s-sD},$$

where t_i is the number of surveys required until the species was detected at site i, with the summation only applied to the s_D sites where the species was detected.

COVARIATE MODELING

So far, all of the methods detailed have effectively assumed that the probability of a site being occupied and the probability of detecting the species in a survey (given species presence) are equal across sites. Often, this is unlikely to be a reasonable assumption and the probabilities will actually vary among sites; that is, the probabilities will be heterogeneous. In fact, in many situations (such as species-habitat modeling; see Chapter 2), the manner in which these probabilities vary in accordance with characteristics of the site may be the primary focus of the study. By using an appropriate link function and the model-based approach of this chapter, it is possible to model the probability of a site being occupied as a function of measured covariates.

There is some choice in the type of link function one might use (McCullagh and Nelder 1989), but here we tend to use the logit link or logistic equation (see also Chapter 3). In doing so, we can draw parallels with logistic regression techniques. In fact, the modeling we discuss in detail could be considered a form of generalized logistic regression analysis where there is some uncertainty as to whether an observed absence equates to a true absence. There are also clear parallels between our modeling and that used for capture-recapture studies, where generalized logistic regression is frequently used in the face of uncertainty (e.g., Lebreton *et al.* 1992; Williams *et al.* 2002).

Using the logit link, we can express the probability of site i being occupied as:

$$\text{logit}(\psi_i) = \beta_0 + \beta_1 x_{i1} + \beta_2 x_{i2} + \ldots + \beta_U x_{iU},$$

which is a function of U covariates associated with site i ($x_{i1}, x_{i2}, \ldots, x_{iU}$) and the $U + 1$ coefficients that are to be estimated: 1 intercept or constant term (β_0) and U regression coefficients for each covariate. While the probability of occupancy can now vary among sites, the actual parameters being estimated (the β s) are still assumed to be constant across all sites. In addition, note that if ψ_i is modeled only as a function of β_0 (i.e., if there are no covariates in the model), then $\psi_i = \psi$ for all sites.

As the sites are assumed to have constant occupancy status within a season, the types of covariates that may be considered appropriate for modeling ψ_i are those that remain constant for the duration of the season. This includes most types of covariates that could be used to characterize a site or its general locality, for example, habitat type, site size, site isolation, elevation, distance from some focal point, and generalized weather conditions. This approach to modeling the occupancy probabilities of sites has been used by MacKenzie et al. (2002), Tyre et al. (2003), and MacKenzie and Bailey (2004).

Using the same principles, MacKenzie et al. (2002) show that the probability of detecting the target species can also be modeled as a function of covariates (this was also noted by Tyre et al. 2003). Here, however, there are two types of covariates that may be considered. The first type are those that remain constant within a season (as for occupancy probabilities), while the second type are those that may vary from survey to survey, such as local environmental conditions, time of day, or surveyor experience. Again, using the logistic equation, the probability of detecting the species at site i during survey j could be expressed as:

$$\text{logit}(p_{ij}) = \beta_0 + \beta_1 x_{i1} + \ldots + \beta_U x_{iU} + \beta_{U+1} y_{ij1} + \ldots + \beta_{U+V} y_{ijv}$$

where $x_{i1}, \ldots, x_{iU}$ denote the U season-constant covariates associated with site i (which may be different from those used to model occupancy probabilities), and $y_{ij1}, \ldots, y_{ijv}$ are the V survey-specific covariates associated with survey j of site i. Note also that analogies of well-known closed-population, capture-recapture models can be specified by way of covariates on detection probability. For example, by specifying the covariate $y_{ij} = 1$ if the target species was detected at site i prior to survey j, $= 0$ otherwise, one could fit analogs of models M_b and M_{tb} (Otis et al. 1978; Williams et al. 2002).

Having the ability to model both occupancy and detection probabilities as functions of covariates enables a large range of models to be investigated with appropriately collected data. This, in conjunction with the ability to accommodate missing observations, is why we believe model-based frameworks such as that presented above, coupled with sound model selection procedures (e.g., AIC), provide a superior structure to ad hoc approaches for making inferences about occupancy-related parameters.

VIOLATIONS OF MODEL ASSUMPTIONS

There are several critical assumptions for the model described above. Briefly, they include: (1) occupancy status at each site does not change over the survey season; that is, sites are "closed" to changes in occupancy; (2) the probability of occupancy is constant across sites, or differences in occupancy probability are modeled using covariates; (3) the probability of detection is constant across all sites and surveys or is a function of site-survey covariates; there is no unmodeled heterogeneity in detection probabilities; and (4) detection of species and detection histories at each location are independent.

If these assumptions are not met, estimators may be biased and inferences about factors that influence either occupancy or detection may be incorrect. There has been limited investigation into the impact of violations of assumptions on parameter estimators in the model described above; however, insights can also be gleaned from assessments of many of these same assumptions in the context of closed population mark-recapture estimators (recall the close relationship between occupancy models and closed capture-recapture models; see Section 4.3).

Kendall (1999) explored several types of violations of the closure assumptions when estimating the size of a population. When individuals exhibited completely random movement in and out of a study area, the estimated population size was larger than the number of individuals present within the study area at any particular sample occasion, but was accurate for the total number of individuals in the "superpopulation" surrounding the study area. For example, suppose the study area is one-half of a forest. On any given day each individual of the target species tosses a coin to determine which half of the forest it will reside in for the day. Only those individuals that are within the study area portion of the forest are, therefore, at risk of capture each day. Assuming that the total number of individuals within the forest does not change during the sampling (i.e., the superpopulation is closed), the resulting closed-population mark-recapture estimates of abundance (and associated variance or standard error) correspond to the total number of individuals within the forest, not the number within the study area at any occasion. Similarly, the estimated capture probabilities represent the probability that an individual is within the study area (i.e., at risk of capture) *and* captured. By analogy, we would thus expect the occupancy estimator to be unbiased if species randomly moved in and out of a sampling unit, though the occupancy estimator should now be interpreted as the proportion of sites "used" by the target species (MacKenzie 2005a). Likewise, given that a site is used by the target species, the detection probability is now a combination of two different and confounded components: the probability that the species was present in the sampling unit and the probability that the species was detected, given that it was present. For example, suppose that a male warbler's breeding territory overlaps, but is not completely enclosed by, the area sampled with a fixed-radius point count. The male may not be within the sampled area during a given survey, but its movement in and out of the area is likely to be random and does not depend on whether the warbler was within the sampled area during the previous survey.

Kendall's (1999) work also suggests that if movement in and out of the sampling unit is not random, the occupancy estimator may be biased. If the species is present during initial surveys and then permanently vacates sampling units during the surveying (i.e., all individuals leave a sampling unit for the remainder of the season; emigration-only movement), then with only two survey occasions per site, the occupancy estimate should reflect the probability that the species was present at a site in the first survey occasion. This does not hold, however, when more than two surveys are conducted, with occupancy at the first survey occasion now being underestimated. Similarly, for immigration-only movement (i.e., when sites are becoming occupied by the species during the course of the surveying), with two survey occasions, the occupancy estimate will reflect the probability that the species is present at a site in the second survey. Again, this no longer holds for more than two surveys, with occupancy in the final survey now being underestimated. We predict the direction of the bias (or lack of bias) from the work of Kendall (1999), assuming that a model with survey-specific detection probabilities has been fit to the data. The bias is less predictable if a model with constant detection probability is used; allowing detection probabilities to vary among survey occasions absorbs some of the effect of the emigration or immigration. In fact, the systematic decrease or increase of detection probability estimates during a season may indicate that emigration or immigration, respectively, is occurring.

Kendall (1999) suggests that the effect of the bias may be reduced by pooling data, such that the modified data set consists of only two survey occasions. If emigration is suspected, then the detection data for the second survey onwards could be pooled, while for immigration the detection data for all surveys prior to the final survey could be pooled (although note carefully how the occupancy estimate should be interpreted in each case). If both types of movements occur within the system (and are not random, as described in the preceding paragraph), bias is likely to remain. Another option is to truncate data to include surveys between only first and last detection of the species, as demonstrated by MacKenzie et al. (2002) with two frog species in Maryland. These scenarios of entry-only or exit-only movement may occur for species that congregate at breeding, wintering, or stopover locations (e.g., neotropical migrants; shorebirds at stopover locations; pond-breeding amphibians; spawning fish). Obviously, investigators should use their knowledge about the phenology of the target species and design their study to try to minimize violations in the closure assumption.

The impact of unmodeled variation in occupancy probability among sites (occupancy heterogeneity) is relatively unknown compared to other model assumptions, and more thorough simulation studies are needed. Moreover, there is no clear parallel assumption in closed-population capture-recapture literature from which we might develop intuition about violations in this assumption. However, we anticipate that in a simple case in which two groups of sites exist with different occupancy probabilities but detection probabilities among all sites are equal, the estimated occupancy value may appropriately reflect the average level of occupancy for the pooled groups, though the reported variance will be too large (i.e., the precision of the estimator would be conservative). Initially this may seem counterintuitive, but consider the example of pooling the data from two independent binomial experiments, each with n_1 and n_2 trials and probabilities of success π_1 and π_2, respectively. The true variance for the total number of successes ($x_1 + x_2 = x_T$) will be:

$$Var(x_T) = Var(x_1) + Var(x_2)$$
$$= n_1\pi_1(1-\pi_1) + n_2\pi_2(1-\pi_2)$$

However, if group membership was unknown, then one might estimate this variance based upon the pooled estimator for the probability of a success, $\hat{\pi}_P = x_T/(n_1 + n_2)$, that is, $\widehat{Var}(x_T) = (n_1 + n_2)\hat{\pi}_P(1 - \hat{\pi}_P)$. Avoiding algebraic details, it can be shown that the bias of $\widehat{Var}(x_T)$ is $[n_1n_2(\pi_1 - \pi_2)^2]/(n_1 + n_2)$ (i.e., $\widehat{Var}(x_T)$ will be too large unless π_1 and π_2 are equal). If there are more than two groups, or if occupancy probability differs among sites as a result of an unmodeled continuous covariate, it may again to reasonable to interpret the parameter estimates as representing an average for the sites from which data were collected. Whether these estimates are reasonable for the population at large depends on whether the distribution of covariates among sampled units is consistent with the distribution of covariates in the population of all sample units (something that should be achieved, on average, with the random selection of sites), although variances may again need adjusting.

Heterogeneity in detection probability will often result in negatively biased occupancy estimates. The problem has been studied extensively in closed-population capture-recapture studies and has been explored within the site occupancy context by Royle and Nichols (2003), MacKenzie and Bailey (2004), MacKenzie et al. (2005), and Royle (2005). Naturally, low detection probabilities and high levels of variation among detection probabilities (either among sites or surveys) increase the potential bias in occupancy estimates (e.g., Royle and Nichols 2003). Bias is further exaggerated in studies involving a small number of sites or a few repeated surveys at each site. Increased detection heterogeneity also causes ambiguity in determining appropriate model structure (Section 5.5; Royle 2005). Anticipating heterogeneity and minimizing its effects, through both study design where possible and collecting relevant covariates to model variation in detection, are essential for good performance of models presented in this book. Detection probabilities may vary among surveys as a result of factors such as environment (e.g., weather conditions), seasonal behavioral patterns, or differences among observers. Detection probabilities may also vary among sites because of habitat features (e.g., forest density). One unique source of heterogeneity in detection probability among sites is the size of the local population at each site. Logically, the probability of detecting a single individual of a target species increases with increasing local population size of the species. Royle and Nichols (2003) suspect that local abundance might be the biggest source of heterogeneity in detection probabilities for many occupancy studies, and they develop a model to accommodate local abundance as the source of heterogeneity in detection. This "abundance" model and others that deal with more generic forms of heterogeneous detection probabilities are discussed in detail in the next chapter.

If detection is not independent among sites, the precision of the occupancy estimates is usually overstated. Nonindependence can arise if sites do not have adequate spatial separation, allowing animals to be detected at multiple sites simultaneously (e.g., an owl hooting at one site is also recorded by an observer at a nearby site). In these instances, the "effective sample size" (the number of independent sites or detection histories) is actually smaller than the number of sites surveyed, and the estimated standard errors obtained from the above model are too small (MacKenzie and Bailey 2004). This is a form of overdispersion (Chapter 3). Again, investigators should attempt to address this issue from a design standpoint, using their knowledge of the target species' movements to distribute sample units to minimize possible independence violations. The assessment of model fit, described next, has some power to detect problems caused by nonindependence, and standard errors can be adjusted with an estimated variance inflation factor. Other forms of dependence (e.g., more likely to redetect the species once it has been detected at a site for the first time) may also create biases in the parameter estimates, but some could be accommodated by defining an appropriate survey-specific covariate (e.g., $y_{ij} = 1$ if species detected at site i prior to survey j, $= 0$ otherwise).

ASSESSING MODEL FIT

Whenever possible, it should be demonstrated that a fitted model adequately describes the observed data; that is, a model should be assessed for lack of fit (McCullagh and Nelder 1989: 8; Lebreton et al. 1992). It should be demonstrated that the models being considered for the data are realistic and capture the important features of the system under study. Substantial lack of fit in a model(s) may lead to inaccurate inferences, either in terms of bias or in terms of precision (e.g., reported standard errors are too small). In order to have some degree of confidence in the inferences resulting from an analysis of real data, it is important that the model fit be assessed.

An increasingly popular approach for analyzing ecological data is to fit a suite or candidate set of models to the data and use a model selection technique such as AIC, or similar measures, for choosing the "best" model(s). The selection of a "best" model(s) does not guarantee the selection of a "good" model. Given the rising popularity of using such techniques in the analysis of ecological data, it is important to realize that they assume that the candidate set contains at least one model that fits the data adequately (Burnham and Anderson 2002), and that they are not a substitute for assessing model fit. A common approach to model selection using AIC and related metrics is to test fit of the most general model in the model set (Burnham and Anderson 2002). If fit of that model is deemed adequate, then model selection proceeds in the usual manner based on AIC, and if fit is not adequate, then a quasi-likelihood overdispersion parameter is estimated and used to modify AIC in the model selection (Chapter 3).

However, while we advocate that models should be assessed at every opportunity, the reality in many ecological studies is that sample sizes may often be too small to detect poor model fit; that is, tests for model fit may have low power. This could lead to a sense of false confidence, in that one may decide that a model is adequate merely because there were insufficient data, not because the model structure was appropriate. This dilemma is one to which we have no solution except to suggest that if using a hypothesis testing procedure, then it may be appropriate to use larger type I error rates than are often used, thus being willing to reject a null hypothesis of adequate fit based upon weaker evidence—in other words, considering rejecting the null hypothesis even when a P-value is greater than 0.05. As noted above, if fit is not viewed strictly in a hypothesis-testing context, then a reasonable approach is to try to estimate the degree to which fit is inadequate. Overdispersion parameter estimates, $\hat{c}$, provide such estimates for use in model selection and inference (see below and Chapter 3).

For many models that are frequently used by ecologists (including those described later in this book), no adequate methods for assessing fit have been developed. For example, there are no tests for lack of fit for the vast majority of mark-recapture models, particularly when individual covariates are used. We believe the reason for this is that developing such techniques is often viewed as a secondary- or tertiary-level problem when compared to the development of more flexible and realistic methods of data analysis. Only once the new models are accepted and receive widespread use (provided they are not quickly superseded) will an impetus be created to develop techniques for assessing the fit of the models. However, we believe that such techniques will be developed over time. Finally, we note that often the best "test" of a model is to compare predictions from the model with an independent data set.

In terms of single-season occupancy models, MacKenzie and Bailey (2004) have recently developed a method for assessing the fit of the model described in this chapter. They suggest a relatively straightforward approach that effectively tests whether the observed number of sites with each particular detection history has a reasonable chance of occurring if the target model (the model which is being assessed) is assumed to be "correct."

Let O_h be the number of sites observed to have detection history h, and E_h be the expected number of sites with history h according to the target model. For example, suppose the target model assumes that occupancy and detection probabilities are constant across sites and time; that is, the model $\psi(\cdot)p(\cdot)$, and parameter estimates are $\hat{\psi} = 0.82$ and $\hat{p} = 0.43$. The expected number of sites with the detection history 101 would be:

$$E_{101} = s \times \widehat{Pr}(h = 101)$$
$$= s\hat{\psi}\hat{p}(1 - \hat{p})\hat{p}$$
$$= s(0.82)0.43^2(0.57)$$
$$= 0.09s$$

More generally, E_h equates to the sum of the estimated probabilities of observing h across all sites, as the occupancy or detection probabilities may be site specific depending upon the model that has been fit to the data; for example:

$$E_{101} = \sum_{i=1}^{s} \widehat{Pr}(h_i = 101)$$
$$= \sum_{i=1}^{s} \hat{\psi}_i \hat{p}_{i1}(1 - \hat{p}_{i2})\hat{p}_{i3} \tag{4.9}$$

This assumes, however, there are no missing observations and equal sampling effort, as an implicit requirement is that the estimated probabilities of observing each possible detection history sum to 1 (i.e., $\sum_h \widehat{Pr}(h = h) = 1$) so that $\sum_h O_h = \sum_h E_h$. To account for missing observations, MacKenzie and Bailey (2004) suggest that sites with each unique combination of missing values be regarded as separate cohorts, as in Table 4.2. Therefore, for each cohort c, the expected number of sites with each detection history becomes:

$$E_{h_r} = \sum_{i=1}^{s_c} Pr(h_i = h_c),$$

where s_c is the number of sites in the cohort.

Once the E_h's have been calculated, MacKenzie and Bailey (2004) recommend the use of a simple Pearson's chi-square statistic to test whether there is sufficient evidence of poor model fit.

$$X^2 = \sum_c \sum_h \frac{(O_{h_c} - E_{h_c})^2}{E_{h_c}}. \tag{4.10}$$

As many of the E_{h_c} are likely to be relatively small (<2) for even moderate values of K (say ≥ 5), the usual distributional arguments used to justify that

TABLE 4.2 Example of Detection Histories and How One Would Consider Them to Be Different Cohorts for the Purpose of Assessing Model Fit

Site	h_i	Cohort
1	1010	1
2	0100	1
3	1101	1
4	−110	2
5	−010	2
6	0–11	3
7	1---	4

X^2 will have a chi-square distribution with df degrees of freedom are unlikely to hold. MacKenzie and Bailey (2004) therefore suggest using a parametric bootstrap procedure to determine whether the observed value of X^2 is unusually large.

Parametric bootstrapping involves assuming that the target model is correct and then generating alternative sets of data subject to the constraints that the s_c's are fixed. As the target model is known to be correct for the generated data, if the observed data appear typical in comparison, then it would seem reasonable to conclude that the target model may be adequate for the observed data. Hence parametric bootstrapping may be an ideal technique for assessing the model's structure. For the single-season occupancy model, MacKenzie and Bailey (2004) implement parametric bootstrapping as follows:

1. Fit target model to the observed data and estimate parameters $\hat{\psi}$ and $\hat{p}_{ij}$ (which may be functions of covariates).
2. Calculate the test statistic for the observed data, X^2_{Obs}, using the model fit in step 1 [i.e., using Eqs. (4.9) and (4.10)].
3. For each site generate a pseudo-random number (r) between 0 and 1. If $r \leq \hat{\psi}_i$ then the site is occupied, hence generate K further pseudo-random numbers (r_j) between 0 and 1. If $r_j \leq \hat{p}_{ij}$ then the species was "detected" and the corresponding bootstrapped observation is a "1"; otherwise "0." If $r > \hat{\psi}_i$, then the site is unoccupied and the bootstrapped observations will all be "0" for that site.
4. Fit a model with the same structure as in step 1 to the bootstrapped data set.
5. Calculate the test statistic for the bootstrapped data, X^2_B, using the model fit in step 4, and store the result.
6. Repeat steps 3–5 a large number of times to approximate the distribution of the test statistic, given the fitted model is correct.

7. Compare X^2_{Obs} to the bootstrap distribution of X^2_B to determine the probability of observing a larger value (the P-value).

If the target model is found to be a poor fit for the data, but it must still be used to make inferences about the system, then an overdispersion parameter ($\hat{c}$) may be used to inflate standard errors (McCullagh and Nelder 1989) and adjust model selection procedures (Burnham and Anderson 2002). Following White et $al.$ (2002), $\hat{c}$ may be estimated as:

$$\hat{c} = \frac{X^2_{Obs}}{\overline{X}^2_B},$$

where $\overline{X}^2_B$ is the average of the test statistics obtained from the parametric bootstrap. If the target model is an adequate description of the data, then $\hat{c}$ should be approximately 1. Values greater than 1 suggest that there is more variation in the observed data than expected by the model, while values less than 1 suggest less variation.

When multiple models are to be considered for the data and some form of model selection procedure is to be used, it is generally recommended that the most general or "global" model (i.e., the most complex model with the greatest number of parameters) be assessed for lack of fit first. The logic is that if the global model fits the data, then any reduced model that has fewer parameters but explains a similar level of variation in the data (i.e., a more parsimonious model) should also provide an adequate description of the data. If the global model is found to be a poor fit, then any adjustments to standard errors or model selection procedures should be made on the basis of the $\hat{c}$ value calculated from the global model (Burnham and Anderson 2002). For example, in an analysis of occupancy data collected on members of the terrestrial salamander complex $Plethodon$ $glutinosus$, MacKenzie and Bailey (2004) found evidence that the global model considered was a poor fit to the data (X^2 = 63.1, P-value = 0.056). QAIC (Chapter 3; Burnham and Anderson 2002) was therefore used for model selection with a value $\hat{c}$ = 1.43 obtained in the manner described above. The standard errors for the parameter estimates were also inflated by a factor of $\sqrt{\hat{c}}$ = 1.20.

MacKenzie and Bailey (2004) found, using simulation, that their procedure had some power for detecting poor model fit for the scenarios considered (at a level of α = 5%), particularly when $K > 5$. Interestingly, when lack of fit was caused by not including an important site-specific covariate for detection probabilities in the target model, their test had greater power when the average detection was high (>0.5), but when the poor model fit was caused by a lack of independence among sites, the test performed better when detection probabilities were lower (<0.5). In terms of assessing the fit of a target model with respect to heterogeneity of occupancy probabilities, MacKenzie and Bailey

(2004) found their test to have no power. They suggest this should not be unexpected, because when the model is misspecified with respect to detection probabilities, some sites will have an unusually large (or small) number of species' detections. However, as occupancy is effectively a single binary observation for each site, there is no such outward indication that the model may be inadequately describing the data. They go on to draw a comparison with the similar problem that exists for logistic regression and that led to the development of the Hosmer-Lemeshow test (Hosmer and Lemeshow 1989:140), which uses the predicted probabilities of a success to classify the observations into k groups, and speculate that such an approach may be modified to assess the fit of a target model with respect to occupancy probabilities. However, point estimates of the occupancy-related parameters may still be considered reasonable as an average for the sites in the study (as discussed in the previous section); therefore, the consequences of not identifying heterogeneous occupancy probabilities will not be as serious as for heterogeneous detection probabilities.

EXAMPLES

To illustrate the above methods we now present two examples. The first is an example from MacKenzie (2006) on pronghorn antelope (*Antilocapra americana*) that was briefly mentioned in Chapter 2, which nicely illustrates that misleading inferences can result from not explicitly accounting for detection probability. In our second example we revisit a data set first considered by MacKenzie *et al.* (2005): occupancy of Mahoenui giant weta (*Deinacrida mahoenui*) within a scientific reserve in New Zealand. The data are analyzed using the software PRESENCE 2.0. In the examples we introduce the idea of formally comparing competing hypotheses or models as part of the inferential procedure. Generally, different hypotheses about the system can be specified by way of covariates that are either collected in the field or constructed after the fact. The different hypotheses of interest can therefore be represented by a suite of candidate models. The candidate models can then be ranked according to some model selection procedure, with models (and hence hypotheses) that have a lot of support being ranked highly, while those less supported by the data will have lower rankings. AIC has been used here to rank the models.

Pronghorn Antelope

MacKenzie (2006) reanalyzed data on pronghorn antelope considered by Manly *et al.* (2002) in a resource selection context. During the northern hemisphere winters of 1980–81 and 1981–82, 256 locations in Wyoming, were surveyed to determine whether the sites were being used by pronghorn antelope (i.e., that antelope were present at a site). A number of site characteristics were also recorded (e.g., distance to water source and grass type) with the intent of identifying factors that affected whether pronghorn antelope used particular sites. Manly et al. (2002) noted that use could be defined in a number of ways, and MacKenzie (2006) opted for a definition in which it was assumed that sites were either used or not used by the antelope over both winters (i.e., the sampling season was defined to encompass both seasons). Hence, the surveys of the sites in each winter represent the repeated surveys of the sites with the "season." MacKenzie (2006) analyzed the data using two approaches in order to illustrate how not accounting for detection probability can result in misleading inferences. The first approach was to use simple logistic regression in which the nondetection of antelope in both winters is assumed to equate to pronghorn antelope not using the site. The second approach was to use the above occupancy models to make inference about habitat selection while explicitly accounting for detection probability. We simply focus on the model selection aspects of the two approaches here to highlight the contrasting results, without specific discussion of the resulting inference on habitat selection. Both approaches considered the effect of four covariates on habitat selection: sagebrush density (Sg), slope (Sl), distance to water (DW), and aspect (A).

The results of using logistic regression are presented in Table 4.3. Note that when used in this context, the implicit assumptions made for the use of logistic regression to be valid is that either sufficient repeat surveys have been conducted such that the probability of a false absence is negligible, or detection probability is constant across all sites, in which case the results should be interpreted as relative rather than absolute measures of occupancy (or "use" in the present context). The summed model weights for the four variables are: distance from water: 86%; slope: 52%; sagebrush density: 35%; and aspect: 16%. Hence, one would likely conclude that distance from a water source is the most important factor for determining whether a site was used by pronghorn antelope.

Quite different conclusions are reached when using the occupancy models described above (Table 4.4). Rather than attempting to perform model selection on both occupancy and detection probabilities simultaneously, MacKenzie (2006) focused on model selection of the occupancy probabilities while maintaining a very general model for detection probabilities (i.e., $p(Sg + Sl + DW + A)$). Now the summed model weights for each factor with respect to use (occupancy) are: slope: 55%; sagebrush density: 41%; distance from a water source: 29%; and aspect: 6%. Distance from a water source has gone from being very strongly supported to having only weak support. MacKenzie (2006) suggests that the difference in the conclusion is caused by these factors

TABLE 4.3 Summary of Model Selection Procedure for Pronghorn Antelope Example Where Factors Affecting Habitat Selection Have Been Investigated Using Simple Logistic Regression

Model	ΔAIC	w	$-2l$	NPar
$\psi(Sl + DW)$	0.00	23%	345.26	3
$\psi(DW)$	0.22	21%	347.48	2
$\psi(Sg + Sl + DW)$	0.82	16%	344.08	4
$\psi(Sg + DW)$	1.18	13%	346.44	3
$\psi(Sl + DW + A)$	2.79	6%	342.05	6
$\psi(DW + A)$	3.08	5%	344.34	5
$\psi(Sl)$	3.81	3%	351.07	2
$\psi(Sg + Sl + DW + A)$	4.05	3%	341.31	7
$\psi(Sg + DW + A)$	4.45	3%	343.71	6
$\psi(Sg + Sl)$	4.67	2%	349.93	3
$\psi(\cdot)$	5.63	1%	354.89	1
$\psi(Sl + A)$	6.11	1%	347.37	5
$\psi(Sg)$	6.65	1%	353.91	2
$\psi(Sg + Sl + A)$	7.45	1%	346.71	6
$\psi(A)$	7.67	1%	350.93	4
$\psi(Sg + A)$	9.13	0%	350.39	5

Given are the relative difference in AIC values compared to the top-ranked model (ΔAIC); the AIC model weights (w); twice the negative log-likelihood ($-2l$); and the number of parameters in the model (NPar).
Source: MacKenzie 2006.

TABLE 4.4 Summary of Model Selection Procedure for Pronghorn Antelope Example

Model	ΔAIC	w	$-2l$	NPar
$\psi(Sl)$	0.00	23%	615.48	9
$\psi(\cdot)$	0.72	16%	618.20	8
$\psi(Sg + Sl)$	0.85	15%	614.33	10
$\psi(Sg)$	1.12	13%	616.60	9
$\psi(Sl + DW)$	1.95	9%	615.44	10
$\psi(DW)$	2.25	7%	617.73	9
$\psi(Sg + Sl + DW)$	2.85	6%	614.33	11
$\psi(Sg + DW)$	2.99	5%	616.47	10
$\psi(Sl + A)$	5.53	1%	615.01	12
$\psi(A)$	6.05	1%	617.53	11
$\psi(Sg + Sl + A)$	6.41	1%	613.89	13
$\psi(DW + A)$	6.60	1%	616.08	12
$\psi(Sl + DW + A)$	6.87	1%	614.35	13
$\psi(Sg + A)$	7.03	1%	616.51	12
$\psi(Sg + Sl + DW + A)$	8.21	0%	613.69	14
$\psi(Sg + DW + A)$	8.34	0%	615.83	13

Factors affecting habitat selection have been investigated using the occupancy models described in this chapter, with a general model for detection probabilities (i.e., $p(Sg + Sl + DW + A)$). Given are the relative difference in AIC values compared to the top-ranked model (ΔAIC); the AIC model weights (w); twice the negative log-likelihood ($-2l$); and the number of parameters in the model (NPar).
Source: MacKenzie 2006.

TABLE 4.5 Summary of Model Selection Procedure for Pronghorn Antelope Example Examining Factors Affecting Detection Probabilities, with a General Model for Occupancy (i.e., $\psi(Sg + Sl + DW + A)$)

Model	ΔAIC	w	$-2l$	NPar
$p(DW)$	0.00	24%	618.54	9
$p(Sl + A)$	1.37	12%	617.913	10
$p(A)$	1.75	10%	616.286	11
$p(Sg + DW)$	1.79	10%	618.325	10
$p(DW + A)$	1.82	10%	614.36	12
$p(\cdot)$	2.80	6%	623.345	8
$p(Sg + Sl + DW)$	3.17	5%	617.707	11
$p(Sl + DW + A)$	3.56	4%	614.1	13
$p(Sg + DW + A)$	3.58	4%	614.116	13
$p(Sg + A)$	3.62	4%	616.157	12
$p(Sl + A)$	3.73	4%	616.274	12
$p(Sl)$	4.76	2%	623.305	9
$p(Sg)$	4.79	2%	623.332	9
$p(Sg + Sl + DW + A)$	5.15	2%	613.691	14
$p(Sg + Sl + A)$	5.61	1%	616.153	13
$p(Sg + Sl)$	6.75	1%	623.295	10

Given are the relative difference in AIC values compared to the top-ranked model (ΔAIC); the AIC model weights (w); twice the negative log-likelihood ($-2l$); and the number of parameters in the model (NPar).

also affecting detection probabilities. Subsequent modeling supports this premise. The results of performing model section on the detection probabilities while maintaining a general model for occupancy (i.e., $\psi(Sg + Sl + DW + A)$) are presented in Table 4.5. The summed model weights for the factors with respect to detection probability are: distance to water source: 68%; aspect: 37%; slope: 29%; and sagebrush density: 27%. Clearly, there is moderately strong support that detection probability is affected by distance to water and less support for the other factors. Interestingly, note that the model with constant detection probability ($p(\cdot)$, the de facto assumption when using simple logistic regression) has only a very low model weight, suggesting little support for this hypothesis.

Mahoenui Giant Weta

Weta are ancient species of the order *Orthoptera* (e.g., grasshoppers, crickets, and locusts), with more than 70 endemic species surviving in New Zealand today. Based upon fossil records, they have remained almost unchanged from their ancestors of 190 million years ago. In the absence of native ground-dwelling mammals, the weta of New Zealand evolved to fulfill the roles rodents play in other ecosystems. As such, the introduction of rats and other small mammals to New Zealand ecosystems with the arrival of the Maori and Europeans decimated weta populations.

The Mahoenui giant weta (*Deinacrida mahoenui*) is endemic to the King Country on the North Island of New Zealand, with only two surviving populations. The main naturally occurring population is restricted to a 240-ha scientific reserve at Mahoenui (near the town of Te Kuiti) maintained by the New Zealand Department of Conservation. The reserve is characterized by steep-sided gullies and is largely covered by dense gorse (*Ulex europaeus*), a perennial pest plant with sharp spiny stems and bright yellow flowers that can form dense thickets, originally introduced to New Zealand as a hedging plant by the early European settlers. The weta use the prickly gorse plants as protection from predators and also as a food source. Goats and cattle are used to browse the gorse, encouraging dense foliage and providing further protection for weta.

As part of a pilot study, in March 2004, 72 circular plots of 3 m radius were surveyed for the Mahoenui giant weta within the more accessible regions of the reserve. Clearly, inference can only be made as to the more accessible parts of the reserve, but this was deemed reasonable given the nature of the pilot study. Each plot was surveyed between three and five times during the five-day period ($K_{average} = 3.6$). Three different observers were used, and the study was designed such that each observer surveyed each site at least once. This was done to avoid introducing heterogeneity in detection probabilities caused by the use of multiple observers (see Chapter 6 for further discussion).

Weta were detected at 35 of the 72 plots (a naïve occupancy estimate of 0.49); however, often weta were only detected in one or two of the repeated surveys, clearly indicating that detection probabilities are less than 1. There conceivably may be a number of plots where weta were indeed present but simply never detected during the surveys.

Here we wish to estimate the probability of occupancy for the weta. Detection probabilities will be allowed to vary by day and also among observers, but simpler models that do not include these effects will be included in our candidate set also in the interests of parsimony. Daily variation in detection probabilities (or more generally, survey-specific detection probabilities) can be easily accommodated with the design-matrix interface used in PRESENCE 2.0 to build models. To allow detection probabilities to vary among the three observers, two survey occasion-specific covariates were defined, *Obs1* and *Obs2*. For site i, survey j, *Obs1* = 1 if the survey was conducted by observer 1, 0 otherwise; and *Obs2* = 1 if the survey was conducted by observer 2, 0 otherwise. Note that if the survey was conducted by observer 3, then *Obs1* = 0 and *Obs2* = 0; thus, the third observer is considered the standard or reference observer against which the other observers are to be compared (Table 4.6).

TABLE 4.6 Coding Used to Define Observer Effects in Mahoenui Giant Weta Example Using the *Obs1* and *Obs2* Covariates

Survey conducted by	*Obs1*	*Obs2*
Observer 1	1	0
Observer 2	0	1
Observer 3	0	0

TABLE 4.7 Summary of Model Selection Procedure for Mahoenui Giant Weta Example

Model	ΔAIC	*w*	*NPar*	$-2l$	*Browse*	SE
$\psi(Browse)p(Day + Obs)$	0.00	0.27	9	239.60	1.17	0.74
$\psi(\cdot)p(Day + Obs)$	0.95	0.17	8	242.55		
$\psi(\cdot)p(Day + Obs + Browse)$	1.81	0.11	9	241.41		
$\psi(Browse)p(Day)$	1.84	0.11	7	245.44	1.24	0.75
$\psi(Browse)p(Day + Obs + Browse)$	2.00	0.10	10	239.60	1.17	0.89
$\psi(\cdot)p(Day)$	3.19	0.06	6	248.79		
$\psi(\cdot)p(Day + Browse)$	3.58	0.05	7	247.18		
$\psi(Browse)p(Day + Browse)$	3.81	0.04	8	245.41	1.15	0.88
$\psi(Browse)p(Obs)$	4.44	0.03	5	252.04	1.18	0.70
$\psi(\cdot)p(Obs)$	5.76	0.02	4	255.36		
$\psi(Browse)p(Obs + Browse)$	6.42	0.01	6	252.02	1.25	0.83
$\psi(Browse)p(\cdot)$	6.66	0.01	3	258.26	1.23	0.72
$\psi(\cdot)p(Obs + Browse)$	6.90	0.01	5	254.50		
$\psi(\cdot)p(\cdot)$	8.19	0.00	2	261.79		
$\psi(Browse)p(Browse)$	8.66	0.00	4	258.26	1.20	0.84
$\psi(\cdot)p(Browse)$	8.85	0.00	3	260.45		

ΔAIC is the difference in AIC value for a particular model when compared with the top-ranked model; *w* is the AIC model weight; *NPar* is the number of parameters; $-2l$ is twice the negative log-likelihood value; *Browse* is the value of the coefficient for the *Browse* variable with respect to its effect on occupancy probability; and SE is the associated standard error (blank entries indicate that the *Browse* variable was not included in the models).

The estimated coefficients related to the covariates *Obs1* and *Obs2* therefore represent the difference (on the logistic scale) between the respective observers and observer 3. While the pilot study was not specifically designed for this purpose, the effect of browsing on occupancy and detection probability will also be assessed. The level of browsing at each plot was assessed by the field crew prior to the pilot study, and a covariate *Browse* has been defined here as = 1 if the plot showed evidence of sustained browsing (based upon shape of bushes and foliage density), 0 otherwise. Therefore, our candidate set contains 16 models without considering interactions between factors (Table 4.7).

Models are denoted with the relevant factors indicated in parentheses following each probability. For example, $\psi(Browse)$ indicates the probability of occupancy being different for browsed and unbrowsed sites, while $p(Day + Obs)$ indicates that detection probability varied by day with additive (on the logistic scale) observer effects.

Testing the global model from the candidate set, $\psi(Browse)p(Day + Obs + Browse)$, does not indicate any evidence of lack of fit using 10,000 bootstrap samples ($X^2 = 154.1$, P-value $= 0.999$, $\hat{c} = 0.35$), although one would perhaps be concerned that the P-value is so close to 1.0 that it may indicate the model "over-fits" the data (i.e., there may be too many parameters in the model). As such, no adjustment has been made to the model selection procedure (AIC) or the standard errors of parameter estimates.

Table 4.7 presents the 16 models ranked according to AIC. The first thing to note is that no single model is demonstrably better than the others; the five top models are separated by less than 2.0 AIC units. As such, the AIC model weight is distributed across a number of models, indicating that a number of models may be reasonable for the collected data. There are, however, a number of common features among the top-ranked models. The eight models where detection probability varied daily are all ranked higher than the models without daily variation. In terms of model weights, the $p(Day)$ models have 91% of the total, providing clear evidence that Day is an important factor in terms of accurately modeling detection probabilities. In terms of comparing hypotheses, the hypothesis that the detection probability varied among days therefore has much greater support than the hypothesis that it was constant. Many of the top-ranked models also contain the factor Obs for detection probability, providing evidence that the observers differed in their ability to find weta in the plots. The combined model weight for $p(Obs)$ models is 73%. There is substantially less support for the hypothesis that the level of browse affects detection probabilities, with a combined model weight of 33%.

In terms of occupancy probability, based upon rankings and AIC model weights, the results are somewhat inconclusive about the effect of browse. The combined weight for the $\psi(Browse)$ models is 58%—in other words, similar levels of support for the hypotheses that occupancy is/is not affected by whether bushes within the plots are browsed. However, there is an important point to note that illustrates that unthinking use of model selection procedures can be misleading. Given the biology of the situation, a priori we would expect browsing to increase the probability of occupancy (by creating better habitat); therefore the parameter estimate associated with the factor $Browse$ for occupancy should be positive. This was in fact the case (Table 4.7). From the respective models, all estimates for the $Browse$ factor were very similar. Yet AIC and similar metrics (and therefore the derived model weights) do not account for the fact that one could specify a priori the direction of a particular relationship (i.e., one could very loosely describe these model selection procedures as "two-sided"). Therefore, one could argue that, as the estimated effect matches our *a priori* expectations, the level of support for these models should be greater than that indicated by the model weights. Unfortunately, we cannot make any firm recommendations at this time for how one might objectively incorporate this idea into an information-theoretic framework. As such, when interpreting the magnitude of the various effects below, rather than taking a model-averaging approach to account for uncertainty about which model(s) provides the most efficient representation of the data, we only consider the parameter estimates from the top-ranked model. We acknowledge that our standard errors do not account for model selection uncertainty (Burnham and Anderson 2002).

From the model $\psi(Browse)p(Day + Obs)$ we have the following equations for estimating occupancy and detection probabilities:

$$\text{logit}(\psi_i) = 0.02 + 1.17 Browse_i \tag{4.11}$$

and:

$$\text{logit}(p_{ij}) = -0.23 Day1 - 0.38 Day2 - 1.17 Day3 - 0.30 Day4 \\ + 0.81 Day5 - 1.07 Obs1_{ij} - 0.34 Obs2_{ij} \tag{4.12}$$

where $Browse_i$ is the value of the variable $Browse$ for plot i (1 or 0); $Day1$–$Day5$ are just indicator variables for the day of the study; and $Obs1_{ij}$ and $Obs2_{ij}$ are used to denote which of the three observers surveyed a given site on a given day (see Table 4.6).

From Eq. (4.11), for an unbrowsed site $\text{logit}(\psi_i) = 0.02$ (as $Browse_i = 0$), which gives odds of occupancy of $e^{0.02} = 1.02$ (:1) and a probability of occupancy of $1.02/(1 + 1.02) = 0.50$. To interpret the effect of browsing on the probability of occupancy, we shall do so in terms of odds ratios (see Chapter 3). The odds ratio for a browsed site being occupied by weta is $e^{1.17} = 3.22$; hence, the odds of occupancy at a browsed site is $3.22 \times 1.02 = 3.28$ (:1; or a probability of occupancy of $3.28/(1 - 3.28) = 0.77$). A confidence interval for the effect of browsing could be calculated on the logit scale, then transformed to the scale of an odds ratio. For example, given that the standard error for the estimated $Browse$ effect is 0.74, an approximate two-sided 95% confidence interval on the logit scale would be $1.17 \pm 2 \times 0.74 = (-0.31, 2.65)$, giving an interval of $(e^{-0.31}, e^{2.65}) = (0.73, 14.15)$ for the odds ratio. As mentioned in Section 3.4, an odds ratio of 1.0 would indicate that the factor has no effect; hence, as 1.0 is included in the confidence interval, we do not have strong evidence that browsing has an effect on the probability of occupancy by weta. Although, as it was expected *a priori* that browsing should have a positive effect on occupancy, it would be more appropriate to consider a one-sided 95%

confidence interval (calculated here by taking the lower limit of a two-sided 90% confidence interval). That is, $(1.17 - 1.65 \times 0.74, \infty) = (-0.05, \infty)$ on the logit scale, or $(0.95, \infty)$ in terms of odds ratios. As 1.0 is only just inside the confidence interval, we would be comfortable in concluding that it does appear browsing has a positive effect on occupancy, but the magnitude of the effect is only poorly known.

In terms of an overall estimate of occupancy, for the plots surveyed we could calculate an average from the estimated occupancy probabilities for the browsed and unbrowsed plots, weighted by the number of plots of each type. Here, 35 plots were classified as browsed and 37 classified as unbrowsed. An overall estimate would therefore be:

$$\frac{s_{Browsed}\hat{\psi}_{Browsed} + s_{Unbrowsed}\hat{\psi}_{Unbrowsed}}{s_{Browsed} + s_{Unbrowsed}} = \frac{35 \times 0.77 + 37 \times 0.50}{35 + 37}$$

$$= \frac{26.95 + 18.5}{72}$$

$$= \frac{45.45}{72}$$

$$= 0.63$$

This is 30% larger than the naïve estimate (the fraction of plots where the species was detected) of 0.49. Clearly, accounting for detection probability has increased the estimated level of occupancy as expected (see below). While not applicable in this example due to the nature of the study design, an alternative method for obtaining an overall estimate of occupancy for the area of interest would be to predict occupancy at each of the potential sampling units. For example, suppose the area could be divided into grid cells (say) of which a sample was surveyed and the resulting data used to build an occupancy model(s). If the variable information used within the occupancy model was available for all grid cells within the regions (e.g., from GIS), then the occupancy probability for each cell could be predicted, and the total level of occupancy would simply be the sum of the occupancy probabilities for all grid cells within the region of interest.

The variance (or standard error) for the overall level of occupancy can be calculated in two steps. First, calculate the variance for a browsed and unbrowsed plot using the delta method (Chapter 3), then combine these variance terms in the normal manner for a weighted average; that is:

$$Var(\hat{\psi}_{Overall}) = \frac{s_{Browsed}^2 Var(\hat{\psi}_{Browsed}) + s_{Unbrowsed}^2 Var(\hat{\psi}_{Unbrowsed})}{\left(s_{Browsed} + s_{Unbrowsed}\right)^2},$$

where the s^2 terms relate the number of plots squared, not a sample variance. It is important to note that the full variance-covariance matrix is for all the βs in the model, both those related to the occupancy probability and those related to detection probability. However, as the logit-link function for occupancy does not involve the βs from the detection probability logit-link function, when applying the delta method to obtain the variance for an estimate of occupancy probability, only the portion of the variance-covariance matrix that relates to the βs in the occupancy logit-link function need to be considered (similarly when estimating the variance for a detection probability). Here then, we have:

$$V = \begin{bmatrix} 0.2113 & -0.1521 \\ -0.1521 & 0.5495 \end{bmatrix}$$

and calculating $Var(\hat{\psi}_{Browsed})$ and $Var(\hat{\psi}_{Unbrowsed})$ proceeds in a very similar manner to the example given in Chapter 3, resulting in values of 0.015 and 0.013, respectively (we leave the derivation of these values as an exercise). Therefore, $Var(\psi_{Overall}) = (35^2 \times 0.015 + 37^2 \times 0.013)/72^2 = 0.007$, or a standard error of 0.08.

A brief examination of the estimated detection probabilities clearly indicates why the overall level of occupancy is estimated to be 30% larger than the naïve estimate (based simply on the number of plots where weta were detected). Estimates for the detection probabilities for each observer on each day are given in Table 4.8, based upon Eq. (4.12). There are clearly a reasonable level of daily variation and substantial differences among observers. The average estimated detection probability is 0.36, which in combination with the average number of surveys per plot (3.6) suggests the expected probability of not detecting weta at a plot where they are present (i.e., the probability of declaring a false absence) is about $(1 - 0.34)^{3.6} = 0.22$, or weta would not be detected at approximately 1 in 5 occupied plots.

TABLE 4.8 Estimated Daily Detection Probabilities for Each Observer in the Mahoenui Giant Weta Pilot Study from the Logit-link Function Given in Eq. (4.12)

Observer	Day 1	Day 2	Day 3	Day 4	Day 5
1	0.21	0.19	0.10	0.20	0.43
2	0.36	0.33	0.18	0.34	0.61
3	0.44	0.41	0.24	0.43	0.69

4.5. ESTIMATING OCCUPANCY FOR A FINITE POPULATION OR SMALL AREA

As mentioned in Section 4.1, the occupancy models developed above (and elsewhere in this book) are based on the view that the sampled locations constitute a random sample from a large (theoretically infinite) population, and the site occupancy parameter ψ is the probability that a site in that infinite population is occupied. While this conceptual formulation may be appropriate, or at least adequate, for many problems, there are inference problems for which this view is insufficient. For example, suppose occupancy metrics are adopted for summarizing the metapopulation status of anurans on a small number of wetland basins in a refuge or park. Conceivably one could sample most or even all of the available wetlands, and in such cases the object of inference might be the occupancy status of the particular wetlands that were sampled or the occupancy status of the larger set of known available wetlands (including some that were not sampled). In such a situation, the distinction between probability of occupancy and the proportion of sites occupied becomes important (Chapter 2). While the basis of the above modeling can still be used to make inference about the number or proportion of sites occupied in a relatively small population, concerns about the asymptotic approximations used by likelihood theory to calculate variances and standard errors may lead us to explore other inferential methods. Furthermore, we know that x (the number of occupied sites in the sample) can only take on integer (whole number) values between s_D and s. Consequently, the proportion of sites occupied in a sample can only have a finite number of possible values, which has not been considered previously (e.g., if $s = 20$, the proportion of sites occupied can only take values of $0, 0.05, 0.1, 0.15, \ldots$). In the remainder of this section, we discuss methods for estimating the number of occupied sites in the population, but note that an estimate of the proportion of sites occupied can be simply derived by dividing the estimated number of sites occupied by the size of the population of interest (s or S). To avoid confusion, we denote the number of occupied sites in the sample as x, and the number of occupied sites in the larger, but finite, population as x_{pop}.

PREDICTION OF UNOBSERVED OCCUPANCY STATE

The essence of estimating the number of occupied sites in a population is predicting the unknown occupancy state of specific sites. The exact occupancy state of a site may be unknown either because of imperfect detection or because the site was not included in the sample where surveys of the species were conducted. The occupancy state of each of these sites can be referred to as a *latent* (i.e., unobserved) state variable. While we have not used this term previously, the concept has been used implicitly in the models described above (and elsewhere) in order to focus estimation on the (population) model parameters governing the latent occupancy state variables (e.g., presence or absence of the species). Nevertheless, many inference problems (such as estimating the number of occupied sites) require that we focus attention directly on the latent state variables (see Dorazio and Royle 2005 for some related estimation problems). To formalize this, let z_i be the binary occupancy state for site i (i.e., presence/absence of the species). If this were observed for every site, then the number of occupied sites that were surveyed is the quantity:

$$x = \sum_{i=1}^{s} z_i.$$

However, due to imperfect detection, the occupancy state is not known for some sites. The logical estimator of x is:

$$\hat{x} = s_D + \sum_{i=s_D+1}^{s} \hat{z}_i,$$

that is, the number of occupied sites in which the species was detected plus the estimated occupancy status of each site where no detections occurred. The expected value of the second term in this expression is $(s - s_D)\psi_{condl}$, where ψ_{condl} is the probability a site is occupied, conditional upon the species not being detected at the site. Recall from previous discussion that ψ_{condl} is related to p and ψ by Bayes' Theorem:

$$\psi_{condl} = \frac{\psi(1-p)^K}{(1-\psi)+\psi(1-p)^K}$$

and so an estimator of ψ_{condl} can be obtained by substituting estimates of ψ and p into this expression. Thus, the estimator of x is:

$$\hat{x} = s_D + (s - s_D)\hat{\psi}_{condl}. \tag{4.13}$$

Note the analogy here with mark-recapture estimators of population size (e.g., Otis $et\ al.$ 1978; Williams $et\ al.$ 2002). The estimated number of animals in a population is the number that were captured, plus the estimated number of animals that were in the population but never captured. Nichols and Karanth (2002; Section 4.3) identified this analogy and exploited it to estimate the probability of occupancy.

For a larger collection of sites (including unsampled sites), indexed by $j = s + 1, s + 2, \ldots S$, where S is the list of potential sample sites, we need to add another term to the estimator of x, so that $\hat{x}_{pop}$ is:

$$\hat{x}_{pop} = s_D + \sum_{i=s_D+1}^{s} \hat{z}_i + \sum_{j=s+1}^{S} \hat{z}_j. \tag{4.14}$$

As these unsampled sites were never surveyed, the expected value of z_j is the unconditional probability of occupancy, ψ_j. When there are no covariates to describe variation in occupancy, the third term could simply be estimated as $(S - s)\hat{\psi}$. However, we would not expect this form of the estimator to be useful when landscape covariates are available for all sites to describe variation in occupancy (e.g., from a GIS database). In such cases, the occupancy status of all potential sample units should be considered in the context of the available covariates, and the general form of the estimator takes this into explicit consideration.

While we expect these estimators to be unbiased, we are less certain about obtaining accurate variance estimates for these quantities. Based purely on heuristics, we suggest the estimator for the number of sites occupied in the sample when detection probability is constant [Eq. (4.13)] will have an asymptotic variance of:

$$Var(\hat{x}) = \frac{s\psi(1-p^*)}{p^* - Kp(1-p)^{K-1}}.$$

This is based upon the second component of $Var(\hat{\psi}_{MLE})$ (in the two-component form) given in Section 4.4. Recall that we interpreted this second component as being associated with the uncertainty in the number of occupied sites in the sample due to imperfect detection of the species. Similarly, in the situation where we seek to estimate the number of occupied sites in a finite population, where some sites were unsampled [i.e., using Eq. (4.14)], and detection probability is constant, we suggest the asymptotic variance may be:

$$Var(\hat{x}_{pop}) = \frac{s\psi(1-p^*)}{p^* - Kp(1-p)^{K-1}} + (S-s)^2 Var(\hat{\psi}_{MLE}).$$

Again, we stress that these variance equations are based purely on the heuristics of the situation and, while we have found them to provide reasonable results in the few situations in which we have used them, we are uncertain whether they have the desired properties or whether they generalize to situations with covariates. Regardless, we would not consider them as accurate variance estimates for two reasons: (1) they rely on the asymptotic properties of maximum likelihood estimates; and (2) they do not account for the discreteness of the estimator that we referred to before. We believe that using a Bayesian approach to estimation overcomes these problems in a manner that is very easy to implement, given the advent of Markov chain Monte Carlo algorithms.

A BAYESIAN FORMULATION OF THE MODEL

A more generic way to resolve issues having to do with inference about occupancy state is to formulate the model explicitly in terms of the latent occupancy state variables, the z_i's introduced previously, and then adopt a Bayesian analysis of the resulting hierarchical model. The important benefit of this approach (in the present context) is that estimation and inference of specific values of z_i and functions of z_i's (such as x) can be achieved directly. In addition, we achieve some generality in terms of addressing these inference problems under model extensions such as when covariates are available. More general extensions are relatively straightforward, including to multi-season situations (Chapter 7) and to multiple-species models (this is the basis of the work by Dorazio and Royle 2005; see Chapter 9). As a final comment, we note that Bayesian inference is not asymptotic. That is, measures of parameter uncertainty, including for functions of parameters (i.e., derived parameters) and predictions of latent variables, are valid for any sample size.

Using previously established notation, we let z_i be the occupancy state of site i, with $z_i = 1$ for an occupied site and $z_i = 0$ for an unoccupied site. In the first stage of the hierarchical model we specify a model for the observed number of detections at each site, conditional on the occupancy state of the site. For $z_i = 1$, this is a binomial distribution when the probability of detection is constant, and for $z_i = 0$ this is a point mass at 0 (i.e., if the site is unoccupied there can only be 0 detections of the species). Thus, the first component of the model is the product-binomial seen previously, except that here, when expressed conditional on the latent state variables z_i, is the product only over those sites for which $z_i = 1$:

$$f(y_1, y_2, \ldots, y_x | z_1, z_2, \ldots, z_x) = \prod_{z_i=1} Bin(y_i; p),$$

where y_i is the observed number of detections. In the second stage of the model we specify the model for the actual latent state variables z_i; $i = 1, 2, \ldots, s$. We suppose that z_i are independent Bernoulli random variables with probability ψ so that we have:

$$g(z_1, z_2, \ldots, z_s) = \prod_{i=1}^{s} \psi^{z_i} (1 - \psi)^{1-z_i}.$$

Thus, the joint distribution of the observed detection frequencies and the latent occupancy state variables is the product $f(y_1, y_2, \ldots, y_{s_1} | z_1, z_2, \ldots, z_s) g(z_1, z_2, \ldots, z_s)$. Note that previously we implicitly integrated the z_i's from this joint distribution to express the likelihood only as a function of the parameters p and ψ. However, for present purposes, there is no need to do so.

Finally, we must specify probability distributions for the two "parameters" in the model, p and ψ, that is, "prior distributions" (see Chapter 3). Lacking formal prior information (e.g., through a previous study) or informed consensus among experts, a natural choice (and also a common choice) of prior distribution in both cases is a uniform $(0,1)$ distribution (i.e., all values for p and ψ between 0 and 1 are equally likely).

These model assumptions yield a completely specified Bayesian model in the sense that calculation of the posterior distribution of any unknown component of the model (e.g., the parameters or the latent state variables) can now be obtained. Obtaining these posterior distributions proceeds with the aid of simulation-based methods known as Markov chain Monte Carlo (MCMC). Presently, we make use of the software package WinBUGS (Spiegelhalter *et al.* 2003) to carry-out the MCMC calculations. However, for most of the occupancy models described thus far, developing MCMC code in common software such as R or SAS is straightforward. In WinBUGS, one describes the model using an intuitive "pseudo-code" representation. The WinBUGS model description for the site-occupancy model with constant p and ψ is given in Fig. 4.2.

For specifics on the syntax used by WinBUGS, readers are referred to WinBUGS supporting documentation. Here we focus on explaining how the WinBUGS code implements the modeling we have described above. In Fig. 4.2, the main workings for the model appear on lines 3–10. Lines 3 and 4 are simply defining the prior distribution for p and ψ (psi) as uniform $(0,1)$. The three lines of code within the "for" loop (lines 6–8) describe the actual model (a zero-inflated binomial). In line 6 a quantity mu[i] is calculated as the product of p and z[i] (i.e., z_i). Recall that z_i is the binary latent state variable of occupancy which $= 1$ if the species is present at site i, and $= 0$ if the species

```
 1:  model SiteOcc;
 2:  {
 3:      p~dunif(0,1)
 4:      psi~dunif(0,1)
 5:      for(i in 1:nsite){
 6:          mu[i]<- p*z[i]
 7:          z[i]~dbern(psi)
 8:          y[i]~dbin(mu[i],K)
 9:      }
10:      x<-sum(z[])
11:  }
```

FIGURE 4.2 Syntax for defining a basic occupancy model in WinBUGS with a constant occupancy probability and constant detection probability (i.e., model $\psi(\cdot)p(\cdot)$). Line numbers on the left-hand side are for ease of interpretation only and are not part of the WinBUGS code.

is absent from the site. Hence mu[i] will $= p$ if the species is present at the site and $= 0$ otherwise. Therefore mu[i] is the probability of detecting the species at a site, given the site's occupancy status. This "trick" will ensure that WinBUGS fits a model in which the probability of detection is zero for unoccupied sites. Line 7 specifies that our latent state variable (presence/absence of the species) is simply a random variable from a Bernoulli distribution with probability ψ, and line 8 specifies that the number of detections at a site (y_i; the data) is a random variable from a binomial distribution with probability mu[i] and K trials (i.e., K surveys at a site). The last line of the model specification (line 10) calculates the number of occupied sites (x) in the sample: the sum of the unknown occupancy state variables. Note that it is not necessary to enforce that z[i] $= 1$ for the sites where the species was detected, as this is performed implicitly within the MCMC algorithms used by WinBUGS (i.e., z[i] must be 1 at those sites where the species was detected).

Developing the pseudo-code description of the model requires most of the effort in using WinBUGS (obviously, not too much effort is required!). In addition to this representation of the model, data must be loaded, initial values selected, and then the MCMC algorithm executed. This is all described in detail in WinBUGS supporting documentation and we avoid providing that level of introduction here, but note this is accomplished with a few lines of code. A good introduction to MCMC and WinBUGS for wildlife biologists is given by Link *et al.* (2002). Once the posterior distribution for each quantity has been approximated using WinBUGS or some other MCMC algorithm, any desired posterior summary such as point estimates (e.g., the posterior mean) or credible intervals (posterior quantiles) can be easily obtained. For example, let $\psi^{(1)}$, $\psi^{(2)}$, ..., $\psi^{(M)}$ be the M posterior samples from the posterior distribution of the probability of occupancy, ψ. We can depict the posterior distribution with a histogram or density plot based on these values, or simply report a point estimate based on the Monte Carlo estimate of the posterior mean:

$$\hat{\psi} = \frac{1}{M} \sum_{m=1}^{M} \psi^{(m)}.$$

As noted above however, the main benefit of Bayesian analysis in the present context is that we have obtained posterior samples of the latent state variables z_i: $i = 1, 2, \ldots, s$ (i.e., whether the species was present/absent at a site), which we may use to estimate functions of those latent state variables (e.g., the number of occupied sites in the sample). In general, the posterior distribution of a function of model parameters can be estimated by applying that function to the posterior samples that have been obtained by MCMC. For example, one can obtain a posterior draw (i.e., one value from the posterior distribution) of the number of occupied sites out of the s sample sites by calculating:

$$x^{(m)} = s_D + \sum_{i=s_D+1}^{s} z_i^{(m)}$$

using the MCMC output for $z_i^{(m)}$: $i = s_D + 1, s_D + 2, \ldots, s$. This can be repeated for all posterior samples of $z_i^{(m)}$ to obtain the posterior distribution of x. In practice, rather than calculating these values after completing the entire MCMC procedure, one would specify that $x^{(m)}$ is to be calculated and stored on each interaction of the procedure (e.g., line 10 of Fig. 4.2). Note that the posterior distribution of x will be discrete (i.e., can only have integer values between s_D and s); hence, variance estimates based upon this posterior, distribution should be accurate for the sampling situation. No asymptotic considerations are required to assert the validity of our inferences. Furthermore, to make inference about the number of occupied sites in the population, we can predict the value of the latent state variables $z_j^{(m)}$: $j = s + 1, s + 2, \ldots, S$ as random values from the Bernoulli distribution with probability ψ. The entire procedure can also be very easily generalized to account for missing observations and allow occupancy and detection probabilities to be functions of covariates.

BLUE-RIDGE TWO-LINED SALAMANDERS REVISITED

We return now to the example of blue-ridge two-lined salamanders considered previously. Suppose it is of interest to estimate the number of the 39 transects that were occupied by the salamanders. Here we do so using the Bayesian approach outlined above with WinBUGS 1.4.

As stated previously, MCMC methods are iterative, simulation-based methods that generate samples of the parameters from their posterior distri-

TABLE 4.9 Summary of Posterior Distributions for Occupancy Model Parameters Fitted to Blue-ridge Two-lined Salamander Data

Quantity	Mean	SD	Central 95% Credible Interval
$\hat{\psi}$	0.61	0.12	(0.39, 0.88)
$\hat{p}$	0.26	0.06	(0.16, 0.37)
$\hat{x}$	24.15	4.07	(19, 34)

Estimated number of occupied transects ($\hat{x}$); estimated probability of occupancy ($\hat{\psi}$); and estimated probability of detection ($\hat{p}$). Given are the mean, standard deviation (SD), and central 95% credible interval (2.5% quantile, 97.5% quantile) of the posterior distributions. Posterior distributions were approximated with 100,000 samples using WinBUGS 1.4, assuming uniform(0,1) prior distributions on ψ and p.

butions. As with any sampling problem, a greater number of samples provides a more accurate view of the sampled "population" (i.e., the posterior distribution). Here we obtained 101,000 samples to approximate the posterior distribution, with the first 1,000 samples being used as the "burn-in" period; thus the results presented below are from the 100,000 latter samples. Fig. (4.3) is the density plot, or histogram, for the estimated posterior distributions of x, ψ, and p, in panels (a), (b), and (c), respectively. Note that the posterior distribution for x is discrete and takes values in the range of 18 (the number of transects where blue-ridge two-lined salamanders were detected) to 39. Various summaries of the posterior distributions are reported in Table 4.9. In particular, note that the posterior means of p and ψ are very similar to those reported in Table 4.1 for the constant-p model. The posterior standard devia-

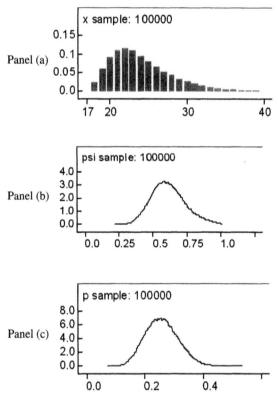

FIGURE 4.3 Posterior distributions of the number of occupied sites (x), and the model parameters ψ and p for the blue-ridge two-lined salamander data. Posterior distributions were approximated with 100,000 samples using WinBUGS 1.4, assuming uniform $(0,1)$ prior distributions on ψ and p.

tion for ψ is also very similar to the asymptotic standard error reported in Table 4.1. As a result, our inference about the underlying probability of occupancy is similar in both cases despite having a relatively small sample of 39 transects (although generally this need not be the case). The last row of Table 4.9 provides posterior summaries for the number of occupied sites (x). A 95% posterior credible interval (i.e., a Bayesian "confidence interval") is reported as (19, 34), and this interval is asymmetric [see Fig. 4.3(a)], as one might expect in small sample situations.

4.6. DISCUSSION

In this chapter we have outlined a general sampling protocol that is used in many occupancy studies and have introduced the concept that to explicitly account for detection probability, repeated surveys of sites are required within a season (the period during which the occupancy state of sites does not change in a completely random manner). We have also reviewed a number of methods for estimating the proportion of area occupied by a target species, and then detailed a model-based framework that provides a great deal of flexibility because of its ability to incorporate missing observations and covariate values into an analysis. Our recent extensions of the modeling to small-area estimation or finite populations is an exciting area of current research that will be very applicable to many situations. The Bayesian implementation of the modeling in this context is very natural and provides valid, non-asymptotic inferences to be made.

As reviewed in Chapter 2, many investigators have quantified occupancy for use in various kinds of ecological investigations. Such investigations include metapopulation studies (e.g., Hanski 1999), species-habitat modeling (including resource selection probability functions; Manly et al. 2002) and monitoring (e.g., Scott et al. 2002). This large body of literature (Chapter 2) generally does not adequately address what we believe to be one of the fundamental characteristics of sampling animal populations, that in most cases a target species will not always be detected at a sampling unit when it is present. By not accounting for imperfect detection within the methods of analysis, the resulting inferences will be less robust, because any apparent change or differences in "occupancy" may actually be due to differences in the detectability of the species (e.g., the pronghorn antelope example).

We have not discussed methods that account for spatial autocorrelation, such as the autologistic model (e.g., Klute et al. 2002; Lichstein et al. 2002). Undoubtedly at some scale, the probability of one sample unit being occupied is likely to be affected by whether a neighboring sampling unit is also occupied, and techniques for investigating the degree of spatial correlation would be a valuable addition to the ecologist's toolbox. However, we believe that such modeling should be done within the context of the models we have described and acknowledge the possibility of false absences. As we stated in the preface, we do not see this book as the last word in occupancy estimation and modeling, but one of the first, and we hope its contents will stimulate others to extend these general methods to address interesting ecological questions.

In the following chapter we extend the above model-based estimation methods to allow for heterogeneous detection probabilities among sites not accounted for by covariates. In Chapter 6, we go on to discuss design issues for single-season occupancy studies. It is vitally important to note that how a study is designed will influence the reliability of resulting estimates and also affect how the estimates should be interpreted. No amount of statistical magic will ever improve poorly collected data.

CHAPTER 5

Single-species, Single-season Models with Heterogeneous Detection Probabilities

In previous chapters, we have formulated site-occupancy models under the assumption that detection probability (p) is constant, or only varies in response to measurable covariates. In this chapter, building on Royle (2005), we consider general classes of site occupancy models that allow for heterogeneity in detection probability among sites. It is natural to consider heterogeneous detection probability models for site occupancy because factors that influence detectability are many and varied, and it may not be possible to identify, much less control for, all of them. For example, variation in detection probability may be induced by covariates that affect detection but that were not measured and hence were omitted from the detection probability model. A crucial factor is that the data upon which site-occupancy models are based are typically observations of detection/nondetection of the species, and variation in site-specific abundance of the species must surely affect the probability of detecting the species (i.e., detecting at least one individual). Detection of at least one member of the species will tend to be higher at sites where abundance is high and lower at sites where abundance is low. Thus, one could only reasonably rule out abundance-induced heterogeneity if abundance could be viewed as being relatively constant across "occupied" sites, which we believe is unlikely

133

to be the case in many animal sampling problems. The phenomenon of abundance-induced heterogeneity in detection probability is more likely to be important when sampled populations are small (e.g., <10) and will diminish in importance as average population size becomes larger, in which case detection probability may be sufficiently well approximated as being constant.

The problem of heterogeneous detection probabilities among individuals has been considered extensively in conventional capture-recapture models. Analogous to the effect of heterogeneity on closed population abundance estimators, unaccounted-for heterogeneity yields biased estimates of site occupancy (Royle and Nichols 2003). The literature on modeling heterogeneity in the context of estimating the size of a closed population is vast (e.g., Burnham and Overton 1978; Norris and Pollock 1996; Coull and Agresti 1999; Pledger 2000; Fienberg *et al.* 1999; Dorazio and Royle 2003; Pledger *et al.* 2003; Link 2003). These existing approaches suppose that p varies by individual (p_i) and that each p_i is a random value from some distribution, commonly referred to as a mixture distribution. Mixture distributions that have been used for p include discrete distributions (i.e., there is only a finite, usually small, number of values p could take; also known as a finite mixture) and continuous distributions such as the beta and the logit-normal (i.e., values are normally distributed on the logit scale). These mixture models can be extended directly to the site-occupancy modeling framework.

Royle and Nichols (2003) suggested an alternative formulation of models for heterogeneity by exploiting the relationship between detection probability and abundance under binomial sampling. Their model also allows for the estimation of abundance from detection/nondetection data. While this model is appealing because it admits an explicit linkage between abundance and occupancy, other (more standard) mixture distributions might also be considered, such as those commonly employed in conventional capture-recapture studies. Such models do not require that parametric assumptions be made about abundance, and so might be viewed as being more general, as they encompass generic heterogeneity due to all mechanisms.

We now provide in Section 5.1 a general formulation of site occupancy models with heterogeneous detection probabilities that is based on the conventional view of p as a random variable, endowed with a probability distribution. We begin by considering the discrete or finite mixture distribution, then advance to continuous or infinite mixtures such as the beta and logit-normal. We also consider the Royle and Nichols (2003) model that arises by considering that heterogeneity in detection probability is derived from variation in abundance. We provide an example of fitting and interpreting these models to avian detection/nondetection data in Section 5.2. In Section 5.3 we generalize these basic heterogeneity models to allow for covariates that are thought to influence detection probability, and we provide an illustration using anuran calling survey data. Finally, we discuss a recent issue raised by Link (2003) in the context of estimating the population size of closed populations using analogous mixture models. He noted that different mixture distributions may produce equivalent (or nearly so) fits to the observed data, but vary widely in their estimates of population size. That is, population size is not identifiable across mixture distributions, and we consider this problem within the context of site occupancy models with heterogeneous detection probabilities.

5.1. SITE OCCUPANCY MODELS WITH HETEROGENEOUS DETECTION

In this section we consider a class of occupancy model in which detection probability may vary among sites but is constant in all other respects (i.e., detection probabilities do not vary in time, or in response to measurable covariates). Such models are analogs of "Model M_h" used in classical abundance estimation problems (Otis *et al.* 1978; Williams *et al.* 2002). The incorporation of time or covariate effects is considered in Section 5.3. A key concept is that now there may be a (potentially infinite) number of possible values for the detection probability at each site, and the likely value for a particular site is not known *a priori*. Therefore, when formulating an expression for the probability of observing a particular detection history, we must account for the fact that the detection probability could take a number of possible values.

GENERAL FORMULATION

Recall from Chapter 4 that under the assumption that detection probability is constant for each survey, the number of detections at each site (Y) is a zero-inflated binomial random variable. The probability of observing y_i detections at site i for K surveys is therefore:

$$\Pr(Y = y_i) = \psi \binom{K}{y_i} p^{y_i} (1-p)^{K-y_i}, \quad y_i > 0$$
$$= \psi(1-p)^K + (1-\psi), \quad y_i = 0$$

Here we use an alternative, more convenient form for expressing the zero-inflated binomial distribution:

$$\Pr(Y = y_i) = \psi \binom{K}{y_i} p^{y_i} (1-p)^{K-y_i} + (1-\psi)I(y_i = 0), \tag{5.1}$$

where $I(y_i = 0)$ is an indicator function that equals 1 if the species was not detected at the site (i.e., $y_i = 0$), 0 otherwise. Expressed in this manner, the probability of observing y_i detections can be considered as the combination of two components, one conditional on occupancy and the other conditional on the site being unoccupied. Conditional on the site being occupied, the probability of observing y_i detections is simply given by the binomial distribution (i.e., $\binom{K}{y_i} p^{y_i} (1-p)^{K-y_i}$). Given the site is unoccupied, the species will never be detected; hence the probability that $y_i = 0$ is 1, and 0 otherwise (as represented by the indicator function). The zero-inflated binomial distribution is then obtained by multiplying each of the conditional probabilities by the probability that a site is either occupied or unoccupied (ψ and $1 - \psi$, respectively; this is known as marginalizing). The joint likelihood based on data from s sites is therefore the product of s versions (one for each site) of Eq. (5.1), that is:

$$L(\psi, p | y_1, y_2, \ldots, y_s) = \prod_{i=1}^{s} \Pr(Y = y_i).$$

The basic strategy to developing models that allow for heterogeneity in p is to view p as what is commonly referred to as a *random effect*. That is, we suppose that p may potentially take on a different value for each sample site and that its potential values are governed by a probability or mixing distribution for p, say $f(p|\theta)$, where θ are the parameters of the mixing distribution to be estimated. In other areas of applied statistics, it is common to choose one of various continuous distributions, such as the beta distribution, or a normal distribution on the logit-transformation of p. Alternatively, discrete mixing distributions in which two or more values of p are allowed may be constructed. Regardless of the form of $f(p|\theta)$, the view of p as a random variable enables standard results to be used to effectively remove site-specific p's from the likelihood. Analysis can therefore focus on the likelihood as a function of ψ, and the parameters of the mixing distribution, θ.

To formalize this notion, recall that the probability of obtaining y_i detections from K surveys, conditional on the site being occupied, is a binomial distribution with parameter p. As p is now considered as a random variable, using the results of Chapter 3, Section 3.2, we can therefore calculate the *expected* probability of obtaining y_i detections. That is, the average probability of observing y_i detections from K surveys, where the averaging is taken over possible values of p.

Let $\pi_f^{(c)}(y_i|\theta)$ denote this expected probability $\Big(\text{i.e.,} \; \pi_f^{(c)}(y_i|\theta) = E_{\tilde{p}}\Big[\binom{K}{y_i} p^{y_i} (1-p)^{K-y_i}\Big]\Big)$, where the superscript "(c)" indicates that the expected probability is conditional on occurrence of the species at a site, and the subscript "f" indicates the mixture distribution for p. To calculate $\pi_f^{(c)}(y_i|\theta)$, note that the probability of obtaining the y_i detections is simply some function of the random variable p $\left(\text{i.e., } g(p) = \binom{K}{y_i} p^{y_i}(1-p)^{K-y_i} \right)$. Hence, when $f(p|\theta)$ is a discrete distribution, such as a finite mixture:

$$\pi_f^{(c)}(y_i|\theta) = \sum_p \binom{K}{y_i} p^{y_i}(1-p)^{K-y_i} f(p|\theta),$$

with the obvious extension to a continuous distribution of:

$$\pi_f^{(c)}(y_i|\theta) = \int_0^1 \binom{K}{y_i} p^{y_i}(1-p)^{K-y_i} f(p|\theta)dp.$$

(Note that integration can be considered as the summation of an infinite number of very small terms; see Appendix.) Finally, we have to marginalize over occupancy state to account for the fact that observed zeros arise from two possible sources (i.e., present but not detected, or not present). This gives rise to the unconditional probability of observing y_i detections from K surveys ($\pi_f(y_i|\theta, \psi)$) of the form:

$$\pi_f(y_i|\theta, \psi) = \psi\pi_f^{(c)}(y_i|\theta) + (1-\psi)I(y_i = 0), \tag{5.2}$$

which, generally, we will refer to as a zero-inflated f = binomial = mixture distribution. It is these probabilities that are used to construct the likelihood of each observation. As above, the likelihood for the observed data from s sites is then given by:

$$L(\psi, \theta|y_1, y_2, \ldots, y_s) = \prod_{i=1}^s \pi_f(y_i|\theta, \psi).$$

This general form [i.e., Eq. (5.2)] holds regardless of the choice of mixture distribution $f(p|\theta)$. For any choice of $f(p|\theta)$, evaluation of the likelihood requires only that the expected probabilities for the $K+1$ possible values of y_i (i.e.; 0, 1, ..., K) be computed. We now consider some special cases of models that permit heterogeneity in p among sites.

FINITE MIXTURES

We begin by considering the simple finite mixture distribution to model heterogeneous detection probabilities. Here it is assumed there are M possible values for p, and the probability that a site has detection probability p_m is f_m. Note that as sites must have one of the M detection probabilities, the sum of

f_m values must be 1.0; that is, $\sum_{m=1}^{M} f_m$. Because of this constraint, only $M-1$ values of f_m are required, and the last can be obtained by subtraction. The parameters to be estimated include ψ, the detection probabilities $p_1, p_2, \ldots, p_M$ and the probability masses $f_1, f_2, \ldots, f_M$. The logic behind the finite mixtures distribution is that the population actually consists of M groups of sites, each with a different detection probability for the species, but group membership is unknown (e.g., the species may be easy to detect at some sites and difficult at others, but looking at the sites prior to the surveying, we could not classify which sites would have a high or low detection probability). Using the previous results, the expected conditional probability for observing the y_i detections will be:

$$\pi_{fm}^{(c)}(y_i|\theta) = \sum_{m=1}^{M} \binom{K}{y_i} p_m^{y_i}(1-p_m)^{K-y_i} f_m, \tag{5.3}$$

where the subscript fm denotes that p is a random variable with a finite mixture distribution. Here, the parameters associated with the distribution (θ) are the M possible values for p (technically known as the support points for the distribution), and the relative frequency for $M-1$ of the possible values for p (i.e., f_m; with the last obtained by subtraction). The unconditional probability of observing the y_i detections can then be obtained from Eq. (5.2).

An equivalent approach to the above when developing an expression for the probability of observing y_i detections is to consider the fact that the site may have any of the M detection probabilities. This is easily incorporated by determining the probability for observing y_i detections assuming the detection probability is p_m (conditional upon occupancy state) and multiplying this component by the probability that the detection probability is p_m (i.e., by f_m). This is repeated for all the M possible values for p_m. This process is similar to that used in the previous chapter for developing the basic occupancy models.

For example, suppose we want to consider a finite mixture model with $M=2$; that is, there are only two possible detection probabilities: p_1 and p_2 (say, high and low). Consider the case in which four surveys were conducted at a site and the species was detected twice; hence $K=4$ and $y_i=2$. From Eq. (5.3), the expected conditional probability of observing two detections will be:

$$\pi_{fm}^{(c)}(2|\theta) = \binom{4}{2} p_1^2(1-p_1)^{4-2} f_1 + \binom{4}{2} p_2^2(1-p_2)^{4-2} f_2,$$

where $f_2 = 1 - f_1$. While this expression has been derived from considering the expected value of a function of a random variable, the expression could also be obtained simply by noting that there are two possible mechanisms that could have produced the observation of two detections, corresponding to binomial draws with probabilities either p_1 or p_2. Applying Eq. (5.2) (note that here $I(y_i = 0)$ evaluates to 0), the unconditional probability of observing two detections is therefore:

$$\pi_{fm}(2|\theta) = \psi \left[\sum_{m=1}^{2} \binom{4}{2} p_m^2 (1 - p_m)^{4-2} f_m \right].$$

Finite mixture models of this type are implemented in PRESENCE 2.0. The user may specify whether two or three groups are used to model the heterogeneity, and maximum likelihood estimates are provided. This finite mixture approach has been used in the closely related situation of mark-recapture abundance estimation where detection probabilities are allowed to vary among individuals (Norris and Pollock 1996; Pledger 2000). Finally, note that in the case when $M = 1$, as should be expected, the formulation for the finite mixture model reduces to that of the models given in Chapter 4.

CONTINUOUS MIXTURES

A natural choice for f is the beta distribution, because this is the conjugate prior distribution for p of the binomial distribution. The beta distribution is defined by:

$$f(y_i|\alpha, \beta) = \frac{\Gamma(\alpha)\Gamma(\beta)}{\Gamma(\alpha+\beta)} y_i^{\alpha+1} (1 - y_i)^{\beta-1}$$

with parameters α and β, and where $\Gamma(\cdot)$ is a gamma function (Williams et al. 2002: 730). Conjugacy is the property that the posterior distribution is of the same parametric form as the prior distribution. The utility of this is not so important in the present context, but conjugate priors are often convenient in Bayesian analysis. Under a beta prior distribution for p, the expected conditional probability of y_i detections has a closed form; it is a zero-inflated beta-binomial with:

$$\pi_{bb}^{(c)}(y_i|\theta) = \frac{\Gamma(K+1)}{\Gamma(y_i+1)\Gamma(K-y_i+1)} \frac{\Gamma(\alpha+y_i)\Gamma(K+\beta-y_i)}{\Gamma(\alpha+\beta+K)} \frac{\Gamma(\alpha+\beta)}{\Gamma(\alpha)\Gamma(\beta)}.$$

This is commonly parameterized in terms of the mean $\mu = \alpha/(\alpha+\beta)$ and precision $\tau = \alpha + \beta$, or the mean and standard deviation μ and $\sigma = \sqrt{\mu(1-\mu)/(\tau+1)}$. The beta mixture is appealing primarily because of its convenient form, although the likelihood must still be maximized numerically. Another natural class of continuous models for describing variation in p is the logit-normal class of models (Coull and Agresti 1999), in which $\text{logit}(p_i)$ is assumed to have a normal distribution with mean μ and standard deviation

σ. In this case, $\pi_{LN}^{(c)}(y_i|\theta)$ does not have a closed form and involves integrals, but the integrals can be evaluated numerically. Therefore the complexity here has no practical consequence, as a computer program can do the calculations.

ABUNDANCE MODELS

The mixture models considered previously are constructed by specifying a mixing distribution on the detection probability parameters p. In contrast, Royle and Nichols (2003) noted that heterogeneity in p can be induced by variation in abundance among sites, and they developed a model of heterogeneity based on this consideration. That is, they placed the mixture distribution on abundance. Specifically, let N_i be the abundance at site i. Then, binomial sampling considerations yield that the "net" probability of detection (i.e., of at least one individual) is $p(N_i, r) = 1 - (1 - r)^{N_i}$, where r is the individual detection probability and the notation $p(N_i, r)$ is used simply to denote that the probability of detecting the species at site i is now a function of N_i and r. That is, the probability of detecting at least one individual of the species is 1 minus the probability of detecting none of the N_i individuals at the site. However as the N_i are unknown, by assuming they are random values from an appropriate discrete distribution, the expected probability of observing y_i detections can be calculated in a manner similar to the above, although here we do not have to condition upon occupancy status, as this is accounted for through the mixture distribution on N_i (i.e., a site is occupied when $N_i > 0$ and unoccupied when $N_i = 0$). The expected unconditional probability of observing y_i detections at a site can therefore be calculated as:

$$\pi_m(y_i|\theta) = \sum_{N_i=0}^{\infty} \left[\binom{K}{y_i} p(N_i, r)^{y_i} [1 - p(N_i, r)]^{K-y_i} \Pr(N = N_i) \right]$$

where $\Pr(N = N_i)$ is the probability of N_i individuals being present at the site according to the mixture distribution for abundance (N). For example, suppose that site-specific abundance has a Poisson distribution with mean λ. Then, the probability of observing y_i detections at site i is:

$$\pi_m(y_i|\theta) = \sum_{N_i=0}^{\infty} \frac{K!}{y!(K-y)!} p(N_i, r)^{y_i} [1 - p(N_i, r)]^{K-y_i} \frac{e^{-\lambda}\lambda^{N_i}}{N_i!}$$

Other discrete probability distributions can be considered for N_i, for example, the negative binomial or other models that allow for more complex mean/variance relationships. See Johnson et al. (1992) for an extensive list of potential discrete distributions.

The appeal of this model is that in some cases it may be reasonable to view λ as density. That is, if individuals are distributed in space according to a Poisson process, and the detection of individuals is independent, then λ is the density of individuals (expected number per sample unit). This interpretation should be considered only within the context of the underlying assumptions, that abundance is Poisson and that detection of individuals is independent. While such assumptions are not likely to be valid in most situations, some model extensions are possible. For example, one might consider alternative abundance distributions such as the negative binomial. However, such models can be difficult to fit given data only on detection/nondetection. In general, there is no reason that N_i must be interpreted as abundance *per se*, nor that λ be interpreted as density. Rather, N_i could be viewed as a generic (integer) random effect that yields variation in p, and thus we might view this model merely as an alternative mixing distribution that accommodates heterogeneity in detection probability. Indeed, it need not even be an integer. Royle (2006) notes a relationship between this model and the logit-normal model. Specifically, note that $p_i = 1 - (1 - r)^{N_i}$ implies that $\log[\log(1 - p_i)] = \log[\log(1 - r)] + \log(N_i)$. This yields a linear model for the complementary log-log link of p with intercept $\log[\log(1 - r)]$ and random effect $u_i \equiv \log(N_i)$, which might as well be assumed normal with mean μ and variance σ^2. The point of this is that, while the model is constructed as a mixture over abundance states, it can be made to resemble a model that is a mixture on (the complement of) p, consistent with the construction of the other models that we have considered previously.

Note that under this model, site occupancy is a derived parameter. For example, under the assumption of Poisson abundance, $\psi = \Pr(N > 0) = 1 - e^{-\lambda}$. Also, even though the model is parameterized in terms of individual detection probability, one can extract a parameter that is analogous to the conditional (on occurrence) detection probability that one usually considers. The more familiar expected conditional detection probability is:

$$p_c = \sum_{N=1}^{\infty} \frac{\Pr(y > 0|N, r)\Pr(N|\lambda)}{\Pr(N > 0|\lambda)}.$$

Finally, we note that the Poisson abundance model presented above and in Royle and Nichols (2003) is implemented in PRESENCE 2.0. Note that these models involve summations over all possible population sizes. Practical implementation of these models requires that these infinite summations be replaced by summations that stop at some large population size expected to exceed any that would be possible in practical situations. This is the approach taken in PRESENCE 2.0.

MODEL FIT

To evaluate model adequacy, we consider the usual deviance statistic based on the expected cell frequencies under the model in question:

$$D_g = 2\sum_{k=0}^{K} n_k \left[\ln(\pi_{sat}(k)) - \ln(\pi_g(k|\hat{\theta})) \right]$$

where $\pi_{sat}(k)$ is the observed proportion of sites with k detections and $\pi_g(k|\hat{\theta})$ are the model-based estimates obtained by plugging the MLEs of model parameters into Eq. (5.3). Asymptotically, and under the hypothesis that the model is correctly specified, we can expect this statistic to have a chi-square distribution on $K - NPar_g$ degrees of freedom where $NPar_g$ is the number of parameters in the heterogeneity model g.

Of course, in practice we often fail to achieve large samples, which renders the null distribution of this statistic invalid. In such cases we would probably rely on the usual strategies for assessing goodness of fit. That is, in small samples we might pool cells where the expected cell count is less than five. Alternatively, we might estimate the small sample null distribution using common resampling strategies such as the parametric bootstrap (see Chapter 4). This would be preferred when the number of cell frequencies is small, so that pooling cells reduces the number of cells to the point of yielding a test with low power. We address some important theoretical issues of assessing model fit in Section 5.5.

5.2. EXAMPLE: BREEDING BIRD POINT COUNT DATA

Here we demonstrate the application of these models for heterogeneous detection probabilities using avian survey data. These data originate from a study conducted to evaluate sources of variation in bird count data (Link et al. 1994) on North American Breeding Bird Survey (BBS) routes (see Robbins et al. 1986 for description of this survey). The specific data used here are from a single route (50 "stops" or sample locations at which the number of detected birds is recorded over a three-minute period). Each sample location was sampled 11 times during an approximately one-month season. The actual count data were reduced to observations of detection/nondetection. We consider data on four species: blue jay (Cyanocitta cristata), common yellow-throat (Geothlypis trichas), song sparrow (Melospiza melodia), and gray catbird (Dumetella carolinensis). The data for each species are given in Table 5.1.

No covariates thought to influence either detection or occurrence are available and so, in the analyses presented here, we have considered only the basic heterogeneity models described in the preceding section. That is, in addition to the constant-p model (described in Chapter 4), we consider heterogeneity models based on the logit-normal ("LN"), beta-binomial ("BB"), two-component finite-mixture ("FM2"), and Royle and Nichols abundance ("RN") models. We note that the sample size is relatively small here ($s = 50$), which

TABLE 5.1 Number of Detections (out of $K = 11$ surveys) for Each of Four Bird Species at Each of the 50 BBS Stops

Species	Number of detections												Naive occupancy
	0	1	2	3	4	5	6	7	8	9	10	11	
JAY	17	9	11	6	5	2	0	0	0	0	0	0	0.66
CYT	14	6	7	5	3	1	4	5	3	2	0	0	0.72
SOSP	24	5	1	5	1	5	3	1	3	1	0	1	0.52
CATB	31	6	4	5	2	0	2	0	0	0	0	0	0.38

Species codes are: blue jay (JAY); common yellow-throat (CYT); song sparrow (SOSP); catbird (CATB). "Naive occupancy" is the observed proportion of occupied sites.

impairs the direct application of the deviance statistic described previously for assessing fit. Rather than undertaking a bootstrap characterization of the proper null distribution for each model, we will present deviance statistics merely as indices of relative fit. Results of fitting these various models to the avian point count data are presented in Table 5.2.

For the blue jay, we note that there does not appear to be much heterogeneity among sites in detection probability because under the logit normal model $\hat{\sigma} = 0.023$, and none of the heterogeneity models yields more than a marginal improvement in log-likelihood. Consequently, we are inclined to favor the constant-p model for this species. Estimates of site occupancy are similar across most of the heterogeneity models (except the RN model). Heterogeneity in detection probability is indicated for the remaining three species. For common yellow-throat, the two-component finite mixture appears favored, with $\hat{\psi} = 0.773$. For song sparrow, the beta or logit-normal models appear favored, with $\hat{\psi} = 0.597$ under the beta model and $\hat{\psi} = 0.578$ under the logit-normal model. Moderate heterogeneity is indicated for the catbird data, but it is less clear which heterogeneity model is to be preferred. All heterogeneity models fit the data equally well, and all have similar negative log-likelihoods. A strict adherent to the use of AIC as a model selection tool would choose the Royle and Nichols model because it wins out on parsimony grounds. Regardless, the four estimates of ψ range from 0.428 (RN) to 0.466 (Beta) and are about 10–15% higher than the estimate from the constant-p model.

Note that we might conduct a formal test of the hypothesis of no heterogeneity based on the results of Self and Liang (1987), who demonstrated the asymptotic distribution of the usual likelihood ratio statistic to be a 50/50 mixture of a χ_0^2 and a χ_1^2 (thus the common advice to "halve the P-value" of the conventional test when used for testing that a variance component is zero). Such a strategy would yield a conclusion of no heterogeneity for the jay,

TABLE 5.2 Parameter Estimates and Summary Statistics from Each Site-occupancy Model Fit to the BBS Survey Data

Species	Model	$\hat{E}[p]$	$\hat{\sigma}_p$	$\hat{\psi}$	$-2l$	NPar	DEV
JAY	const	0.199	0	0.723	164.08	2	1.88
	LN	0.197	0.023	0.728	164.07	3	1.87
	BB	0.197	0.024	0.728	164.07	3	1.87
	FM2	0.191	0.046	0.753	164.01	4	1.81
	RN	0.178	0.085	0.806	165.45	2	3.25
CYT	const	0.385	0	0.723	250.55	2	41.22
	LN	0.341	0.218	0.809	218.26	3	8.93
	BB	0.330	0.221	0.835	218.26	3	8.93
	FM2	0.360	0.218	0.773	212.40	4	3.07
	RN	0.374	0.147	0.765	222.77	2	13.44
SOSP	const	0.419	0	0.521	215.08	2	37.89
	LN	0.377	0.237	0.578	188.43	3	11.24
	BB	0.365	0.243	0.597	188.18	3	10.99
	FM2	0.398	0.206	0.549	192.18	4	14.99
	RN	0.403	0.132	0.567	196.25	2	19.06
CATB	const	0.219	0	0.407	132.11	2	8.05
	LN	0.196	0.102	0.454	130.08	3	6.02
	BB	0.191	0.106	0.466	130.04	3	5.98
	FM2	0.206	0.095	0.433	129.99	4	5.92
	RN	0.209	0.075	0.428	130.33	2	6.27

Model designations are: constant-p model ("const"), logit-normal ("LN"), beta ("BB"), two-point finite-mixture ("FM2"), and Royle and Nichols (2003) abundance model ("RN"). $\hat{E}[p]$ and $\hat{\sigma}_p$ are the mean and standard deviation of the estimated heterogeneity distribution, $\hat{\psi}$ is the estimated proportion of sites occupied, $-2l$ is twice the negative log-likelihood. NPar is the number of parameters for each model, and DEV is the model deviance, used as a relative measure of model fit for each species.

significant heterogeneity for the common yellow-throat and song sparrow, and a more equivocal conclusion for the catbird.

5.3. GENERALIZATIONS: COVARIATE EFFECTS

In many problems, the possibility that p varies temporally or in relation to measurable covariates might be considered, perhaps in addition to heterogeneity among sites. For such purposes, we require a detection history formulation of the likelihood. For example, let $\mathbf{h}_i = 0101$. The probability of observing this particular detection history at an occupied site (hence ψ does not appear in the probability statement) is $Pr(\mathbf{h}_i = 0101|\text{site is occupied}) =$

$(1 - p_{i1})p_{i2}(1 - p_{i3})p_{i4}$. A convenient manner in which to parameterize heterogeneity in p among sites is to assume a linear logit-link function between p_{it} and covariates according to:

$$\text{logit}(p_{ij}) = \alpha + \beta x_{ij} \tag{5.4}$$

where x_{ij} is a covariate measured at site i during survey j, and α and β are the parameters to be estimated (Chapter 3; where $\alpha = \beta_0$ and $\beta = \beta_1$, which is done to avoid confusion with the notation below). The extension of this basic model to accommodate heterogeneity among sites is straightforward, and that is to replace α with α_i, where α_i is a site-specific random effect endowed with a suitable mixing distribution. For example, we might suppose that α_i has a normal distribution, $g(\alpha_i) = \text{Normal}(\mu, \sigma^2)$, or there may be a small number of discrete values that α_i could take, as under the finite mixture models described previously. Under this detection history formulation of the likelihood, parameter estimates may be obtained by integrating the likelihood contribution of each detection history over the specified mixing distribution and then zero-inflating the resulting marginal probabilities. For example, under a two-point finite mixture model in which there are two values of α_i, say α_1 and α_2 with masses f_1 and $f_2 = 1 - f_1$, then the likelihood contribution for a site with the detection history $\mathbf{h}_i = 0101$ is:

$$\pi_{fm}(\mathbf{h}_i | \alpha_1, \alpha_2, f_1, \beta) = \psi\pi_{fm}^{(c)}(\mathbf{h}_i | \alpha_1, \alpha_2, f_1, \beta) + I(\mathbf{h}_i = 0)(1 - \psi)$$

where

$$\pi_{fm}^{(c)}(\mathbf{h}_i | \alpha_1, \alpha_2, f_1, \beta) = \sum_{m=1}^{2} f_m[(1 - p_{m,1})p_{m,2}(1 - p_{m,3})p_{m,4}]$$

and $p_{m,j}$ is obtained by substituting each α_m into Eq. (5.4).

Under the logit-normal model, the conditional (on occurrence) probability of obtaining capture history $\mathbf{h}_i$ is computed by evaluating the integral $\pi_{ln}^{(c)}(\mathbf{h}_i | \beta, \mu, \sigma^2) = \int_{-\infty}^{\infty} \text{Pr}(\mathbf{h}_i | \alpha, \beta)g(\alpha_i | \mu, \sigma^2)d\alpha_i$ for which the unconditional likelihood contribution is obtained by zero-inflating this probability:

$$\pi_{ln}(\mathbf{h}_i | \beta, \mu, \sigma^2) = \psi\pi_{ln}^{(c)}(\mathbf{h}_i | \beta, \mu, \sigma^2) + I(\mathbf{h}_i = 0)(1 - \psi)$$

The full model likelihood of observing the detection histories for all sites is the product (over i) of each site's likelihood contribution, that is,

$$\prod_{i=1}^{s} \pi_{ln}(\mathbf{h}_i | \beta, \mu, \sigma^2).$$

The Royle and Nichols (2003) formulation is somewhat different. In this case, covariates are modeled on the parameter r (individual detection probability) so that, for example, $\text{logit}(r_{ij}) = \alpha + \beta x_{ij}$ is substituted into the expression for net detection probability $p(N_i, r_{ij}) = 1 - (1 - r_{ij})^{N_i}$ and then used in constructing the likelihood of each detection history as before. As with finite mixture models, the likelihood of each detection history is averaged over all possible values of the detection probability parameter—in other words, values of N_i. Under this model, the intercept, α, remains constant because the heterogeneity is induced by mixing over the abundance distribution.

5.4. EXAMPLE: ANURAN CALLING SURVEY DATA

To illustrate the application of models with covariates on detection, we consider data from a study conducted to evaluate sources of variation in detection probability in North American Amphibian Monitoring Program (NAAMP) survey data. The particular data consist of observations of anuran detection/nondetection at 220 roadside "stops" in Maryland. Each stop is associated with the potential breeding habitat of anurans. See Weir et al. (2005) for further details on this study and results. Between 3 and 14 visits were made to each of the 220 stops between early March 2002 and the end of July 2002. We focus here on data for the gray treefrog (Hyla versicolor).

We consider two covariates that are likely to influence detection probability. Most importantly, the breeding phenology of all North American anurans is strongly seasonal, and can be expected to vary even within the putative breeding season. In particular, while one might detect a given species over a relatively long period, there should be an increase in breeding activity subsequent to the onset of calling activity, with a distinctive period of peak calling behavior, followed by a gradual decline. Here we choose to model this seasonality in detection that derives from breeding behavior as a quadratic on (integer) sample day, defining March 1 to be day = 1. The only other covariate on detection probability considered here is air temperature at the time of survey, and we suppose that the response to temperature may also be quadratic.

Models containing these covariates, in addition to the null heterogeneity model without covariates and one containing no heterogeneity, were fitted to the gray treefrog data. The parameter estimates for each model, ordered by AIC, are presented in Table 5.3. First, we note that moderate heterogeneity is indicated (comparing the null model without heterogeneity to that with heterogeneity), with $\hat{\sigma} = 0.76$ under the best-fitting model, which contains a quadratic response to temperature and a linear day effect.

The fitted quadratic temperature indicates a convex response with detection probability achieving a maximum at about 20.4°C. Also, the fitted response to day indicates increasing detection probability as the season pro-

TABLE 5.3 Parameter Estimates for Detection Probability Covariate Models with Heterogeneity Fit to the Gray Treefrog Data

Model	$\hat{\mu}$	$\log(\sigma)$	$\hat{\beta}_0$	$temp$	$temp^2$	day	day^2	$-2l$	NPar	AIC
$p(het, temp, temp^2, day)$	-2.32	-0.28	0.21	0.73	-0.59	0.68		747.41	6	759.41
$p(het, temp, temp^2, day, day^2)$	-2.24	-0.26	0.25	0.68	-0.51	0.84	-0.24	745.83	7	759.83
$p(het, temp, day, day^2)$	-2.47	-0.17	0.32	0.29		1.10	-0.48	754.61	6	766.61
$p(het, day, day^2)$	-2.50	-0.15	0.38			1.28	-0.47	757.44	5	767.44
$p(het, day)$	-2.75	-0.19	0.28			1.00		764.78	4	772.78
$p(het, temp, temp^2)$	-2.03	-0.45	0.10	1.23	-0.69			766.95	5	776.95
$p(het, temp)$	-2.50	-0.33	0.15	0.84				787.77	4	795.77
$p(het)$	-2.33	-0.45	0.26					842.81	3	848.81
$p(\cdot)$	-2.01		-0.09					845.04	2	849.04

The logit-normal model (intercept μ and standard deviation σ) with covariates was used to model heterogeneity. The estimated coefficients for each covariate are indicated in the column labeled with the corresponding covariate. β_0 is the logit-transform of ψ. No entry dash indicates that the particular covariate is absent from the model. $-2l$ is twice the negative log-likelihood, NPar is the number of model parameters, and AIC is Akaike's information criterion.

gresses, consistent with the late onset of breeding in this species. Finally, the estimated occupancy rate of the species at stops along NAAMP routes in Maryland is $\hat{\psi} \times \exp(0.21)/[1 + \exp(0.21)] = 0.55$.

5.5. ON THE IDENTIFIABILITY OF ψ

Link (2003) demonstrated that in closed population models for estimating abundance, abundance is not identifiable in the presence of heterogeneity, in the sense that different mixture distributions may give rise to identically distributed data (or nearly so), yet produce substantially different inferences about abundance. Although site occupancy models with heterogeneous detection probabilities are trivial to construct, as demonstrated previously, it is natural to consider whether Link's result applies to these classes of models. In closed population models for estimating abundance, the inability to observe the "zero frequency" (number of uncaptured individuals) is the genesis of the nonidentifiability problem reported by Link 2003). However, in occupancy models, the zero frequency is observed. Unfortunately, site occupancy models introduce uncertainty about nondetection and absence of the species (by introduction of the site occupancy parameter) that may largely mitigate any ability to partition the observed zeros into those due to nondetection and those due to non-occurrence of the species in the presence of heterogeneity.

Royle (2005) considers this problem in some detail and demonstrates that the same basic phenomenon should be of some concern in the context of site occupancy models. It can be established empirically that, in certain instances, one cannot reasonably expect to distinguish between alternative mixture distributions from data. To demonstrate this, he chose g (e.g., logit-normal) and f (e.g., beta) to minimize twice the Kullback-Liebler distance (Burnham and Anderson 1998: 37) between π_g to π_f:

$$KL_f = 2\sum_{k=0}^{K} \pi_g(k)\ln\left(\frac{\pi_g(k)}{\pi_f(k)}\right).$$

In some cases, KL_f is close to zero, yet ψ_f and ψ_g differ markedly (note that Royle (2005) provides an example of a g and f where in fact $KL_f = 0$, but this is not generally the case). Link (2003) noted that s (number of sampled sites) times KL_f is the noncentrality parameter that can be used to assess the power of a goodness-of-fit test of f against the g alternative. Consequently, there will be little power to distinguish between f and g, and hence to make the correct inference about ψ, when the KL distance is small.

For example, suppose that data are collected on $K = 5$ sampling occasions, that the logits of p_i have a normal distribution with $\mu = -2$ and $\sigma = 1$, and that

TABLE 5.4 Multinomial Cell Probabilities and Kullback-Liebler Distance Between Models Under the Specified Logit-normal Model for Heterogeneity

Model	ψ	Cell probabilities						$2 \times KL$
		$k = 0$	$k = 1$	$k = 2$	$k = 3$	$k = 4$	$k = 5$	
LN ($\mu = -2, \sigma = 1$)	0.75	0.630	0.222	0.097	0.037	0.011	0.002	0
Const. $p = 0.23$	0.51	0.630	0.205	0.122	0.037	0.005	0.000	0.017
BB $\mu = 0.13, \tau = 5.87$	0.91	0.630	0.223	0.096	0.037	0.011	0.002	2.8e-05
FM2 $p = 0.15, 0.48$; $f = 0.87$	0.62	0.630	0.222	0.097	0.036	0.012	0.002	6.6e-05
RN $r = 0.15, \lambda = 0.84$	0.57	0.629	0.213	0.111	0.037	0.008	0.001	0.0048

The mixture models are logit-normal ("LN"), beta ("BB"), two-component finite mixture ("FM2"), and the Poisson abundance model of Royle and Nichols ("RN").

$\psi = 0.75$. The closest (in the Kullback-Liebler sense) models in each of the other classes are given in Table 5.4.

The marginal cell probabilities are very similar across the five models, and the noncentrality parameter is small in all cases, indicating low power to choose among them, consistent with the results reported by Link (2003). For example, if one obtained a size sample of 200 sites, the power to correctly reject the constant-p model is 0.274, but the heterogeneity models are all very similar. In the free software package R (Ihaka and Gentleman 1996), this calculation is done by issuing the command:

$$R > 1 - \text{pchisq}(\text{qchisq}(0.95, df=6-2), df=6-2, ncp=200*0.17)$$

These results are consistent with Link's (2003) conclusion that "there is virtually no power to distinguish the beta and logit-normal models, except with very large samples." The same can be said about the set of four heterogeneity models considered in Table 5.2. Importantly, the models imply very different values of ψ.

Consider this issue in the context of some of the data analyzed in Section 5.2. The results for the song sparrow suggested the presence of fairly extreme heterogeneity. Furthermore, the LN or BB models appeared to be favored, and they provided nearly identical fits (as measured by deviance) to the data. If we suppose the LN is truth (with parameters given by those in Table 5.2), and carry out the calculation of the KL distance, we obtain the results given in Table 5.5.

Note that while the differences between the logit-normal and several of the mixture models are substantial (in particular, between the LN and constant and RN models), the LN and BB models are nearly indistinguishable. Their expected cell frequencies (the expected data) are nearly identical, and in rea-

TABLE 5.5 Multinomial Cell Frequencies and Kullback-Liebler Distance ("KL") of Closest Heterogeneity Models to the Best-fitting Logit-normal Model to the Sosp Data (K = 11 visits)

| Model | ψ | Expected cell frequencies | | | | | | | | | | | KL |
		0	1	2	3	4	5	6	7	8	9	10	11	
LN	0.578	0.480	0.074	0.074	0.069	0.062	0.056	0.049	0.042	0.036	0.028	0.020	0.010	0.000000
const	0.521	0.480	0.011	0.039	0.083	0.119	0.119	0.086	0.044	0.016	0.004	0.001	0.000	0.563191
BB	0.595	0.480	0.075	0.072	0.068	0.062	0.056	0.050	0.043	0.035	0.028	0.019	0.010	0.000176
FM2	0.535	0.480	0.054	0.088	0.087	0.061	0.040	0.039	0.048	0.049	0.035	0.015	0.003	0.040050
RN	0.568	0.439	0.031	0.071	0.101	0.104	0.087	0.066	0.047	0.030	0.016	0.006	0.001	0.146583

Models are: constant-p ("const"), beta ("BB"), two-point finite mixture ("FM2"), Poisson abundance model ("RN").

150

sonable sample sizes there will be no power to reject the BB model if the LN model is the correct model. However, this ambiguity between the LN and BB is mitigated to some extent by the fact that the estimated occupancy rates under the two models are not very different.

The fact is, having $K = 11$ and a moderate detection probability partially mitigates the problem, either by enunciating differences between the models (in terms of fit) or by minimizing differences between ψ under different models. Some might argue that this is a situation wherein statistical methods are not useful because most of the occupied sites have been detected. However, we note that we are not interested in making inferences about the *apparent* level of occupancy, and we know that the naive estimator used to obtain $\neq \psi = 0.52$ in Table 5.1 is biased low. Consequently, the adjustment of ψ upwards by about 15% as indicated by the LN or BB models probably seems reasonable in light of the evidence resulting from investigation of a suite of what are generally regarded as being reasonable models.

Next, consider the catbird data, for which a lower mean detection probability and moderate heterogeneity were indicated. For this species, the naive estimate (Table 5.1) of occupancy is 0.38. All heterogeneity models are fairly similar in terms of fit, and the KL distances between LN and const, BB, FM2, and RN models are (0.04179, 0.00012, 0.00094, 0.00434). In this case, the models are much more similar, and the implied level of occupancy rate estimates range from about 0.43 to 0.47. Pragmatically, we might feel comfortable thinking that truth is in the vicinity of 0.43 to 0.47. However, the basis for making a formal inference is muddled at best in light of the ambiguity among these heterogeneity models.

5.6. DISCUSSION

In this chapter, we extended site occupancy models to allow for heterogeneity in detection probability among sites. Under the simplest heterogeneity model in which there are no covariates that influence detection probability, we considered several continuous and discrete mixture distributions for p. We also presented a model based on the consideration that heterogeneity in detection probability is due to variation in abundance. Several of these models extend easily to the situation when covariates thought to influence detection are available. Similar extensions can be considered to other classes of site occupancy models, for example to the multiseason models described by MacKenzie *et al.* (2003; see Chapter 7).

We considered the problem raised by Link (2003). He demonstrated that in closed population models for estimating abundance, abundance is not identifiable in the presence of heterogeneity, in the sense that different mixture distributions may give rise to identically distributed data (or nearly so), yet produce substantially different inferences about abundance. In closed population models for estimating abundance, the inability to observe the "zero frequency" (number of individuals never captured) is the genesis of the nonidentifiability problem reported by Link (2003). However, in occupancy models, the zero frequency is observed. Unfortunately, site occupancy models introduce uncertainty about nondetection and nonoccurrence (by introduction of the site occupancy parameter) that largely mitigates any ability to partition the observed zeros into those due to nondetection and those due to nonoccurrence in the presence of heterogeneity. We believe that the problem raised by Link (2003) should be considered when attempting to estimate site occupancy because the same general phenomenon can occur in these models.

The effect of misspecification of heterogeneity models is related to the degree of heterogeneity and the mean detection probability. As Link (2003) noted, differences among mixtures are more pronounced as the mass of $g(p)$ is concentrated near zero (i.e., when most sites have a very low detection probability). One might view low mean detection or high levels of heterogeneity as suggesting that the species of interest cannot be reliably, or effectively, sampled. This should be viewed as a biological sampling issue to consider in survey design prior to data collection, and not as a statistical issue to rectify after the fact by considering complex models of the detection process. The results described in Section 5.5 suggest that the latter may not be a viable option in some situations.

When faced with discrepant estimates of ψ, obtained from models that all appear to fit the data, it is not clear what conclusion (if any) can be drawn. Certainly, if the heterogeneity models yield consistent results, we might feel comfortable with an estimate that is based on one or several of the competing models. However, there is no formal basis for inference in this situation. How does one decide what constitutes consistency among the models? While a 10% difference may seem negligible, on the occupancy scale of 0 to 1 such model differences are substantial and serve to complicate the central inference problem. One could argue that the estimates are only similar because there are models that were not considered that might yield more substantial discrepancies.

The practical consequence of this issue is that monitoring programs that emphasize site occupancy as a metric of population status must consider the possibility that the existence of heterogeneity may diminish the utility of such metrics, and they should take steps to minimize heterogeneity or to increase mean detection probability. For example, establishing rigorous sampling protocols or identifying covariates that affect detection probability may reduce heterogeneity to the point of being unimportant in terms of selecting among possible mixture models. While this may be difficult in the context of estimating population size, in surveys for estimating site occupancy it is often possible to measure a number of covariates about the site being sampled and the conditions under which sampling occurs, and sampling methods often are more flexible. A key problem in dealing with heterogeneous capture probabilities of individual animals is that covariates cannot be measured on the animals that are never captured. In contrast, as noted above, occupancy studies do permit covariate measurement at all sites, including those at which detections never occur. This provides a substantive advantage for occupancy studies and argues for careful consideration of covariates affecting detection probability.

In closed population models, high p may be unrealistic or, if achieved, suggests that most of the population was captured, hence reducing the need for complex estimators of abundance in the first place. This is not the case in site-occupancy models wherein there will always be ambiguous zeros, at least for species for which site occupancy is a useful metric.

For site occupancy models that are based on detection/nondetection data, heterogeneity may be due to variation in abundance, and it is less clear how to deal with this issue in survey design. In sampling of populations in which there is thought to exist considerable variation in abundance, hence site occupancy is a less useful summary of demographic state, demographic summaries that focus on abundance should be considered more explicitly by collecting data that are more informative than simple detection/nondetection (e.g., counts of individuals detected per site; see Chapter 10). Alternatively, one might consider restricting attention to models of heterogeneity that are derived from models of variation in abundance, if that is thought to be the primary source of heterogeneity. While biological arguments to restrict the classes of models under consideration may be appealing to biologists, they may be less appealing when faced with a wide array of seemingly reasonable models for describing heterogeneity (e.g., due to the environment or habitat), and so *a priori* exclusion of such models is limiting.

In Chapter 6 we consider the design of single-season occupancy studies. As noted above, some sources of heterogeneity can be controlled through the careful design of occupancy studies and many suggestions made in the following chapter are given to achieve this end.

Design of Single-season Occupancy Studies

In Chapters 1 and 3 we briefly described some of the basic principles and reasons for carefully designing a study. A poorly designed study may not yield the type and quality of information required to achieve study objectives. In a worst-case scenario, a poorly designed study may yield no useful information and may have only succeeded in wasting precious time, effort, and money. There is no statistical magic wand for poorly collected data that will improve their inferential power. Statisticians may occasionally pull a veritable rabbit out of the hat by developing new methods for such data sets, but there is no substitute for a well-designed (and executed) study. Generally the principle of "Garbage In, Garbage Out" stands.

In this chapter we focus on the issues involved when designing a single-season study for estimating occupancy for a target species, based upon the analysis methods covered in the previous two chapters. Particularly, we consider issues related to how to define a "site," how to select a site, how to define a "season," how to incorporate repeated surveys, and how to allocate effort. This is not an instruction book for designing *your* study, but a "playbook" with different ideas and possible approaches that may be appropriate in some situations and not in others. We draw from our own and others' practical experi-

155

ences, computer simulation studies, and careful thought and intuition. Many of the issues covered in this chapter also apply to other occupancy-related problems that are addressed later in the book.

First, we argue very strongly that, for most species, it is important that data be collected in such a manner that the probability of detecting the species in a survey can be estimated. Generally we suggest that this requires repeated surveys be conducted at, at least, a subset of sites. At sites where the species is detected, the repeated surveys provide the necessary information about the chances of detecting the species at an occupied site in any single survey—in other words, the detection probability. Without repeated surveys it is impossible to disentangle false and genuine absences, unless there is some other form of auxiliary information. It is important to realize that in order to perform repeated surveys, sites do not have to be visited more than once. There are a number of practical options for how repeated surveys can be conducted in the field, some of which will be covered later in this chapter.

We stress that studies generally need to be designed on a case-by-case basis. In most situations, there are unique aspects that must be addressed. Often the best studies arise when biologists, statisticians, and other relevant experts work in unison as a team, with each person examining a proposed study design from his or her own perspective and looking for potential problems either with the practicalities of the design or with the adequacy of the information to be collected. We strongly suggest that input should always be sought from people with the relevant expertise.

We also stress the overarching importance of a clear objective when designing a study (Chapter 1). Subtle differences in the study objective may have a major impact on how a study should be conducted in the field. For example, suppose that there are two possible objectives for an occupancy study within a national park: (1) to compare occupancy for two specific habitat types, and (2) to obtain an overall estimate of occupancy for the entire national park. For both objectives, one potential design would be to randomly select sites from across the national park and to collect the relevant data including habitat type at each site. To address Objective 1, one could use the techniques detailed in Chapter 4 and determine whether there was evidence that occupancy varied by habitat type by using "habitat type" as a covariate (although note there may be more than just the two habitat types of interest represented in the data). A second design, tailored to address Objective 1, would be to first identify all areas within the park comprising the two habitat types and then randomly select sites solely from within those two habitats. Again, the methods of Chapter 4 could be applied to determine whether there was evidence that occupancy differed for the two habitat types. The latter design is likely to be much more efficient for addressing Objective 1, but would not be appropriate for Objective 2 as the sample is not drawn from the entire park. Obviously, these are not the only two possible study designs that could be used, and with others it may be possible to address both objectives (e.g., by stratifying the entire park into three habitat types [the two of direct interest and "others"], then randomly selecting sites within each stratum). However, by attempting to adequately address both objectives simultaneously, a greater level of field effort may be required (i.e., surveying more sites) than would be needed to achieve a single objective.

6.1. DEFINING A "SITE"

We use *site* simply as a generic term to represent the sampling units from the population, or area, of interest. These may be naturally occurring (e.g., ponds or habitat remnants) or defined arbitrarily (e.g., five-hectare blocks within a forest); however, at the spatial scale of the "site," the intent is to establish either species presence or species absence. That is, the observed outcome of surveying a "site" (briefly ignoring the issue of detection probability) could only take two values, presence or absence (i.e., 1 or 0). Only by combining the data from a number of sites do we obtain a measure of occupancy that could take values between 0 and 1. So, clearly, the first consideration with respect to defining a "site" should be the spatial scale at which an observed 0 or 1 is meaningful. For example, suppose a landscape contains a number of remnant forest stands, and management wants to monitor for a rare species within these stands based on the occupancy state variable. Does determining whether the species is present within the stand (yes/no) provide the manager with sufficient information to make informed decisions, or is information required at a finer level of resolution (i.e., the fraction of each stand that is occupied)? In the former case, it may be appropriate to consider the individual stands as "sites," while in the latter case, each stand would have to contain multiple sites and a "site" thus would have to be defined at some smaller scale. Which option would be appropriate in any given situation clearly depends upon a number of additional factors such as management objectives (i.e., what level of information managers require) and the size and number of the stands within the landscape.

A second point with respect to defining a "site" is realizing that measures of occupancy can be very scale dependent, particularly for arbitrarily defined sites within a contiguous habitat. In a given situation, a larger "site" is likely to have a higher probability of occupancy (i.e., contains at least one individual of the target species) than a smaller "site." For example, consider Fig. 6.1, which represents a one-hectare (100 m × 100 m) area of interest with the black cells representing the location of the species; sites have been defined at two different scales: (a) 50 m × 50 m; or (b) 20 m × 20 m. Sampling and detectabil-

(a)

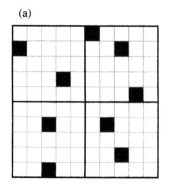

(b)

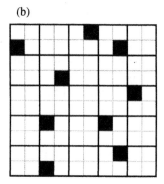

FIGURE 6.1 Graphical representation of a one-hectare (100 m × 100 m) area with the location of the target species indicated by the black cells. "Sites" have been defined at two different scales: (a) 50 m × 50 m; and (b) 20 m × 20 m.

ity issues aside, the true proportion of sites occupied is 1.0 if the first definition is used (i.e., the species is present in all four "sites") and 0.36 using the latter definition (or even 0.09 if a site was defined to be 10 m × 10 m). Generally we suggest that a site should be large enough to have a reasonable probability of the species being there (we suggest that the probability should be 0.2–0.8 in most situations), but small enough so that any measure of occupancy is meaningful and the site can be surveyed with a reasonable level of effort.

6.2. SITE SELECTION

In earlier chapters we briefly mentioned various probabilistic sampling schemes that one might use to select a sample of sites from the population or area of interest (e.g., simple random sampling, stratified random sampling, etc.). The importance of having a probabilistic sampling scheme is to generalize the results from the analysis of data collected from the specific study sites to the wider population of interest. If there is no intent to generalize the results beyond the actual study sites sampled, a probabilistic sampling scheme is not required (i.e., the study sites represent the entire population of interest). However, if no probabilistic sampling scheme is used (i.e., sites are selected purely in a haphazard manner or because of convenience), there is no statistical basis for generalizing the results to beyond the specific study sites.

The methods detailed in this book implicitly assume that the study sites represent a simple random sample from the wider population. If a stratified random sample is used, then one could analyze the data for each stratum separately (or within a single analysis with covariates defined to represent each stratum), and then one could combine the estimates for each stratum using standard results (e.g., Cochran 1977; Thompson 2002). For other sampling schemes, in which random sampling occurs at multiple levels (e.g., first a random sample of forest stands within a region, and then a random sample of sites from within the selected stands), the exact method of analysis depends upon the level at which inference is to be made. No additional structure to the modeling is required if the results are not to be generalized beyond the random sample of stands (i.e., the random sample of units at the level immediately above sites), as the sample of stands represent the entire population of interest. However, if inference is to be generalized beyond the initial sample of units (e.g., to all stands within the region), then the method of analysis should reflect that random sampling has occurred at multiple levels. We believe the easiest way to accomplish this would be through the use of hierarchical models using a Bayesian form of analysis, and we see this as a potential extension of the methods that are presented in this book. If an adaptive sampling scheme (Thompson and Seber 1996; Thompson 2002) is used (i.e., neighboring sites are included in the sample once the species is detected at a site), then the methods in this book would again need to be extended. This is because the probability that any given site will be included in the final sample now depends upon the probabilities that are the very focus of the models of Chapters 4 and 5 (i.e., sites may be included in the final sample when the species is present and detected at a neighboring site). We believe that by extending the models to account for the probability of inclusion in the final sample, unbiased estimates of occupancy could be obtained.

While there are some sampling schemes that are not directly compatible with the methods discussed in this book at present, we do not discourage the use of such schemes. We are confident that in the future, techniques will be developed to estimate and model occupancy data collected using these schemes. In addition, it is frequently possible to obtain model-based occupancy estimates for small spatial units using the methods described in this book and then to use these estimates in replication-based estimation approaches (e.g., Skalski 1994). Although this two-step approach may not be optimal, it provides a reasonable approach until more inclusive modeling is completed. However, we urge people to think very carefully about why they wish to use a more complicated scheme if similar results could be obtained using a much simpler design.

A final comment with respect to selecting sites is that generally we advise against only selecting sites based upon knowledge of their likely occupancy state (e.g., sites that were known to be occupied by the species in the recent past, or based upon casual observations) when the population of interest consists of sites about which such knowledge is and is not available. Unless this group of sites actually represents the population of interest, estimates of occupancy for the entire population may be biased. For example, suppose occupancy is to be estimated within a stream system for a particular salamander species. Within this system, there are locations where the salamanders have been reported as present, based on sightings in the past by members of the public and local herpetologists. If these locations were selected as study sites, then the estimated level of occupancy is likely to be higher than for a random sample of study sites from throughout the stream system. Alternatively, if interest lies both in locations where the salamanders have been reported and in all other places within the stream system, then these regions could be treated as separate strata within the population (the stream system), and a random sample of sites selected from each stratum. We return to this point in Chapter 7, where we consider design issues for multiple-season occupancy studies. However, one situation in which this type of design may be appropriate is when the fraction of sites that are still occupied is now of direct interest (e.g., as a "survival" probability for the species over the intervening time period). In this case, the population of interest consists only of sites that were known to be occupied during the past (e.g., based on museum records, field notes of naturalists and explorers, etc.), and "occupancy" may be interpreted as a measure of "survival" accordingly. Asking questions about whether current "occupancy" (of previously occupied sites) is lower in human-developed sites than less developed sites (for example) may be worthwhile study objectives.

6.3. DEFINING A "SEASON"

In Chapter 4 we considered how violation of the closure assumption influenced an estimate of occupancy. Provided changes in occupancy within a season are random (i.e., the probability of the species being physically present within a site at one point in time does not depend upon whether it was physically present at an earlier point in time), then occupancy estimates are unbiased, provided they are now interpreted in terms of sites that are "used" (places at which the species is *sometimes* physically present during the season) rather than "occupied" (places at which the species is *always* physically present during the season). For other kinds of changes in occupancy within a season (e.g., Markovian changes depending on previous occupancy state, sites becoming permanently occupied [immigration], or sites becoming permanently unoccupied [emigration; see Chapter 4]), the estimated occupancy parameter may not relate to any relevant aspect of the population. In terms of deciding what time period should be defined as a "season" to provide a meaningful interpretation of the resulting occupancy estimate, these results suggest that a "season" should be a time frame of sufficient length such that the species is either always present or always absent from the sites (i.e., sites are completely closed to changes in occupancy), or changes occur at random.

It is also vitally important to consider the study objectives when defining a "season." When changes in occupancy do occur at random over some time frame, then at a smaller timescale it is likely that sites are effectively closed with respect to occupancy. For example, over a five-day period, a wide-ranging carnivore may be physically present at random within a given site as it roams throughout its territory, but on any single day the carnivore would be either present or absent at the site. If the objective of the study was to investigate what habitats are used by the carnivore (perhaps to identify habitats to prioritize for conservation), then it may be appropriate to define a week (or perhaps a much longer period) as a "season," but if the objective was to provide more of a "snapshot" of the carnivore population at a given point in time (say, occupancy is being used as a surrogate for population size), then a "season" should be defined as a much shorter time interval.

Definition of a "season" can also depend upon how a "site" is defined, particularly when "site" is an arbitrary unit. Depending upon the species, the closure assumption may be more reasonable for a large "site" than a small "site." Thus in some instances it may be possible to satisfy the closure assumption by increasing the size of a "site," provided that the relevant information (according to the study objectives) will still be collected. Recall that the closure assumption is at the species level, and not at that of the individual animals (although there may be little distinction for low-density species).

Finally, we stress that as the estimate of occupancy is to apply to the entire population, consideration of the closure assumption not only applies at the time frame during which surveying is conducted at a single site but also to the time frame applied to all sites within the population. Therefore, if a week is thought to be a reasonable definition of a "season" at the site level, yet it takes a month to complete the surveying for the entire population, the "season" should be considered as a month-long period if the occupancy estimate is to be applied to all the sites surveyed within that month.

6.4. CONDUCTING REPEAT SURVEYS

Once a meaningful definition of a "season" has been determined, there are a number of practical options for how repeated surveys could be conducted at a site. These include:

1. visit the site multiple times and conduct a single survey on each visit;
2. conduct multiple surveys within a single visit, where surveys are separated by sufficient time so that they can be considered independent (i.e.,

the probability of detecting the species in a subsequent survey does not depend upon whether the species was detected in a previous survey);

3. have multiple surveys conducted simultaneously by independent surveyors (or by using independent sampling methods) during a single visit;

4. within a larger site, conduct surveys at multiple smaller plots (e.g., conduct multiple transect surveys within a five-hectare block).

There are various advantages and disadvantages to each approach, but the choice of which approach is most appropriate for a given situation really depends upon the nature of the biological question, the timings of the surveys in relation to the biology of the species in question, and the factors that may affect the probability of detecting the species.

It is worthwhile to try to ensure (as much as possible) that each survey is independent. That is, the probability of detecting the species in one survey should not depend upon the outcome (detection or nondetection) of another survey. In options 1 and 2 above, this may occur, for example, if once the species is detected at a nest or den (say), the probability of redetecting is increased because the observer knows where to look for the species or its sign in a subsequent survey. However, this could be easily accounted for during the design of the study either by having a different observer visit the site each time or by using a "removal" design in which surveys halt once the species is first detected. The effect could also be easily modeled with a survey-specific covariate that allows the probability of detection to be different after the first detection of the species (e.g., the survey-specific covariate equals 1 for all surveys after the species is first detected, and 0 otherwise). When conducting multiple surveys simultaneously, as in options 3 and 4, a lack of independence may be introduced if the number of opportunities for detecting the species is limited, such that it would not be possible for each of the surveys to detect the species. For example, suppose two observers are surveying a site for blue jay (*Cyanocitta cristata*) nests. They conduct a 10-minute search of the site, each searching one-half of the site. Now suppose that sites have been defined such that there will only be either a single nest or no nest within the site. Given that there is a nest at the site, as each observer only surveys one-half of the site, it will be impossible for them both to find the nest (it cannot be in both halves); yet with no prior knowledge of the nest's location, each observer has some non-zero probability of finding the nest (the probability that the nest is located within either half, multiplied by the probability of each observer finding the nest if it is in his or her half). That is, the *a priori* probabilities of observer 1 and observer 2 finding the nest are p_1 and p_2, respectively, which are both greater than zero. Assuming independence of the surveys, the probability of both observers finding the nest will be $p_1 p_2$, which again will be greater than zero. Yet because of the practical situation we know this probability to be zero; therefore, the surveys could not be considered independent. If there are limited opportunities for detecting the species, this again can be accommodated either through design (each observer randomly searches the entire site for 20 minutes, or each tosses a coin to determine which of the two halves he or she will survey, such that both observers may survey the same or different halves), or via modeling of the data. However, in this instance the modeling would have to make some assumptions about the number of opportunities available for detecting the species, which may be unlikely in practice.

When determining which of the four options above may be appropriate, the key is to consider how the biology of the species may affect a surveyor's ability to detect it and the factors that may affect the detection probabilities (e.g., animal activity patterns, observer experience, weather conditions). The choice of whether to conduct repeated surveys during a single visit or multiple visits largely rests with the degree of variability in detection probabilities and the timescale over which it varies. For example, suppose detection probability is thought to be relatively constant over a certain time frame (e.g., a week). If all sites can be surveyed within a week, it is probably more efficient (logistically) to visit each site once and conduct the repeated surveys within that single visit than to visit sites multiple times and conduct a single survey each time.

However, if the detection probability varied daily (say), and sites were only visited once during a week-long study (i.e., a timescale during which the detection probability may have varied), a form of heterogeneity in detection probabilities may be introduced. The detection probability associated with each site may be different, as sites were surveyed on different days (Fig. 6.2). If a moderate number of sites (more than 20) are visited each day, then it may be possible to estimate a daily detection probability for each group of sites. Alternatively, when the mechanisms that drive changes in detection probability are thought to be known (e.g., cloud cover), these could be measured as covariates and then included in the occupancy models of Chapters 4 and 5. However, in this case our resulting inference could be heavily dependent upon the appropriateness of that assumption about the mechanisms underlying variation in detection probability. Where practical, we suggest a more prudent course would be to conduct multiple visits, so that each site can be surveyed under a range of conditions.

There are a number of other ways in which characteristics of the study design may unintentionally introduce heterogeneity into the detection probabilities. For example, MacKenzie et al. (2004a) discuss design aspects of the Mahoenui giant weta pilot study, the data for which were used as an example in Chapter 4. Two potential sources of heterogeneity they discuss are due to use of multiple observers and different times of doing the surveying. In many

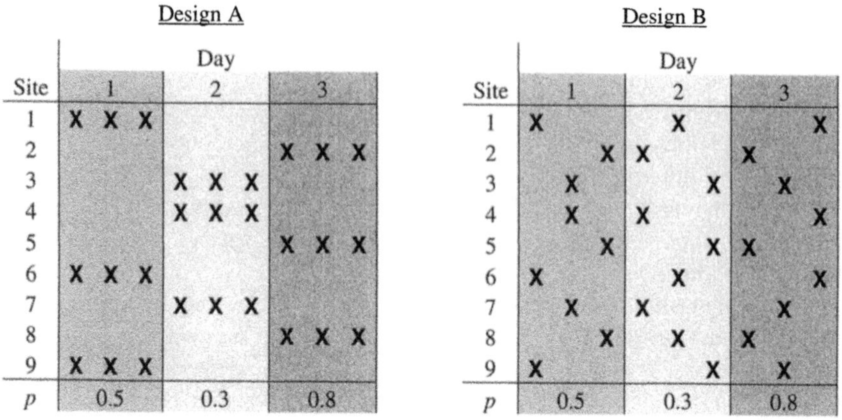

FIGURE 6.2 Example of study design–induced heterogeneity when detection probability varies at some time scale. In Design A, all three surveys of each site are only conducted on a single day; hence, the detection probability is different for sites surveyed on different days. In Design B, each site is surveyed on different days, thereby removing the heterogeneity.

situations, observers will vary in their ability to detect the target species because of such factors as differences in experience, eyesight, and hearing. As a result, the detection probabilities among sites will be heterogeneous if the same observers always survey the same sites. Again, if each observer surveys a moderate number of sites, then perhaps the effects could be estimated. MacKenzie et al. (2004) suggest a simple design-based solution that enables better estimation of potential observer effects: rotating observers amongst the sites so that no site is ever surveyed by only a single observer. In a similar manner, heterogeneity could be introduced by always surveying the same sites at the same time of day (e.g., due to the observers always surveying sites in the same sequence). Many species exhibit different behavioral patterns at different times of the day that may result in their being more or less detectable at certain times [e.g., reptiles sunning themselves early in the day; morning song of breeding birds; temperature-related inactivity of butterflies early in the day (Casula and Nichols 2003)]. If surveying for the species is conducted over a period of the day where detection probabilities change appreciably, and different sites are being surveyed during that period, heterogeneity may be the result. While a "time of day" covariate could be used to account for such variation, there is the potential for such an effect to be confounded with other site-specific factors. MacKenzie et al. (2004a) again suggest a simple design-based solution that would provide more robust information about any potential "time of day" effect, which is to change the order in which sites are surveyed each day.

In summary, we suggest that when considering the exact manner in which repeat surveys are to be conducted, careful thought must be given to the timescale and likely factors that may impact upon the probability of detecting the target species. If it is not possible to conduct the required number of repeated surveys at a moderate number of sites within the timescale over which detection probability is believed to be relatively constant, a prudent course of action would be to visit the sites repeatedly, with a single survey of each site per visit. One should also design a study to avoid inducing heterogeneity in detection probabilities with factors that could (partially) be controlled (e.g., conducting all repeat surveys in a single day when only a small number of sites are surveyed per day).

6.5. ALLOCATION OF EFFORT: NUMBER OF SITES VS. NUMBER OF SURVEYS

Once the investigator has some idea of how repeat surveys should be conducted, given the biology of the species and the practicalities of attempting to detect it at a site, the next obvious questions are "How many repeated surveys are required?" and "How many sites should be surveyed?" Ultimately, increasing the number of sites surveyed will improve the precision of the occupancy estimate. However, this should not be to the detriment of conducting repeated surveys; otherwise the variance component related to the uncertainty due to the species' imperfect detection (see Chapter 4) will increase and potentially negate any beneficial effect of surveying more sites. In effect, this means a delicate balancing act is required to appropriately allocate resources between the sites to survey and the number of repeated surveys per site.

In this section we shall begin by considering a general design in which all sites are surveyed an equal number of times, and then continue to consider other designs. While in practice this general and simple design may not be the most efficient design, it is useful to consider such a design to establish some general principles.

In order to determine how resources should be allocated, three pieces of information are required:

1. the level of acceptable precision for an occupancy estimate;
2. initial estimates of the probability of occupancy and detection;
3. an indication of the maximum number of surveys (where a survey denotes a single search of a single site) that could be conducted.

The level of acceptable precision or uncertainty should be clearly stated as part of the study's objective. Often such a statement will specify that the standard error (say) for the estimated proportion of area occupied by the species is less than some desired value. Without an *a priori* inclination of the level of acceptable uncertainty for the study, it is difficult to judge whether the resources that will be consumed during the fieldwork will yield sufficient information to be useful for the intended purpose. Commonly used metrics for specifying the level of acceptable uncertainty are the variance $(Var(\hat{\psi}))$, standard error $(\sqrt{Var(\hat{\psi})})$, and coefficient of variation $(\sqrt{Var(\hat{\psi})}/\psi)$.

As in all study design situations, initial estimates or guesses of the key population parameters are required in order to determine the number of samples needed to achieve the study's objective. This is because the measures of uncertainty depend on these values. Here, this relates to the occupancy and detection probabilities. Often a range of potential values should be suggested, so that the sensitivity of the required sample size to the values that are estimated can be examined.

Finally, it is important to have some idea of the maximum number of surveys that could be conducted in practice (remembering that in some cases it may be appropriate to conduct multiple surveys within a single visit to a site). This is determined by the amount of resources that are available and the total sampling effort that could be used. There is little point in designing a study if there are insufficient resources to implement it (although it would highlight that more resources may be needed in order to achieve the objective!). A good study design must have all the specified statistical properties, but still be practical to implement in the field.

Once all this information has been obtained, a useful starting point is to consider the (unrealistic) situation in which species are always detected at sites without error, that is, $p = 1$. This represents a best possible case and so establishes the minimum number of sites that would be required to obtain an occupancy estimate with the desired level of uncertainty (Chapter 4). If it is not possible to survey the minimum number of sites within a season, then clearly the study's objectives will need to be reconsidered.

For example, suppose it is thought that the probability of a site being occupied by the target species within an area is 0.8 ($\psi = 0.8$), and suppose that we would like the standard error of the estimate to be 0.05 (i.e., an approximate 95% confidence interval would be $\pm$ 0.1). Therefore, assuming $p = 1$, the number of sites that should be surveyed can be determined from:

$$0.05 = \sqrt{Var(\hat{\psi})}$$
$$= \sqrt{\frac{\psi(1-\psi)}{s}},$$
$$= \sqrt{\frac{0.8 \times 0.2}{s}}$$

which gives:

$$s = \frac{0.16}{0.05^2}.$$
$$= 64$$

Hence, based upon the above information, a minimum of 64 sites must be surveyed in order to achieve the desired level of precision. If it is not feasible to sample 64 sites, then it may be necessary to reconsider the study's objective. If 64 sites is an acceptable number, then the next issue is how many repeated surveys and additional sites should be surveyed (given that $p < 1$) in order to achieve a standard error of 0.05.

Below we consider three general sampling schemes that are compatible with the model described in Chapter 4: (1) a "standard design" in which s sites are each surveyed K times; (2) a "double sampling design" in where s_K sites are surveyed K times and s_1 sites surveyed once; and (3) a "removal design" in which s sites are surveyed up to a maximum of K times, but surveying halts at a site once the species is detected. We draw heavily on the work of MacKenzie and Royle (2005), which, to the best of our knowledge, is the only published paper to make a detailed examination of the important issue of allocating effort between the number of sites and the number of surveys per site in a single season, particularly when the total available effort may be limited. They did so under the assumption that one aspect of the study objective was based upon the level of uncertainty in an occupancy estimate (i.e., the variance of $\hat{\psi}$, $Var(\hat{\psi})$). MacKenzie and Royle (2005) found that for each of the three sampling schemes listed above, there was an optimal number of repeated surveys that should be conducted regardless of whether the objective was to design a study to: (i) achieve a desired level of precision for minimal total survey effort; or (ii) minimize the variance of the occupancy estimator for a given total number of surveys.

STANDARD DESIGN

No Consideration of Cost

Under a standard design in which detection probability is assumed constant, $Var(\hat{\psi})$ can be expressed as:

$$Var(\hat{\psi}) = \frac{\psi}{s}\left[(1-\psi) + \frac{(1-p^*)}{p^* - Kp(1-p)^{K-1}}\right], \tag{6.1}$$

where $p^* = 1 - (1 - p)^K$ is the probability of detecting the species at least once during K surveys of an occupied site (Chapter 4). Further, the total number of surveys (TS) will be:

TABLE 6.1   Optimum Number of Surveys to Conduct at Each Site (K) for a Standard Design Where All Sites Are Surveyed an Equal Number of Times with No Consideration of Survey Costs, for Selected Values of Occupancy (ψ) and Detection Probabilities (p)

	ψ								
p	0.1	0.2	0.3	0.4	0.5	0.6	0.7	0.8	0.9
0.1	14	15	16	17	18	20	23	26	34
0.2	7	7	8	8	9	10	11	13	16
0.3	5	5	5	5	6	6	7	8	10
0.4	3	4	4	4	4	5	5	6	7
0.5	3	3	3	3	3	3	4	4	5
0.6	2	2	2	2	3	3	3	3	4
0.7	2	2	2	2	2	2	2	3	3
0.8	2	2	2	2	2	2	2	2	2
0.9	2	2	2	2	2	2	2	2	2

Source: MacKenzie and Royle (2005).

$$TS = s \times K. \tag{6.2}$$

Table 6.1 (constructed from the results of MacKenzie and Royle, 2005) indicates the number of surveys that should be conducted at each site (without consideration of survey costs). For given values of occupancy (ψ) and detection probabilities (p), these values for K will result in the most efficient standard design possible (given all model assumptions are met). There are a number of notable features about the values in Table 6.1 to which we wish to draw attention. The first is that the optimal course of action whenever there is a chance that the species may go undetected at an occupied site is never to conduct only one survey at all the sites. Second, when detection probability is moderately high (greater than 0.5), conducting only two to three surveys per site results in a reasonably efficient design, but many more surveys may be required when detection probability is lower. That is, as detection probability decreases, the optimal number of surveys increases. Further, the optimal number of surveys also increases as the occupancy probability increases, the reason for which may not be immediately obvious. However, consider a species that is rare on the landscape (in terms of occupancy). A randomly selected site will have a low probability of occupancy; hence, expending a large amount of effort trying to detect a species that is often not present may not be an efficient use of resources available to the study. For a common species, on the other hand, expending reasonable effort to confirm that the species really is present at a site may be more worthwhile than being relatively uncertain about occupancy status and moving on to another site where the species is also likely to be present. From these results, MacKenzie and Royle (2005) suggest a crucial general principle for the design of occupancy studies. For a rare species, one should survey more sites less intensively (i.e., with fewer repeated surveys), while for a common species, one should survey fewer sites more intensively (i.e., with more repeated surveys).

In terms of designing a particular study, the required value for K can be simply obtained based upon the values of ψ and p that are assumed for the species. Then the number of sites to survey to obtain a desired level of precision, or for a fixed total level of survey effort, can be simply calculated from Eqs. (6.3) or (6.4) respectively.

$$s = \frac{\psi}{Var(\hat{\psi})}\left[(1-\psi) + \frac{(1-p^*)}{p^* - Kp(1-p)^{K-1}}\right] \tag{6.3}$$

$$s = \frac{TS}{K} \tag{6.4}$$

Oftentimes one should consider a range of values for ψ and p if there is uncertainty about what values may be reasonable, and consider the resulting impact upon the suggested design because of that uncertainty.

For example, previously the desired standard error was 0.05 ($\sqrt{Var(\hat{\psi})}$ = 0.05) and the probability of occupancy was assumed to be about 0.8 ($\psi \approx 0.8$). Now suppose that the probability of detecting the species (given presence at a site) in a single survey is thought to be about 0.4 ($p \approx 0.4$). From Table 6.1 we find that the optimal number of repeated surveys per site in this situation is six. This gives the expected probability of detecting the species at least once (p^*) as approximately $p^* = 1 - (1-0.4)^6 = 0.953$. To achieve a standard error of 0.05, the number of sites to survey is:

$$s = \frac{0.8}{0.05^2}\left[0.2 + \frac{0.047}{0.953 - 6(0.4)(0.6)^5}\right]$$
$$= \frac{0.8}{0.05^2}[0.2 + 0.061]$$
$$= 83.5$$

Therefore, approximately 84 sites should be surveyed (with $504 = 84 \times 6$ total surveys) to give the desired level of precision (note that the 84 sites required is greater than the 64 sites required if the detection probability is 1).

However, suppose that there are only enough available resources to conduct 300 total surveys. What is the best standard error that could be achieved? Even though the criterion for the design has changed, the optimal number of repeated surveys per site is still six; hence, there are only enough resources to survey 50 sites. Substituting these values into Eq. (6.1) and taking the square root gives a standard error of 0.065. A decision should then be made about whether a standard error that is 30% higher than what was originally desired can be tolerated, or if the objective of the study should be reassessed. As noted earlier, it would be wise to check how sensitive these results are to the assumed values of ψ and p before reconsidering the study's objective. In addition, there may be other designs in which sites are surveyed an unequal number of times that may yield a standard error close to the desired level for a reasonable level of effort. We will discuss this later in the chapter.

Naturally, in some instances it may not be practical to conduct the optimal number of surveys at each site (e.g., Table 6.1 suggests that 20 is the optimal number of surveys when $\psi = 0.6$ and $p = 0.1!$). Eqs. (6.1) and (6.2) could then be used explore how much worse a suboptimal design would perform, enabling an assessment to be made as to whether the study's objectives can be practically achieved. In such situations (and sometimes more generally), the field methods and protocols being used to detect the species should also be reassessed to see if alternative methods could be used to improve detection probabilities.

Including Survey Cost

So far we have implicitly assumed that the "cost" (as measured in some meaningful unit) of conducting each survey is constant, or at least that cost is immaterial to the design of a study. However, the study's budget will often be one of the greatest limiting factors in many situations, and the cost associated with each survey may vary. For example, if it is appropriate to conduct multiple surveys within a single site visit, most of the cost will be associated with traveling to the site and conducting the first survey. The cost of conducting subsequent surveys during the same visit may be relatively small. Another example is where the surveying of additional sites becomes more costly as the number of sites (s) increases, which could be due to increased travel cost, the need for hiring of additional staff, and so on.

The general techniques used by MacKenzie and Royle (2005) for finding the optimal allocation of resources can be easily extended to account for costs by defining a cost function. A study can then be designed either in terms of minimizing cost to obtain a desired level of precision or by seeking the best design to minimize the variance of the occupancy estimator for a fixed cost [note that Eq. (6.2), which calculates the total number of surveys in a standard design, could be considered as a simple cost function, where the cost for each survey is 1]. The cost function itself could take a great variety of forms. For instance, if the cost of subsequent surveys is different from the cost of the first survey, then the cost function could be of the form:

$$C = c_0 + s[c_1 + c_2(K - 1)],$$

where c_0 is a fixed overhead cost; c_1 is the cost of conducting the first survey of a site; and c_2 is the cost of conducting subsequent surveys. Alternatively, in a situation where the cost of surveying additional sites continues to increase, the cost function may take the form:

$$C = c_0 + \sum_{i=1}^{s} K[c_1 + c_2(i-1)]$$

where c_0 is the overhead cost; c_1 is the cost of conducting each survey at the first site; and c_2 is the extra cost per survey for each additional site (i.e., the cost per survey at site 2 is $c_1 + c_2$, at site 3 the cost per survey is $c_1 + 2c_2$, etc.).

In the first case, in which the cost of an initial survey may be different than that of a subsequent survey of a site, MacKenzie and Royle (2005) used analytic techniques to show that similar results to those presented above hold. That is, regardless of whether the study is being designed to achieve a desired level of precision with minimal cost or to minimize uncertainty for a fixed cost, the optimal value for K is the same. This is similar to what happened for the situation in which cost was not considered. However, the optimal value for K does depend upon the relative cost of an initial and subsequent survey (c_1 and c_2, respectively, and on ψ and p), although K is relatively stable when the relative cost is in the range $0.5c_2 < c_1 < 2c_2$. Table 6.2 presents the optimal value of K when $c_1 = 10c_2$. Note that the same general trends occur (K decreases as p increases but increases as ψ increases), although the optimal value for K tends to be larger by one or two surveys than when cost was not considered. This result is intuitive since, if it is relatively cheap to conduct subsequent surveys, then it would be reasonable that the best use of resources is to reduce

TABLE 6.2 Optimum Number of Surveys to Conduct at Each Site for a Standard Design Where All Sites Are Surveyed an Equal Number of Times and the Relative Cost of Conducting Initial (c_1) to Subsequent (c_2) Surveys Is $c_1 = 10c_2$, for Selected Values of Occupancy (ψ) and Detection Probabilities (p)

p	ψ								
	0.1	0.2	0.3	0.4	0.5	0.6	0.7	0.8	0.9
0.1	18	19	20	21	22	24	26	30	37
0.2	10	10	11	11	12	13	14	15	18
0.3	7	7	7	8	8	9	9	10	12
0.4	5	5	6	6	6	6	7	8	9
0.5	4	4	4	5	5	5	5	6	7
0.6	3	3	4	4	4	4	4	5	5
0.7	3	3	3	3	3	3	3	4	4
0.8	2	2	2	2	2	3	3	3	3
0.9	2	2	2	2	2	2	2	2	2

Source: MacKenzie and Royle (2005).

uncertainty related to the potential presence of the species at a site rather than incur the higher costs associated with sampling a new site.

Returning to our example, above we found that 504 was the minimum number of surveys required to obtain a standard error of 0.05, with $s = 84$ and $K = 6$, and that if there were only enough available resources to conduct 300 surveys, the best design of $s = 50$ and $K = 6$ would give a standard error of 0.065 (given the assumed values of ψ and p). Now suppose that for the species in question it would be appropriate to conduct multiple surveys within a single site visit; hence, the cost of subsequent surveys (\$20) is much less than the cost of the first survey (\$200). Let us assume that the overhead cost is \$3,000, hence the cost function takes the form:

$$C = \$3,000 + s[\$200 + \$20(K - 1)].$$

The issue here is, what design would minimize the total cost while achieving a standard error of 0.05? The relative cost of an initial to subsequent survey is $c_1 = 10c_2$; hence, from Table 6.2 the optimal value for K is 8. Setting $K = 8$ and $Var(\hat{\psi}) = 0.05^2$ in Eq. (6.1) and solving for s suggests that $s \approx 70$ should give the best design to minimize the total cost at \$26,800 (given $\psi \approx 0.8$ and $p = 0.4$). Compare this to the design for which cost was not considered, where $s = 84$ and $K = 6$. Although 11% more surveys would be conducted (560 here vs. 504), the design that accounts for the cost of surveying would be \$1,400 cheaper (assuming the same cost function is used).

For more complicated cost functions, the optimal choice for K may no longer be constant with respect to s or total cost. The techniques used by MacKenzie and Royle (2005) can be easily replicated for the new cost function using the numerical routines available in spreadsheet software such as Microsoft Excel. Most spreadsheet packages have a numerical routine that can be used to find a set of values that provide a solution to a particular optimization problem. Such routines can be used to find the best allocation of resources given some design criteria. The general procedure would be:

1. Specify values for ψ, p, s, and K in a series of cells.
2. Enter a formula to calculate $Var(\hat{\psi})$ (or the appropriate uncertainty measure being used) based upon these cell entries.
3. Enter a formula corresponding to the study's cost function.
4. Run the numerical optimization routine where the values in the corresponding cells for s and K are to be changed. The exact procedure used here depends upon the design criteria. If $Var(\hat{\psi})$ is to be minimized for a fixed (maximum) cost, then the routine should be run such that the cell corresponding to $Var(\hat{\psi})$ is minimized subject to the constraint that the value in the cost function cell equals (or is less than) the available budget. If the study is designed to achieve a specific $Var(\hat{\psi})$ while min-

imizing cost, then the cell corresponding to the cost function should be set as the target with a constraint imposed upon the $Var(\hat{\psi})$ cell.

5. Numerical routines can sometimes be sensitive to starting values. Step 4 should be repeated with different starting values for s and K to verify that the best combination has been found.

6. The resulting values for s and K are likely to be non-integer; therefore, the values will need to be rounded and adjusted to ensure the constraints are still met (e.g., above the numerical routine suggests the values of $s = 71.6$ and $K = 7.6$).

DOUBLE SAMPLING DESIGN

As an alternative to surveying all sites an equal number of times, repeated surveys could be conducted at a subset of sites, with all other sites surveyed only once (MacKenzie *et al.* 2002, 2003, 2004b; MacKenzie 2005a). When a site is only surveyed once, then the probability of occupancy and detection is confounded unless there is additional information. The intent behind the double sampling design is that the additional information required to separate occupancy from detectability at those sites only surveyed once comes from the sites surveyed repeatedly. Detection probabilities can be estimated (and modeled) from the data collected at sites where repeated surveys were conducted, and that information is then applied to the sites only surveyed once. Data collected in this manner are completely compatible with the model detailed in Chapter 4, as occasions when surveys were not conducted could be considered as missing values. Using such an approach, the entire estimation procedure can be conducted within a single framework. However, using a double sampling scheme does require the assumption that detection probabilities (or the model used to describe them) is the same at the sites surveyed repeatedly and those surveyed once.

One of the motivations for the suggested use of such a design is efficiency (MacKenzie *et al.* 2002, 2003, 2004b; MacKenzie 2005a). Intuitively, it might seem reasonable that at some point sufficient repeated surveys will have been conducted across all sites to allow precise estimation and modeling of detection probabilities. Conducting additional repeated surveys of sites could, therefore, be inefficient, in that the resources used on repeat surveys might be better utilized by surveying additional sites, increasing the spatial replication of the study. While this may sound reasonable in theory, MacKenzie and Royle (2005) have recently shown that this idea does not generally hold in practice.

They computed the fraction of the total survey effort that should be used to survey sites only once, which minimized the asymptotic variance for $\hat{\psi}$

TABLE 6.3 Optimal Fraction of Total Survey Effort that Should Be Devoted to Surveying Sites Only Once Using a Double Sampling Design, for Selected Values of Occupancy (ψ) and Detection Probabilities (p)

p	ψ								
	0.1	0.2	0.3	0.4	0.5	0.6	0.7	0.8	0.9
0.1	0	0	0	0	0	0	0	0	0
0.2	0	0	0	0	0	0	0	0	0
0.3	0	0	0	0	0	0	0	0	0
0.4	0	3	0	0	0	0	0	0	0
0.5	6	1	0	0	0	0	0	0	0
0.6	0	0	0	12	4	0	0	0	0
0.7	9	5	0	0	0	0	0	0	0
0.8	33	30	26	21	14	5	0	0	0
0.9	56	54	51	48	44	39	31	17	0

Source: MacKenzie and Royle (2005).

under such a design (assuming p is constant). Only when $p \geq 0.8$ was there a suggestion that a substantial fraction of the total surveys available should be devoted to surveying a number of sites once. In most other cases, the entire survey effort should be used to survey sites repeatedly (i.e., $\hat{\psi}$ had a smaller variance under a standard design than a double sampling design; Table 6.3). Some entries in Table 6.3 may look unusual; for example, when $\psi = 0.4$ and $p = 0.6$, 12% of the total survey effort should be used to survey sites only once, while the entries in nearby cells are very small. This result (and others) is simply because splitting the survey effort in that manner gives an average number of surveys per site that is very close to the *non-integer* optimal value for K under the standard design. As the numerical search used only considered optimal *integer* values for K with the optimal design, the variance in "unusual" situations tended to be smaller. However, even in the situations where it is suggested that double sampling may be feasible, the reduction in the standard error tended to be minor unless ψ was small and p large (Table 6.4). When MacKenzie and Royle (2005) also considered survey costs, only if the cost of an initial survey was less than the cost of a subsequent survey (a situation that is difficult to conceive of in practice) did the double sampling scheme consistently perform better than an optimal standard design.

The likely reason why a double sampling scheme does not (generally) provide more precise estimates of $\hat{\psi}$ than an optimal standard design is that the component of $Var(\hat{\psi})$ related to the uncertainty in an estimate of p is generally much smaller than the component associated with the imperfect detection of the species [see Eq. (4.8)]. That is, even if p is estimated precisely, the uncertainty related to being unable to confirm that a species is absent from a site may still be relatively large.

TABLE 6.4 Percent Reduction in Standard Error of the Occupancy Estimator ($\hat{\psi}$) Using an Optimal Double Sampling Design Compared to an Optimal Standard Design, for Selected Values of Occupancy (ψ) and Detection Probabilities (p)

p	ψ								
	0.1	0.2	0.3	0.4	0.5	0.6	0.7	0.8	0.9
0.1	0	0	0	0	0	0	0	0	0
0.2	0	0	0	0	0	0	0	0	0
0.3	0	0	0	0	0	0	0	0	0
0.4	0	0	0	0	0	0	0	0	0
0.5	0	0	0	0	0	0	0	0	0
0.6	0	0	0	0	0	0	0	0	0
0.7	0	0	0	0	0	0	0	0	0
0.8	4	3	2	2	1	0	0	0	0
0.9	12	11	10	9	7	5	3	1	0

While the results of MacKenzie and Royle (2005) indicate that a double sampling scheme may not perform as well as an optimal standard design, there may be some situations in which such a design could still be useful. For example, if some potential sites are very remote and difficult to access, it may not be possible to implement a standard design, as resurveying the remote sites may be prohibitively expensive. If it can be assumed that occupancy and detection probabilities are similar at the remote and more accessible sites, then it may be reasonable to use a double sampling type of design.

REMOVAL SAMPLING DESIGN

Another type of design that has been used in the field is one in which sites may be surveyed a maximum of K times in a season, but once the species is initially detected, no further surveys of that site are conducted. This is the type of design considered by Azuma et al. (1990) when they developed their occupancy estimation procedure for the monitoring of northern spotted owls in the Pacific Northwest of the United States. Here we refer to such a design as a "removal" design, as sites are removed from the pool of sites being actively surveyed once the species is detected (also because of the analogy with removal experiments used in mark-recapture studies, e.g., Otis et al. 1978; Williams et al. 2002).

In addition to enabling the estimation of detection probabilities, the repeated surveying of sites reduces the chance of the species being declared as falsely absent from a site. Once the species' presence has been confirmed, additional surveys only provide further information about detection probabilities, not about occupancy, which is often the main focus of the type of studies considered here. The logic behind using a removal design is that the main piece of information required is confirmation that the target species is present at a site. The number of surveys required until the species is first detected provides the relevant information allowing detection probabilities to be estimated.

Recently MacKenzie and Royle (2005) have shown that an optimal removal design can provide more precise estimates of $\hat{\psi}$ than an optimal standard design with the same total number of surveys. Effectively, the resources not used to resurvey sites following the first detection of the species can be used to sample additional sites, increasing the spatial replication without increasing the total level of survey effort.

Using a removal design, the variance of the occupancy estimator is:

$$Var(\hat{\psi}) = \frac{\psi}{s}\left[(1-\psi) + \frac{p^*(1-p^*)}{(p^*)^2 - K^2 p^2 (1-p)^{K-1}}\right]. \tag{6.5}$$

As for a standard design, MacKenzie and Royle (2005) found that, regardless of whether a study was being designed to minimize total survey effort to obtain a desired level of precision or to minimize uncertainty in the occupancy estimate for a fixed level of effort, the optimal value for K (now the maximum number of surveys to be conducted) is consistent for given values of ψ and p (Table 6.5). They also found that incorporating differential costs for initial and subsequent surveys with a removal design has a similar effect to that for the standard design (e.g., increasing K by one or two surveys if $c_1 \geq 5c_2$). Note that under a removal design, the total number of surveys required will not be a fixed number because now there is an element of chance involved with the number of surveys that will be required before detecting the species for the

TABLE 6.5 Optimal Maximum Number of Surveys to Conduct at Each Site for a Removal Design (K) Where All Sites Are Surveyed Until the Species Is First Detected, for Selected Values of Occupancy (ψ) and Detection Probabilities (p)

p	ψ								
	0.1	0.2	0.3	0.4	0.5	0.6	0.7	0.8	0.9
0.1	23	24	25	26	28	31	34	39	49
0.2	11	11	12	13	13	15	16	19	23
0.3	7	7	7	8	8	9	10	12	14
0.4	5	5	5	6	6	6	7	8	10
0.5	4	4	4	4	4	5	5	6	8
0.6	3	3	3	3	3	4	4	5	6
0.7	2	2	2	3	3	3	3	4	5
0.8	2	2	2	2	2	2	3	3	4
0.9	2	2	2	2	2	2	2	2	3

Source: MacKenzie and Royle (2005).

first time at an occupied site. However, the expected number of surveys required can be calculated.

To give an indication of the relative efficiency of an optimal removal design to the optimal standard design, MacKenzie and Royle (2005) compared the expected standard errors of the estimators under the two designs with the same (expected) total number of surveys (Table 6.6). Values of the ratio less than 1 indicate situations in which the optimal standard design is more efficient in terms of obtaining a smaller standard error, which only occurs when the level of occupancy is less than 0.3. Therefore, generally an optimal removal design will be more efficient than an optimal standard design, although one must be prepared to conduct a greater maximum number of surveys in order to fully realize the gain in efficiency. For example, if $\psi = 0.8$ and $p = 0.3$, the standard error of an optimal standard design with eight repeat surveys per site (Table 6.1) will be 42% greater than that of an optimal removal design, but sites would have to be surveyed up to a maximum of 12 times (Table 6.5).

Removal studies are likely to be most useful in situations in which, once a species is detected at a site, it is more (or less) likely to be detected again in subsequent surveys. For example, once a surveyor locates the species at its den or nest within the site, the surveyor may simply return to the den or nest in future surveys of the site, resulting in a higher detection probability (or lower detection probability if the surveyor disturbed the species, causing it to vacate the den or nest). By examining the model likelihood, it is easy to see why the removal design can be used without a loss of efficiency compared to using a covariate to model the change in detection probability. Consider again the likelihood equation for a standard design with constant detection probability given in Eq. (4.5):

TABLE 6.6 Ratio of Standard Errors for Optimal Standard and Removal Designs

					ψ				
p	0.1	0.2	0.3	0.4	0.5	0.6	0.7	0.8	0.9
0.1	0.90	0.94	0.98	**1.04**	**1.10**	**1.18**	**1.30**	**1.46**	**1.74**
0.2	0.91	0.94	0.99	**1.04**	**1.10**	**1.18**	**1.28**	**1.44**	**1.71**
0.3	0.92	0.95	0.99	**1.04**	**1.10**	**1.17**	**1.27**	**1.42**	**1.68**
0.4	0.93	0.96	0.99	**1.03**	**1.09**	**1.17**	**1.26**	**1.40**	**1.64**
0.5	0.93	0.96	1.00	**1.04**	**1.08**	**1.16**	**1.24**	**1.37**	**1.60**
0.6	0.94	0.97	**1.01**	**1.06**	**1.09**	**1.15**	**1.22**	**1.35**	**1.55**
0.7	0.95	0.96	0.97	**1.01**	**1.07**	**1.13**	**1.22**	**1.31**	**1.48**
0.8	1.00	**1.02**	**1.04**	**1.07**	**1.09**	**1.11**	**1.15**	**1.25**	**1.45**
0.9	**1.02**	**1.05**	**1.07**	**1.10**	**1.13**	**1.17**	**1.20**	**1.24**	**1.31**

Values greater than 1 (in bold) indicate situations where an optimal removal design has a smaller standard error than the optimal standard design.
Source: MacKenzie and Royle (2005).

$$L(\psi, p|h_1, h_2, \ldots, h_s) = \left[\psi^{s_D} p^{\sum_{j=1}^{K} s_j} (1-p)^{Ks_D - \sum_{j=1}^{K} s_j} \right] \left[\psi(1-p)^K + (1-\psi) \right]^{s-s_D}.$$

This could be factored and written in two components:

$$L(\psi, p|h_1, h_2, \ldots, h_s) = \left\{ \left[\psi^{s_D} p^{s_D} (1-p)^{\sum_{i=1}^{s_D} t_i - s_D} \right] \left[\psi(1-p)^K + (1-\psi) \right]^{s-s_D} \right\}$$

$$\times \left\{ p^{\sum_{j=1}^{K} s_i - s_D} (1-p)^{(K+1)s_D - \sum_{j=1}^{K} s_i - \sum_{k=1}^{s_D} s_i t} \right\}.$$

The first component models the data up to and including the survey where the species was detected for the first time (t_i), and is the likelihood for data collected under the removal design with constant detection probability (Chapter 4). The second component models the detection/nondetection of the species in surveys conducted after the first detection and does not involve ψ. When the detection probability is the same before and after the first detection of the species, the second component does contribute some information towards estimating occupancy, as it helps reduce uncertainty associated with having to estimate detection probability from the data [i.e., the third component of $Var(\hat{\psi}_{MLE})$ in Eq. (4.8)]. However, when the detection probability is different after first detection of the species, then the second component will contribute no information towards estimation of the occupancy probability. (This is also true of closed capture-recapture models in which individuals exhibit a behavioral response to capture; Pollock 1974; Otis et al. 1978.) Therefore, additional modeling of the detection probabilities after first detection of the species may not greatly improve the estimation of ψ; in such situations, it may thus be advantageous to simply halt the surveys once the species is first detected at a site.

However, a disadvantage of the removal design is that there is less flexibility in how the resultant data can be analyzed. For example, when data are collected according to a standard design, models can be fit where detection probability is either survey specific or constant for all surveys. However, for data collected from a removal design, models with fully survey-specific detection probabilities cannot be fit (recall from Chapter 4 that the parameters are not identifiable when p is survey specific under a removal design). However, an equality constraint for even two detection probabilities (e.g., $p_{K-1} = p_K$) will permit estimation under a removal design. If such constraints are not reasonable (i.e., if it is not likely that detection probabilities for multiple surveys will be constant), then it may not be appropriate to use a removal design, because the model cannot reflect the reality of the situation and the resulting estimates of occupancy may be biased. If the changes in detectability are thought to occur in relation to some covariate, then the data from both types of designs should be able to fit appropriate models, although data from a standard design may have more success at differentiating between the effect of covariates that vary in space versus those that vary in time. In an effort to combine efficiency and robustness, MacKenzie and Royle (2005) suggest a hybrid design in which a standard design protocol is conducted at some sites and a removal design protocol at the remainder. To the best of our knowledge, however, the properties of such a design have not been investigated to date.

MORE SITES VS. MORE SURVEYS

A final point highlighted by MacKenzie and Royle (2005) is the realization that increasing the number of sites surveyed at the expense of decreasing the number of repeat surveys may not result in a better design (in terms of precision of the occupancy estimate). They offer an example in which, if it is assumed $\psi \approx 0.4$ and $p \approx 0.3$, the asymptotic standard error of the occupancy estimator for a standard design where 200 sites are each surveyed twice is 0.11. However, by surveying only 80 sites five times, the standard error is reduced to 0.07. Allocating the same total number of surveys in a more efficient manner resulted in a 36% reduction of the standard error. To achieve the same gain in precision with only two surveys of each site, 500 sites would need to be surveyed and total survey effort increased by 250%!

We believe the same holds true more generally when, for example, the objective of the study is to identify important habitat preferences for the species. Without a sufficient number of repeat surveys, the probability of a "false absence" $[= (1 - p)^K]$ at a site may be sufficiently large that it is difficult to identify any important factors associated with occupancy. In our experience, a key design consideration often seems to be reducing the probability of a false absence down to an acceptable level in order to make robust inferences about occupancy patterns in the population. We suggest that inference is best when the probability of a false absence is in the range 0.05–0.15. Any smaller and one may be expending too much effort on repeat surveys that could be better utilized by surveying more sites; any larger and one may not be expending enough effort on repeat surveys, resulting in a greater level of uncertainty about the occupancy status of sites where the species was never detected.

6.6. DISCUSSION

Designing a good study is often as much an art as a science. The science plays a role in giving guidance on what may be an optimal design, given a list of assumptions for specific situations, and also in providing knowledge about the biology of the species. The art comes in taking the information provided by the science and turning it into a practical design that can be used in the field so that the required data can be collected to meet the study's objectives. The art can be viewed as subjective ways of taking account of additional considerations and constraints where the science can provide little insight. This may involve generalizing results, lateral thinking, and even the occasional leap of faith. The science should be used to take as much of the guesswork out of the design of studies as possible, but sometimes intuition (both statistical and biological) has an important role to play, as there may well be a number of unknown factors on which the current science can shed little light. In such cases, however, small pilot studies can often be helpful to collect additional information or at least to test the proposed design.

In this chapter, there has been a lot of discussion related to "optimal" designs. These designs are optimal in the sense of providing the most precise estimate of the occupancy probability, given certain constraints on resources and relatively simple models (i.e., occupancy and detection probabilities are constant). However, an optimal design may not be a robust design. What if there is heterogeneity in occupancy or detection probabilities among sites? What if detection probabilities are affected by some factor that has not been considered? We generally believe the standard design will be the most robust, as it will provide the greatest degree of flexibility in the modeling of the data. Whether an "optimal" standard design should be used depends very much on the application; what may be optimal in one set of circumstances may be suboptimal in another. As such, the values given in Tables 6.1, 6.2, and 6.5 for the "optimal" number of repeat surveys should only be considered as a bare minimum. Generally we would suggest that at least three surveys per site be conducted, but even then there may be insufficient data to identify important sources of variation such as heterogeneity in detection probabilities (either with covariates or the mixture models of Chapter 5), in which case a greater number of surveys (5+) may have to be conducted. Some other general principles that may be useful in the design of single-season occupancy studies are:

1. A removal design is likely to be more efficient than a standard design if detection probability is relatively constant, but may provide less flexibility for modeling.
2. Sampling more sites with fewer surveys for a rare species versus fewer sites with more surveys for a common species with a similar detection probability should hold as a general strategy.
3. For all designs, there is likely to be some optimal value for K that gives the most efficient design for given values of ψ and p. However, if ψ and p are thought to vary among sites (in some known manner that can be

modeled with a covariate or used as a basis for stratification), then the choice of K may also vary among sites.

To combine robustness and efficiency for a fixed level of resources, a combination of study designs could be used. For example, half of the total survey effort may be used to construct a standard design, and the other half of the survey effort is used to construct a removal design. The standard design provides additional flexibility for modeling, and the removal design increases spatial replication for the same level of effort. Given that the same physical sampling methods are used in each, the combined data could be analyzed within a single framework.

As was noted at the beginning of this chapter, it is important to consider how the assumptions of modeling techniques match with the biology of the species and the reality of collecting the data from the field. For example, the basic models of MacKenzie *et al.* (2002) assume no unmodeled heterogeneity in detection probability among sites, yet it can be very easy to accidentally introduce such heterogeneity when designing a study (e.g., through observer effects). The purpose of this chapter, as mentioned earlier, is not to provide a recipe for designing your specific study, but to give some guidance on the various aspects of study design that need to be considered, along with some general principles.

A good study design can make or break a project and thus deserves careful consideration. Studies should usually be designed on a case-by-case basis, as the details of each design (goals, species, environment, etc.) will often be different. Once a satisfactory design has been developed, simulation and pilot studies can be useful tools for assessing whether the design will provide the type and quality of information required to meet the study's objectives. Pilot studies can also be invaluable for testing field protocols and resolving "teething troubles" prior to implementing a full-scale design. Often there may be pressures from various quarters to begin data collection as soon as possible, but without a sound study design in place, the resulting data may be inappropriate or insufficient, and all that has been achieved is the wasting of precious economic resources.

CHAPTER 7

Single-species, Multiple-season Occupancy Models

In the preceding chapters we have considered the problem of estimating the probability of occupancy (or proportion of sites occupied) in a single season. The methods we have detailed may provide some indication of the current patterns in occupancy within that season, a snapshot of the population at a single point in time. However as discussed in Chapter 1, despite the popularity of doing so, it is not always appropriate to attempt to infer process from an observed pattern. Often there are many processes that could result in the same pattern being observed at any given time (e.g., Pirsig 1974; Romesburg 1981; Nichols 1991a; Williams *et al.* 2002).

The only reliable approach to understanding the processes occurring within a system is to observe how the system behaves over a longer time frame. This should not be at all surprising. As an analogy, suppose that you are given a randomly selected photograph from a stack of photographs taken throughout a football game. You are then asked to comment on the current state of the game and how the game has progressed up to that point. It would be possible to tell something about the current state of play, such as which team has the ball and possibly the score; however, it would be impossible to make further comment on how the game has progressed. Not until you are able to go through the entire stack of photographs (in order) would you be able to get some idea of how the game progressed. It is the same situation in ecological studies, where processes of population dynamics can only be fully understood by observing the population at systematic points in time, noting how the patterns change, and modeling these changes in terms of relevant rate parameters. As emphasized in Chapter 1, strong inferences arise when system behavior (e.g., estimated changes in rate parameters) is compared against predictions of *a priori* hypotheses, especially when system dynamics are generated by experimental manipulations within the context of experimental design. The models of this chapter were developed to provide the estimates needed for such investigations.

In this chapter we turn our attention to the problem of estimating occupancy over multiple seasons and, in particular, to understanding the underlying population dynamics that may cause changes in the occupancy state of a site. These dynamic parameters are of interest in many areas of ecology, including metapopulation studies in which the processes of local extinction and colonization (often hypothesized to be functions of patch size and isolation from neighboring patches, respectively) produce an "incidence" function (e.g., Diamond 1975a; Hanski 1994a&b). However, most of the methods used to study these parameters do not explicitly account for detection probability. Moilanen (2002) found false absences to be the greatest contributor of bias to the estimation of the incidence function. In monitoring programs, often the rate of change in occupancy may hold as much or greater interest than the absolute level of occupancy at any point in time. Changes in the use of different habitats over time will also be of interest in many species-habitat studies. For example, are the same habitats used by a species in summer and winter? What effect has a change in the habitat had on the species patterns of use?

We consider two general approaches: (1) a model in which underlying dynamics are implied but not explicitly accounted for (effectively combining several single-season models); and (2) explicitly modeling potential changes in the occupancy state of a site over time with colonization and local extinction probabilities.

7.1. BASIC SAMPLING SCHEME

We assume a situation in which s sites are selected from an area of interest with the intent of establishing the presence or absence of a species, as in single-season studies, although now the assessment is for multiple points in time.

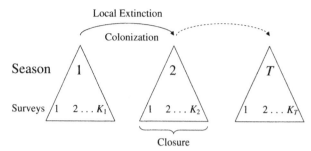

FIGURE 7.1 Graphical representation of the sampling situation for a multi-season occupancy study. Each triangle represents a season (t), with multiple (K_t) surveys within seasons. Sites are closed to changes in occupancy within seasons, but changes may occur between seasons through the processes of colonization and local extinction.

Sites may constitute naturally occurring sampling units such as discrete ponds or patches of vegetation, investigator-defined monitoring stations, or quadrats chosen from a predefined area of interest.

The time frame of the study can now be considered at two scales. First, at the larger scale, the study is conducted over multiple (T) seasons (e.g., years or breeding seasons). Each season is common to all sites, with the occupancy state of sites able to change between seasons, but not within seasons. Within each season, the smaller time scale, appropriate sampling methods are used to survey sites K_t times (Fig. 7.1). Such a design is similar to Pollock's robust design (Pollock 1982) used in mark-recapture studies, where seasons represent the primary sampling periods and surveys within seasons represent secondary sampling periods. Effectively, the general design considered here is a sequence of single-season studies conducted at (usually) the same sites for multiple seasons.

From each survey of a site, the target species is detected (1) or not detected (0) and is never falsely detected when absent. The resulting sequence of detections and nondetections for site i, conducted during season t, is denoted as the detection history $\mathbf{h}_{t,i}$. The complete detection history for site i is denoted as $\mathbf{h}_i$, and is the sequence of the T single-season detection histories. For example, the detection history $\mathbf{h}_i = 110\ 000\ 010$ represents a three-season study (with three surveys per season) in which the target species was detected in the first and second surveys in season 1, was never detected in season 2, and detected only in the second survey in season 3. Similar to the single-season situation, due to imperfect detectability we do not know whether the species was present but undetected in season 2 or instead was genuinely absent.

7.2. AN IMPLICIT DYNAMICS MODEL

One approach to modeling detection/nondetection data from multiple seasons is to effectively apply a single-season model to the data collected in each of the T seasons. Under this approach, occupancy in one season is considered to be a random process in the sense that occupancy status of the previous season is not relevant. Regardless of the underlying processes of changes in occupancy, only the resulting pattern or level of occupancy each season is modeled. Here, let ψ_t be the probability a site is occupied in season t, and $p_{t,j}$ be the probability of detecting the species in the jth survey of a site during season t (given the species was present at the site in season t). Using the model-based approach of MacKenzie et al. (2002; as detailed in Chapter 4), the likelihood for the observed data in season t would be:

$$L_t(\psi_t, \mathbf{p}_t | \mathbf{h}_{t,1}, \mathbf{h}_{t,2}, \ldots, \mathbf{h}_{t,s}) = \prod_{i=1}^{s} \Pr(\mathbf{h}_{t,i}),$$

with the full likelihood evaluated for the T seasons being the product of the seasonal likelihoods, that is,

$$L(\psi, \mathbf{p} | \mathbf{h}_1, \mathbf{h}_2, \ldots, \mathbf{h}_s) = \prod_{i=1}^{T} L_t(\psi_t, \mathbf{p}_t | \mathbf{h}_{t,1}, \mathbf{h}_{t,2}, \ldots, \mathbf{h}_{t,s}).$$

This same model can also be developed directly from the detection histories using the same techniques as have been used in the previous chapters: taking a verbal description of the detection histories and translating them into a mathematical equation. Consider again the detection history $\mathbf{h}_i = 110\ 000\ 010$. A verbal description of these data would be:

In season 1: The site was occupied with the species being detected in the first and second surveys, but not in the third.

In season 2: The site was either occupied with the species not being detected in any of the three surveys, or the site was unoccupied.

In season 3: The site was occupied with the species being detected in the second survey, but not in the first or third survey.

Translating these statements into mathematical equations using the defined model parameters, we have:

Season 1: $\psi_1 p_{1,1} p_{1,2}(1 - p_{1,3})$,
Season 2: $\psi_2(1 - p_{2,1})(1 - p_{2,2})(1 - p_{2,3}) + (1 - \psi_2)$,
Season 3: $\psi_3(1 - p_{3,1})p_{3,2}(1 - p_{3,3})$.

Therefore, the probability of observing the entire detection history would be:

$$Pr(\mathbf{h}_i = 100\ 000\ 010) = \psi_1 p_{1,1} p_{1,2}(1 - p_{1,3})$$
$$\times \left\{ \psi_2 \prod_{j=1}^{3}(1 - p_{2,j}) + (1 - \psi_2) \right\} \quad (7.1)$$
$$\times \psi_3(1 - p_{3,1})p_{3,2}(1 - p_{3,3}).$$

This procedure can be used to obtain the probability statement for each of the s observed detection histories, and the model likelihood would be calculated as:

$$L(\psi, p|\mathbf{h}_1, \mathbf{h}_2, \ldots, \mathbf{h}_s) = \prod_{i=1}^{s} Pr(\mathbf{h}_i).$$

As in Chapter 4, this model can be easily generalized so that the probabilities of occupancy and detection are functions of covariates, and to allow for missing observations. Models can also be considered in which there is some structural relationship among probabilities in different seasons. For example, Field et al. (2005) modeled a systematic decline in occupancy over time by defining seasonal occupancy probabilities with a linear trend on the logit scale—in other words, $\text{logit}(\psi_t) = \beta_0 + \beta_1 t$.

Finally, we note that although the above modeling may appear to be relatively phenomenological, in the sense that vital rates (probabilities of local extinction and colonization) governing the dynamic process do not appear explicitly in this model, it actually makes fairly restrictive assumptions about these vital rates. In Section 7.4 we show that the implicit dynamics model is based on the assumption that the probability of the species not going locally extinct at a previously occupied site is equal to the probability of colonization of a previously unoccupied site. In the next section, we discuss a more general explicit model of occupancy dynamics, from which the above implicit dynamics model can be obtained as a special case.

7.3. MODELING DYNAMIC CHANGES EXPLICITLY

As noted in the previous section, the dynamic processes governing changes in the occupancy state variable are the colonization of an unoccupied site by the species and the local extinction of the species at an occupied site. In this section we consider models that directly incorporate these dynamic processes, as they are often of direct interest. They are somewhat analogous to the birth and death processes of the abundance state variable, and as such supply information relevant to the long-term sustainability of a population. As the drivers of the system (with respect to occupancy), understanding how these dynamic processes are affected by changes of habitat or climatic conditions (for example) may be important for the successful management of ecological systems.

For the remainder of this chapter we consider the dynamic changes in occupancy as a first-order Markov process. That is, the probability of a site being occupied in season t depends upon the occupancy state of the site in the previous season, $t - 1$. In some situations, higher-order Markov processes (e.g., occupancy probability at t depends upon state of occupancy at both $t - 1$ and $t - 2$) may be biologically reasonable to represent long-term "memory" about the occupancy state of a site. For example, from mark-resight data, Hestbeck et al. (1991) modeled transition probabilities between different wintering grounds for Canada geese (*Branta canadensis*) as a second-order Markov process to represent long-term fidelity of individual birds to each region. However, at present we do not know of any data sets where it would be reasonable to hypothesize that a higher-order Markov process is occurring; hence, we have made no attempts to develop such models.

Modeling changes in occupancy as a Markov process also accounts for a form of temporal autocorrelation. When observations on the same sampling unit are positively correlated, values close in time are more similar than those separated by longer periods (i.e., the sampling variance for a short time series will tend to be less than that for a longer time series). In the occupancy context, this equates to the expectation that a site that is occupied now may be more likely to be occupied again in the near future than one that is currently unoccupied. A Markov process adequately achieves this purpose.

Markovian changes in occupancy can also be considered as inducing a form of heterogeneity in occupancy probabilities in the sense that the probability of a site being occupied in season t will be different for sites that were occupied and sites that were unoccupied in the previous season.

Formally, we define colonization (γ_t) and local extinction (ε_t) probabilities to be:

γ_t = the probability that an unoccupied site in season t is occupied by the species in season $t + 1$; and

ε_t = the probability that a site occupied in season t is unoccupied by the species in season $t + 1$.

These dynamic processes represent the probabilities of a site transitioning between the occupied and unoccupied states between consecutive seasons (Fig. 7.2).

Below we detail three approaches to modeling multiple-season occupancy data that explicitly account for the processes of colonization and local extinction. First, we briefly discuss some historical approaches that were developed for situations in which the species is (assumed to be) always detected when present at a site (i.e., detection probability equals 1). We then focus on two

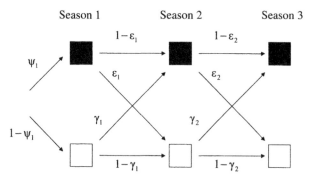

FIGURE 7.2 Representation of how the occupancy state of a site may change between seasons in terms of the processes of occupancy (ψ), colonization (γ), and local extinction (ε). Filled boxes indicate that the site is occupied (species present) in that season, while empty boxes indicate that the site is unoccupied (species absent).

methods that allow for the imperfect detection of the species: a "conditional" and an "unconditional" approach. The "conditional" approach exploits the similarities between the type of data collected in the current context and capture-recapture data collected from individuals. This approach is "conditional" in the sense that the detection history for a site is only modeled from the season in which the species is first detected, that is, the modeling conditions upon the first detection of the species. This is due to the fact that in capture-recapture, an individual is unknown to the researchers prior to the first time it is caught and marked. As a result, it is not generally possible to obtain seasonal estimates of occupancy, only estimates of the dynamic processes themselves. The second approach (upon which we largely focus) is an "unconditional" approach, in which the modeling makes full use of the detection histories. Here it is possible to estimate both occupancy and the dynamic parameters.

MODELING DYNAMIC PROCESSES WHEN DETECTION PROBABILITY IS 1

In the late 1960s and throughout the 1970s, the ecological literature contained a number of studies of animals on islands in which the presence or absence of breeding populations (frequently of birds) was assessed over a number of different years (see Diamond and May 1977 and papers cited therein). This work was motivated largely by the models of MacArthur and Wilson (1967), which suggested that species richness on islands reflected a dynamic equilibrium between rates of local extinction and colonization. Diamond and May

(1977) recommended that such data be viewed as having resulted from a stationary Markov process defined by corresponding rates of extinction and colonization, a recommendation that had been anticipated by Simberloff (1969). Diamond and May (1977) focused on the implications of this model for (1) detection-nondetection data collected at varying time intervals and (2) computation of various turnover statistics.

Clark and Rosenzweig (1994; also see Rosenzweig and Clark 1994) considered the problem of estimating extinction and colonization rates for such a Markov model from detection-nondetection data. They provided maximum likelihood estimates assuming both a stationary process with rate parameters constant over time and detection probabilities of 1. Erwin et al. (1998) expanded this general modeling approach, considering reduced-parameter non-Markovian models, as well as Markovian models permitting time-specific rates of local extinction and colonization. These models only provide reasonable estimates in the situation in which presence and absence can be ascertained with certainty (Clark and Rosenzweig 1994; Erwin et al. 1998) and are thus of limited usefulness.

CONDITIONAL MODELING OF DYNAMIC PROCESSES

Barbraud et al. (2003) considered models for colony site dynamics, the same problem considered by Erwin et al. (1998), but wanted to relax the assumption of detection probabilities equal to 1. They considered the estimation problem by focusing on the analogy between occupancy dynamics of colony sites and population dynamics of individual animals. The simplest form of colony detection history data for multiple seasons consists of 1's and 0's denoting detection or nondetection, respectively, at each study site. These data are analogous to capture history data for individual animals in animal populations open to gains and losses. For example, the capture history 1 0 1 0 would indicate an animal caught in periods (analogous to one survey per season here) 1 and 3, but not in periods 2 and 4. The usual approach to modeling such data uses parameters for survival from one sample period to the next and capture probability at each period (e.g., see Lebreton et al. 1992). Interior 0's (followed and preceded by one or more 1's; e.g., period 2) are usually viewed unambiguously as "present but not captured" and thus modeled with the complement of capture probability. However, this analogy is not especially useful for site occupancy studies in which an interior 0 can reflect either "present but not detected" or "absent, but followed by recolonization."

Barbraud et al. (2003) recognized that there is a close analogy between site occupancy studies and capture-recapture studies with temporary emigration (Kendall et al. 1997; Kendall 1999; Williams et al. 2002). In the case of temporary emigration, an interior 0 can result from either "presence without capture" or "temporary emigration" of the animal. Unfortunately, the probability of being a temporary emigrant is confounded with capture probability in standard models for open populations (Burnham 1993; Kendall et al. 1997). However, Kendall and Nichols (1995) and Kendall et al. (1997) recognized that the robust design (described above, also see Pollock 1982) provides the information needed to estimate capture probability conditional on presence in the sampled area and thus allows separate estimates of this conditional capture probability and the probability of being a temporary emigrant.

Barbraud et al. (2003) viewed the Markovian temporary emigration model of Kendall et al. (1997) as potentially most useful for estimation in the context of site occupancy dynamics. This temporary emigration model contains two parameters for the probability of being a temporary emigrant (i.e., the probability that an animal was not in the study population, but still part of the "superpopulation") at any sampling period: (1) for animals that were not temporary emigrants the previous period, $t - 1$; and (2) for animals that were also temporary emigrants the previous period. The temporary emigration parameter for animals that were not temporary emigrants the previous period (i.e., within the study population at $t - 1$, outside the study population at t) was viewed as local extinction probability in the context of occupancy studies (i.e., site was occupied at $t - 1$, unoccupied at t). The complement of the second temporary emigration parameter, $1 - Pr$ (an animal outside the study population at $t - 1$ was also outside the study population at t), was viewed as a probability of colonization in occupancy studies. Under this analogy, the usual survival probability of open capture-recapture models was set equal to 1, as it reflected the probability that a site always "survived" (i.e., sites will never "die," in the sense that the species could always recolonize the site a later time. Note that one situation where this might not be a reasonable assumption is where a change in the habitat or local environment of a site renders it uninhabitable for the species, in which case the joint modeling of habitat and occupancy may be useful; see Chapter 10.) In order to complete the analogy between the temporary emigration problem and site occupancy dynamics, we note that the random temporary emigration model of Kendall et al. (1997) is equivalent to the implicit dynamics model for site occupancy (Section 7.2).

The advantage of recognizing this analogy between the modeling of temporary emigration and occupancy dynamics involved software and computations. Software had been developed by Kendall and Hines (1999) and White and Burnham (1999) to obtain parameter estimates under the temporary emigration models of Kendall et al. (1997) using robust design data. Barbraud et al. (2003) thus used these programs (MARK: White and Burnham 1999; RDSURVIV: Kendall and Hines 1999) with data from two survey flights per year over the Camargue delta in southern France to estimate local rates of extinction and colonization for purple heron (*Ardea purpurea*) and grey heron (*Ardea cinerea*) breeding colonies at reed bed sites. Of particular interest biologically was the modeling of time-specific local colonization in one area of the Camargue as a function of local extinction probability in a neighboring disturbed area. This ultrastructural modeling dealt explicitly with spatial dependencies in occupancy and provided indirect inference about animal movement without using marked individuals (Barbraud *et al.* 2003).

We do not present the actual temporary emigration model here, as it is best viewed as a special case of the model of MacKenzie *et al.* (2003), which is presented in detail below. Readers especially interested in the details of the use of the temporary emigration modeling approach for site occupancy studies are directed to Barbraud *et al.* (2003). The primary difference between the two approaches (Barbraud *et al.* 2003; MacKenzie *et al.* 2003) is the conditional nature of the temporary emigration modeling. In capture-recapture studies of animal populations, most models (exceptions include the temporal symmetry models of Nichols *et al.* 1986, 2000; Pradel 1996; Williams *et al.* 2002) condition on the release of individual animals at their periods of first capture. Stated differently, the 0's occurring in a capture history before an animal's first capture are not typically modeled. However, in site occupancy studies in which potential sites are identified at the beginning of the study, such conditioning is not needed, and initial 0's can be modeled. We thus refer to the temporary emigration modeling of Barbraud *et al.* (2003) as "conditional," and contrast this with the "unconditional" approach of MacKenzie *et al.* (2003). Both approaches should provide approximately unbiased estimates of the dynamic processes, but the approach of MacKenzie *et al.* (2003) should be more efficient and leads more readily to estimates of site occupancy for each season of the study.

Unconditional Modeling of Dynamic Processes

MacKenzie *et al.* (2003) used the colonization and local extinction probabilities defined above (γ_t and ε_t) to extend the single-season model of MacKenzie *et al.* (2002). Once occupancy state (the probability of occupancy) is established in the first season (ψ_1), potential changes in the occupancy state of a site between seasons are simply incorporated using the dynamic parameters. To construct their model, MacKenzie *et al.* (2003) used the now familiar approach of taking a verbal description of a detection history and translating it into a mathematical equation, giving the probability of observing the detection history. For example, consider once again the detection history $h_i =$ 110 000 010, where the occupancy status of the site in the second season is unknown. A verbal description of these data that incorporates the concepts of colonization and local extinction would be:

In season 1: The site was occupied with the species being detected in the first and second surveys, but not in the third.

From the end of season 1 to the start of season 3 (immediately before surveying commenced): Either the species did not go locally extinct between seasons 1 and 2, was not detected in any of the three surveys within season 2 and did not go locally extinct between seasons 2 and 3, *or* the species went locally extinct between seasons 1 and 2, then recolonized the site between seasons 2 and 3.

In season 3: The species was detected in the second survey, but not in the first or third survey.

Translating these statements into mathematical equations using the model parameters defined earlier, we have:

Season 1: $\psi_1 p_{1,1} p_{1,2} (1 - p_{1,3})$,

From the end of season 1 to the start of season 3:

$$(1 - \varepsilon_1) \prod_{j=1}^{3} (1 - p_{2,j})(1 - \varepsilon_2) + \varepsilon_1 \gamma_2,$$

Season 3: $(1 - p_{3,1}) p_{3,2} (1 - p_{3,3})$.

The probability of observing the complete detection history would be:

$$\Pr(\mathbf{h}_i = 110\,000\,010) = \psi_1 p_{1,1} p_{1,2} (1 - p_{1,3})$$
$$\times \left\{ (1 - \varepsilon_1) \prod_{j=1}^{3} (1 - p_{2,j})(1 - \varepsilon_2) + \varepsilon_1 \gamma_2 \right\} \qquad (7.2)$$
$$\times (1 - p_{3,1}) p_{3,2} (1 - p_{3,3})$$

Note the differences between Eqs. (7.2) and (7.1), where the dynamic processes are not explicitly modeled. Here the model incorporates a mechanistic process for how the occupancy state of a site may change between seasons, whereas in the former model only the state of occupancy each season is considered.

Generally, there may be a number of different possible pathways that could result in an observed detection history. MacKenzie *et al.* (2003) therefore suggest it might be most convenient to describe the model using matrix notation (see the Appendix for a brief introduction to matrices). Let ϕ_t be the 2×2 matrix denoting the probability of a site transitioning between occupancy states from season t to $t + 1$. Rows of ϕ_t represent the occupancy state of the site in season t (state 1 = occupied; state 2 = unoccupied), and columns represent the occupancy state at $t + 1$, that is:

$$\phi_t = \begin{bmatrix} 1 - \varepsilon_t & \varepsilon_t \\ \gamma_t & 1 - \gamma_t \end{bmatrix}.$$

Further, let ϕ_0 be the row vector:

$$\phi_0 = [\psi_1 \quad 1 - \psi_1],$$

where ψ_1 is the probability the site is occupied in the first season. This vector models whether a site was occupied or unoccupied in the first season. Next, let $p_{h,t}$ be a column vector where each entry denotes the probability of observing the detection history $h_{t,i}$ for site i in season t, conditional upon occupancy state. For instance, if the site was occupied by the species, the probability of observing the detection history $h_{t,i} = 101$ would be $p_{t,1}(1 - p_{t,2})p_{t,3}$, while if the site was unoccupied, the probability would be zero (i.e., it is impossible to detect the species at a site if that site is unoccupied). Therefore, $p_{101,t}$ could be defined as:

$$p_{101_t} = \begin{bmatrix} p_{t,1}(1 - p_{t,2})p_{t,3} \\ 0 \end{bmatrix}.$$

The second element of $p_{101,t}$ will always be 0 whenever the species is detected at least once at the site during season t. Using similar reasoning, if the species is never detected at a site during season t ($h_{t,i} = 000$), the second element will always be 1, as this is the only observable detection history for a site that is unoccupied, that is:

$$p_{000_t} = \begin{bmatrix} \prod_{j=1}^{3}(1 - p_{t,j}) \\ 1 \end{bmatrix}.$$

Using this matrix notation, the probability for an observed detection history could be calculated as:

$$\Pr(h_i) = \phi_0 \prod_{t=1}^{T-1} D(p_{h,t})\phi_t p_{h,T}, \qquad (7.3)$$

where $D(p_{h,t})$ is a diagonal matrix with the elements of $p_{h,t}$ along the main diagonal (top left to bottom right), zero otherwise. Diagonalizing the vector is required merely for the matrix algebra to work out correctly. Initially this equation may look somewhat confusing, but stepping through the various components, we see that it does have an intuitive interpretation. ϕ_0 establishes the probability that a site is either in the occupied or unoccupied state immediately prior to surveys commencing in season 1. The term $D(p_{h,t})\phi_t$ next calculates the probability of observing the particular sequence of detections and nondetections in season t (conditional upon occupancy state), and then the probability of the site transitioning to the occupied or unoccupied state immediately before surveying begins in season $t + 1$. This is done recursively from season 1 to immediately before the final season of surveying (season T), hence the product term $\prod_{t=1}^{T-1} D(\mathbf{p}_{\mathbf{h},t})\boldsymbol{\phi}_t$. At this stage, the equation has calculated the probability of observing the particular detection history up to the second-to-last season of surveying, and the site being in either the occupied or unoccupied state immediately prior to the surveying in season T. Therefore, to complete the probability statement, the probability of observing the sequence of detections and nondetections in the final season (conditional upon occupancy state) is required—in other words, $\mathbf{p}_{\mathbf{h},T}$. Some examples of observed detection histories and their probability statements, according to the above model, are given in Table 7.1. We encourage readers to take the time to work through these examples to cement their understanding of the model. From the probability statements for each observed detection history, the model likelihood can be calculated in the usual manner (assuming independence of detection histories), that is:

$$L(\psi_1, \gamma, \varepsilon, \mathbf{p}|\mathbf{h}_1, \ldots, \mathbf{h}_s) = \prod_{i=1}^{s} \Pr(\mathbf{h}_i).$$

Note that if $T = 1$, that is if the study is only conducted for a single season, then the above equation reduces to $\Pr(\mathbf{h}_i) = \boldsymbol{\phi}_0 \mathbf{p}_{\mathbf{h},1}$. This is an equivalent formulation for calculating the probability of observing a detection history for the single-season model of MacKenzie et al. (2002, and described in Chapter 4).

MISSING OBSERVATIONS

Missing observations can be easily accounted for when using this type of modeling approach, as in the single-season case. If the missing observations occur within season t, then the vector $\mathbf{p}_{\mathbf{h},t}$ is adjusted by removing the corresponding $p_{t,j}$ parameter(s). For example, if the history 11– is obtained at primary period t (where "–" indicates a missing observation), then:

$$\mathbf{p}_{11-,t} = \begin{bmatrix} p_{t,1} p_{t,2} \\ 0 \end{bmatrix}.$$

This represents that fact that no information on either detection or nondetection has been collected about the parameter $p_{t,3}$ from the site with this detection history.

TABLE 7.1 Examples of Detection Histories (h_i) and the Associated Probabilities of Observing Them [$Pr(h_i)$] Using the Unconditional Explicit Dynamics Model

h_i	$Pr(h_i)$
11 01 10	$= \phi_0 D(p_{1,1})\phi_1 D(p_{01,2})\phi_2 p_{10,3}$
	$= [\psi_1 \quad 1-\psi_1]\begin{bmatrix} p_{1,1}p_{1,2} & 0 \\ 0 & 0 \end{bmatrix}\begin{bmatrix} 1-\varepsilon_1 & \varepsilon_1 \\ \gamma_1 & 1-\gamma_1 \end{bmatrix}\begin{bmatrix} (1-p_{2,1})p_{2,2} & 0 \\ 0 & 0 \end{bmatrix}$
	$\begin{bmatrix} 1-\varepsilon_2 & \varepsilon_2 \\ \gamma_2 & 1-\gamma_2 \end{bmatrix}\begin{bmatrix} p_{3,1}(1-p_{3,2}) \\ 0 \end{bmatrix}$
	$= \psi_1 p_{1,1}p_{1,2}(1-\varepsilon_1)(1-p_{2,1})p_{2,2}(1-\varepsilon_2)p_{3,1}(1-p_{3,2})$
00 10 00	$= \phi_0 D(p_{00,1})\phi_1 D(p_{10,2})\phi_2 p_{00,3}$
	$= [\psi_1 \quad 1-\psi_1]\begin{bmatrix} \prod_{j=1}^{2}(1-p_{1,j}) & 0 \\ 0 & 1 \end{bmatrix}\begin{bmatrix} 1-\varepsilon_1 & \varepsilon_1 \\ \gamma_1 & 1-\gamma_1 \end{bmatrix}$
	$\times\begin{bmatrix} p_{2,1}(1-p_{2,2}) & 0 \\ 0 & 0 \end{bmatrix}\begin{bmatrix} 1-\varepsilon_2 & \varepsilon_2 \\ \gamma_2 & 1-\gamma_2 \end{bmatrix}\begin{bmatrix} \prod_{j=1}^{2}(1-p_{3,j}) \\ 1 \end{bmatrix}$
	$= \left\{\psi_1\prod_{j=1}^{2}(1-p_{1,j})(1-\varepsilon_1)+(1-\psi_1)\gamma_1\right\}p_{2,1}(1-p_{2,2})\left\{(1-\varepsilon_2)\prod_{j=1}^{2}(1-p_{3,j})+\varepsilon_2\right\}$
00 00 00	$= \phi_0 D(p_{00,1})\phi_1 D(p_{00,2})\phi_2 p_{00,3}$
	$= [\psi_1 \quad 1-\psi_1]\begin{bmatrix} \prod_{j=1}^{2}(1-p_{1,j}) & 0 \\ 0 & 1 \end{bmatrix}\begin{bmatrix} 1-\varepsilon_1 & \varepsilon_1 \\ \gamma_1 & 1-\gamma_1 \end{bmatrix}$
	$\times\begin{bmatrix} \prod_{j=1}^{2}(1-p_{2,j}) & 0 \\ 0 & 1 \end{bmatrix}\begin{bmatrix} 1-\varepsilon_2 & \varepsilon_2 \\ \gamma_2 & 1-\gamma_2 \end{bmatrix}\begin{bmatrix} \prod_{j=1}^{2}(1-p_{3,j}) \\ 1 \end{bmatrix}$
	$= \psi\prod_{j=1}^{2}(1-p_{1,j})\left\{\begin{matrix}(1-\varepsilon_1)\prod_{j=1}^{2}(1-p_{2,j})\left[(1-\varepsilon_2)\prod_{j=1}^{2}(1-p_{3,j})+\varepsilon_2\right] \\ +\varepsilon_1\left[\gamma_2\prod_{j=1}^{2}(1-p_{3,j})+(1-\gamma_2)\right]\end{matrix}\right\}$
	$+ (1-\psi_1)\left\{\begin{matrix}\gamma_1\prod_{j=1}^{2}(1-p_{2,j})\left[(1-\varepsilon_2)\prod_{j=1}^{2}(1-p_{3,j})+\varepsilon_2\right] \\ +(1-\gamma_1)\left[\gamma_2\prod_{j=1}^{2}(1-p_{3,j})+(1-\gamma_2)\right]\end{matrix}\right\}$

Similarly, the model can be adjusted to allow for situations in which a site was not surveyed for an entire season. Consider the following detection history, where the site was not surveyed at all in the second season, 10 -- 11. Again, no information has been collected regarding either the detection or nondetection of the species, although here the occupancy state of the site at season 2 is also unknown; hence, all possibilities must be allowed for. This can be achieved by (effectively) omitting $p_{h,2}$ entirely; in other words, the probability of this detection history is:

$$\Pr(h_i = 10 -\!\!- 11) = \phi_0 D(p_{10,1})\phi_1\phi_2 p_{11,3}.$$

By having the ability to accommodate missing observations, the unconditional model of MacKenzie *et al.* (2003) provides a great deal of flexibility in the way the data can be collected in the field and still be analyzed using this technique. Not only can there be unequal sampling effort across sites within seasons but, potentially, not all sites have to be surveyed each season (within reason). However, it is important to note that even though no data were collected from this site during the second season, the associated colonization and local extinction probabilities still appear in the probabilistic statement (within ϕ_2). As such, it is assumed by the model that these probabilities are the same at both the sites that are and are not surveyed within that season. This assumption of the model must be carefully considered if a study design is proposed that intentionally avoids surveying all sites each season.

INCLUDING COVARIATE INFORMATION

Thus far in this chapter, the implicit assumption has been made that all model parameters are constant across all sites. Failure of this assumption results in heterogeneity in model parameters, which could result in inferences that are inaccurate if the heterogeneity is not accounted for. As already discussed in Chapters 4 and 5, one approach to dealing with potential heterogeneity is the inclusion of information on variables that may affect the value of one or more parameters, or covariates. Indeed, the relationship between the covariates and certain parameters of interest may often be the primary motivation for conducting the study (e.g., habitat variables in habitat modeling, or measures of isolation and patch size in metapopulation studies). As noted in Chapters 3 and 4, covariate information can be included in the model by use of an appropriate link function, such as the logit link (Chapter 3). The mechanics for doing so are identical to those presented in Chapter 4, so we do not cover this material again here.

Similar to the single-season case, occupancy, colonization, and local extinction probabilities could all be functions of variables that have a single, constant value for the duration of the season (season-specific covariates). These may be variables that characterize sites during each sampling season (e.g., habitat type, elevation, or patch size) or variables that represent the level of change in some quantity between seasons (e.g., changes in habitat composition). Detection probabilities can be functions of season-specific covariates, and also of variables that may change with each survey of a site (e.g., rainfall in preceding 24 hours, air temperature, or observer).

Having the ability to incorporate covariate information of these types provides a great deal of flexibility in the models that could be considered as reasonable descriptions of the processes that give rise to the data. Moreover, different hypotheses about the system can often be expressed as models that involve different sets of covariates for each parameter type. The strength of evidence for each hypothesis can then be determined by fitting the suite of models with the different sets of covariates and making a formal comparison of the models (e.g., by using the AIC model selection criterion). For example, in metapopulation studies, local extinction probabilities are frequently assumed to be decreasing functions of patch area (e.g., Moilanen 1999). That is, the species is more likely to go extinct from small patches than large patches. This may be reasonable for some species, but perhaps not in every case. Furthermore in some situations, variation in the areas of the sampled patches may be insufficient to discern such an effect on local extinction probabilities. Therefore, a second hypothesis would be that local extinction probabilities are constant with respect to patch area. These competing hypotheses could be formulated as two models with different sets of covariates for ε_t. To represent the area hypothesis, a model could be fit to the data where "patch area" is included as a covariate for extinction probability, and a second model without the "patch area" covariate for ε_t (but identical in all other respects) could be fit to the data to represent the second hypothesis. The level of support for each of the two models would then reflect the degree of support of each hypothesis. Note that one could also use a similar approach to determine the functional form of such a relationship (e.g., linear or quadratic), or even to compare link functions. However, we caution that such comparisons should only be done on the basis of sound biological reasoning, not in the pursuit of a "best" model.

ALTERNATIVE PARAMETERIZATIONS

MacKenzie *et al.* (2003) noted that in some situations, quantities other than the probability of occupancy in the first season, seasonal colonization, and local extinction probabilities may be of interest. They suggested that these quantities could be derived from the estimated parameters, or that the model could be reparameterized so that the quantities are estimated directly.

Seasonal estimates of occupancy are one such quantity. In some applications (e.g., monitoring), the processes of colonization and local extinction may not be of direct interest, with the main focus of the study being how occupancy changes over time. The three probabilities are simply related by the recursive equation:

$$\psi_{t+1} = \psi_t(1 - \varepsilon_t) + (1 - \psi_t)\gamma_t, \tag{7.4}$$

in other words, sites occupied next season are a combination of those sites occupied this season where the species does not go locally extinct ($\psi_t(1 - \varepsilon_t)$) and the sites that are currently unoccupied that are colonized by the species before next season $(1 - \psi_t)\gamma_t$. This equation can be rearranged to make either of the dynamic processes the subject, that is:

$$\gamma_t = \frac{\psi_{t+1} - \psi_t(1 - \varepsilon_t)}{(1 - \psi_t)}$$

or

$$\varepsilon_t = 1 - \frac{\psi_{t+1} - (1 - \psi_t)\gamma_t}{\psi_t}.$$

The same model as described above would be used except, rather than estimate the γ_t and ε_t parameters directly, one would directly estimate, for example, the seasonal occupancy and local extinction probabilities. The value for γ_t could then be derived using the above formula and used in the model to evaluate the likelihood.

There may be a temptation to use the recursive occupancy equation above in association with the implicit dynamic modeling approach described in Section 7.2 as a means of incorporating colonization and local extinction probabilities into a multi-season occupancy model. However, doing so does not yield the explicit dynamics approach described above. For example, consider the simple detection history $\mathbf{h}_i = 10\ 01$. Using the unconditional approach of MacKenzie et al. (2003), the probability of observing this history would be:

$$\Pr(\mathbf{h}_i = 10\ 01) = \psi_1 p_{1,1}(1 - p_{1,2})(1 - \varepsilon_1)(1 - p_{2,1})p_{2,2},$$

while using the implicit dynamics model from Section 7.2, the probability would be:

$$\Pr(\mathbf{h}_i = 10\ 01) = \psi_1 p_{1,1}(1 - p_{1,2})\psi_2(1 - p_{2,1})p_{2,2}.$$

Substituting the expression for ψ_2 given by Eq. (7.2) into the implicit dynamics model does not give the equivalent of the unconditional explicit dynamics model, for example:

$$\begin{aligned}
\Pr(\mathbf{h}_i = 10\ 01) &= \psi_1 p_{1,1}(1 - p_{1,2})\psi_2(1 - p_{2,1})p_{2,2} \\
&= \psi_1 p_{1,1}(1 - p_{1,2})[\psi_1(1 - \varepsilon_1) + (1 - \psi_1)\gamma_1](1 - p_{2,1})p_{2,2} \\
&\neq \psi_1 p_{1,1}(1 - p_{1,2})(1 - \varepsilon_1)(1 - p_{2,1})p_{2,2}.
\end{aligned}$$

Another quantity suggested by MacKenzie $et\ al.$ (2003) is the rate of change in occupancy. By analogy with population size (where the comparable measure is known as the "finite rate of change" or "growth rate"), they suggest it could be defined as:

$$\lambda_t = \frac{\psi_{t+1}}{\psi_t}.$$

Using the recursive occupancy equation as an intermediate step, the unconditional model could be reparameterized so that λ_t is estimated directly. However, there are some practical problems that limit the usefulness of this parameterization. First, there are bounds on the allowable values of λ_t that vary with ψ_t. For example, suppose that currently the probability of occupancy is 0.5 (i.e., $\psi_t = 0.5$). Then the maximal rate of change in occupancy, as defined above, must be 2; otherwise the probability of occupancy in the next season will exceed 1.0. However, if currently $\psi_t = 0.2$, then the maximal rate of change in occupancy will be 5. Second, you cannot have a constant, long-term, rate of change greater than 1, as eventually it will result in an estimate of $\psi_{t+1} > 1$. For instance, suppose the probability of occupancy in season 1 is 0.2. A long-term rate of change of 1.2 (i.e., occupancy probability increases by 20% each season) would suggest that in season 10 occupancy is greater than 1.0 (0.20, 0.24, 0.29, . . . , 0.72, 0.86, 1.03).

An alternative definition for the rate of change in occupancy is to use odds ratios, that is:

$$\lambda_t' = \frac{\psi_{t+1}/(1-\psi_{t+1})}{\psi_t/(1-\psi_t)}.$$

While it may seem more complicated to interpret, this has the advantage of not suffering from the restrictions of the above definition. Also, the general concept is similar to that of using a logit or log-odds link function (recall from Chapter 3 that the odds ratio is the amount by which the odds of occupancy in season t is multiplied to get the odds of occupancy in season $t + 1$). Further, if λ_t' is constant across time (i.e., $\lambda_1' = \lambda_2' = \ldots = \lambda_{T-1}'$), then $\ln(\lambda')$ will correspond to the trend parameter when modeling occupancy as a linear function of time on the logit scale:

$$\text{logit}(\psi_t) = \beta_0 + \beta_1 t$$
$$= \beta_0 + \ln(\lambda')t.$$

In each of the above cases, the parameter of interest in the reparameterized version of the model could still be a function of covariates, although a link function other than the logit-link may have to be used.

The choice of which parameterization may be most appropriate in a given situation depends on the goals of the study and scientific questions being addressed. If the main focus is on the underlying dynamic processes and the factors that may affect them, then the original parameterization should be used. However, in many management scenarios, occupancy estimates and changes in occupancy over time may be the primary focus, suggesting that one of the alternative parameterizations may be more appropriate (although note the successful application of management actions may require a deeper understanding of occupancy dynamics beyond whether the overall trend is increasing or decreasing). The important thing to note is that the results from fitting different parameterizations of the model to the same data are comparable, including the comparison of model selection metrics such as AIC. There is nothing inherently wrong with the comparison of multiple parameterizations of the model; however, we suggest that choice of parameterization should be generally governed by the study objective, rather than fitting models with all possible parameterizations and using model selection criteria to differentiate among them. Finally, we have often found the original parameterization to be the most numerically stable, particularly when a model contains a large number of covariates. As colonization and local extinction probabilities must take values in the 0–1 interval, there are constraints on allowable values for the occupancy probability. Enforcing these constraints when using a reparameterized version of the unconditional model can make the computer algorithms unstable.

EXAMPLE: HOUSE FINCH EXPANSION IN NORTH AMERICA

House finches (*Carpodacus mexicanus*) are native to western North America, but are not native to the mid-continent or eastern North America. However, in 1942 a small population was released in the east on Long Island, New York, and house finches have exhibited an impressive westward expansion since that time. The magnitude of this expansion is such that it is obvious in the raw data of the North American Breeding Bird Survey (BBS) (Robbins *et al.* 1986). Here we subject the BBS data to the probabilistic modeling of this chapter in an effort to draw formal inferences about this expansion. The North American Breeding Bird Survey has been conducted annually since the mid-1960s by volunteer observers. The counts are conducted during the peak of the breeding season, usually during June. Observers follow a route along roads for ~39.2 km, stopping every 0.8 km for 50 consecutive stops. At each stop, a point count is conducted for three minutes with observers counting all birds detected within a 400 m radius. There are now >4,000 BBS routes throughout North America, so the geographic coverage is impressively extensive.

The BBS protocol specifies that the route be run once each breeding season, so the data do not contain the temporal replication that we typically use for occupancy modeling. We thus take a different approach and view each of the 50 stops as a replicate count from the area covered by the route. This is far from ideal. For example, under the view that the area covered by each stop is a random selection from the area covered by all stops, we would ideally be sampling with replacement. Nevertheless, given the survey design and protocol of the BBS, we view our approach as not only reasonable, but better than most available approaches.

In order to investigate the westward expansion, we defined twenty-six 100 km distance bands from the Long Island point of release and analyzed data from within each distance band every five years for the period 1976–2001. We used a relatively phenomenological kind of modeling in which we focused on time $\times$ distance interactions. Thus, distance band (d) and year ($year$) were covariates in our analysis. The only other covariate used an ad hoc approach to dealing with relative abundance of birds that was possible because of the large number of stops per route. Specifically, we created a categorical variable for observed frequency of occurrence (f), indicating whether house finches were detected on more than 10 stops in the route in any previous time period. This covariate was used to model detection probability, along with distance and time. Only distance and time were used to model local rates of extinction and colonization, and only distance was used to model the initial occupancy level ($\psi_{76}(d)$). Our prediction was that there would be an increase in rate of colonization with distance as time increased. We had no real expectation about rate of extinction. We expected occupancy to increase with distance as time progressed, in much the same manner as colonization (although note that occupancy probabilities were derived using Eq. (7.4) for 1981 onwards).

We fit a number of occupancy models in PRESENCE 2.0, and the top eight ranked models appear in Table 7.2. The model with low AIC ($\psi_{76}(d)\gamma(year \times d)\varepsilon(d)p(year \times d + f)$) received a model weight of >0.75, indicating a good degree of support. Under this model, the year $\times$ distance interaction for rate of colonization followed the primary prediction of the analysis. Tabled estimates of occupancy (Table 7.3) and rate of colonization (Table 7.4) provide a clear view of the westward expansion from the Long Island release point. Estimates of occupancy were >0.01 for distances up to only 400 km in 1976, whereas occupancy estimates were approximately 0.33 for the most distant band (>2600 km) in 2001 (Table 7.3). Rates of colonization showed a similar pattern (Table 7.4). Recall that the rates of colonization are indexed to the initial year of the five-year interval to which they pertain. For example, the estimate of colonization for the first distance band (within 100 km from

TABLE 7.2 Summary of Model Selection Procedure Results for House Finch Example

Model	ΔAIC	w	$-2l$	$NPar$
$\psi_{76}(d)\gamma(year \times d)\varepsilon(d)p(year \times d + f)$	0.00	78%	44415.64	27
$\psi_{76}(d)\gamma(year \times d)\varepsilon(year + d)p(year \times d + f)$	2.50	22%	44410.14	31
$\psi_{76}(d)\gamma(year \times d)\varepsilon(year)p(year \times d + f)$	13.61	0%	44423.24	30
$\psi_{76}(d)\gamma(year \times d)\varepsilon(year \times d)p(year \times d + f)$	14.70	0%	44422.34	31
$\psi_{76}(d)\gamma(year \times d)\varepsilon(\cdot)p(year \times d + f)$	15.64	0%	44433.27	26
$\psi_{76}(d)\gamma(d)\varepsilon(year \times d)p(year \times d + f)$	18.67	0%	44434.31	27
$\psi_{76}(d)\gamma(year)\varepsilon(year \times d)p(year \times d + f)$	27.19	0%	44436.82	30
$\psi_{76}(d)\gamma(\cdot)\varepsilon(year \times d)p(year \times d + f)$	52.56	0%	44470.2	26

Factors affecting occupancy, colonization, and extinction probabilities include: distance band (d) and year ($year$). Detection probabilities may be affected by these same factors and a categorical variable for observed frequency of occurrence (f). Given are the relative difference in AIC values compared to the top-ranked model (ΔAIC), AIC model weights (w), twice the negative log-likelihood ($-2l$), and the number of parameters in the model ($NPar$). Results are only presented for the top eight ranked models.

the release site) for 1976 ($\hat{\gamma}_{76}(d = 0) = 0.576$) indicates that of the routes in this distance interval that were not occupied in 1976, about 58% were occupied by house finches five years later in 1981. The estimates of Tables 7.3 and 7.4 provide a quantitative description of the changes in occupancy and colonization through time with distance.

We believe that the general topic of range expansion and contraction will become increasingly important in the future with the spread of invasive species and range changes induced by climate change. Monitoring programs designed to permit estimation of occupancy will be ideally suited to study these changes. We would like to extend these methods to more mechanistic models of range expansion. In particular, we would like to use an approach similar to that of Wikle (2003) to model colonization as a function of occupancy of nearby sample units, rather than simply as a time × distance interaction.

7.4. INVESTIGATING OCCUPANCY DYNAMICS

In this section we discuss how these multi-season models can be used to address interesting hypotheses about the species biology and ecology through occupancy dynamics, specifically, hypotheses about the mechanistic processes of changes in occupancy and the hypothesis that the population is at some form of equilibrium with respect to occupancy dynamics. Such hypotheses

TABLE 7.3   Estimated Probability of Occupancy ($\hat{\psi}$) by House Finch Within Each of twenty-six 100-km Distance Bands (d) Every Five Years for the Period 1976–2001

	$\hat{\psi}$					
Band (d)	1976	1981	1986	1991	1996	2001
0	0.57 (0.09)	0.80 (0.04)	0.96 (0.01)	0.97 (0.01)	0.96 (0.01)	0.96 (0.01)
1	0.31 (0.06)	0.66 (0.04)	0.93 (0.02)	0.96 (0.01)	0.95 (0.01)	0.95 (0.01)
2	0.13 (0.04)	0.53 (0.05)	0.89 (0.03)	0.95 (0.01)	0.94 (0.02)	0.94 (0.02)
3	0.05 (0.03)	0.44 (0.04)	0.83 (0.04)	0.94 (0.01)	0.93 (0.02)	0.93 (0.02)
4	0.02 (0.01)	0.37 (0.04)	0.75 (0.04)	0.91 (0.02)	0.92 (0.03)	0.92 (0.03)
5	0.01 (0.01)	0.31 (0.04)	0.66 (0.04)	0.89 (0.02)	0.90 (0.04)	0.91 (0.03)
6	0.00	0.26 (0.04)	0.56 (0.05)	0.85 (0.03)	0.88 (0.05)	0.89 (0.04)
7	0.00	0.22 (0.04)	0.46 (0.06)	0.80 (0.03)	0.86 (0.06)	0.87 (0.05)
8	0.00	0.18 (0.04)	0.37 (0.06)	0.75 (0.03)	0.82 (0.07)	0.85 (0.06)
9	0.00	0.15 (0.04)	0.29 (0.06)	0.69 (0.04)	0.79 (0.08)	0.82 (0.08)
10	0.00	0.13 (0.04)	0.22 (0.05)	0.62 (0.04)	0.75 (0.08)	0.79 (0.09)
11	0.00	0.10 (0.04)	0.17 (0.04)	0.56 (0.04)	0.72 (0.09)	0.76 (0.10)
12	0.00	0.08 (0.04)	0.13 (0.04)	0.50 (0.04)	0.68 (0.10)	0.73 (0.11)
13	0.00	0.07 (0.03)	0.09 (0.03)	0.44 (0.04)	0.64 (0.10)	0.69 (0.13)
14	0.00	0.06 (0.03)	0.07 (0.02)	0.38 (0.04)	0.60 (0.10)	0.66 (0.14)
15	0.00	0.04 (0.03)	0.05 (0.02)	0.33 (0.04)	0.56 (0.10)	0.63 (0.15)
16	0.00	0.04 (0.02)	0.04 (0.01)	0.28 (0.04)	0.53 (0.10)	0.59 (0.17)
17	0.00	0.03 (0.02)	0.03 (0.01)	0.24 (0.04)	0.49 (0.10)	0.56 (0.18)
18	0.00	0.02 (0.02)	0.02 (0.01)	0.20 (0.03)	0.46 (0.10)	0.53 (0.19)
19	0.00	0.02 (0.02)	0.02 (0.01)	0.17 (0.03)	0.44 (0.09)	0.50 (0.20)
20	0.00	0.01 (0.01)	0.01	0.14 (0.03)	0.41 (0.09)	0.47 (0.21)
21	0.00	0.01 (0.01)	0.01	0.12 (0.03)	0.39 (0.09)	0.44 (0.22)
22	0.00	0.01 (0.01)	0.01	0.10 (0.03)	0.36 (0.08)	0.41 (0.23)
23	0.00	0.01 (0.01)	0.01	0.08 (0.03)	0.34 (0.08)	0.38 (0.23)
24	0.00	0.01 (0.01)	0.00	0.06 (0.02)	0.32 (0.08)	0.35 (0.24)
25	0.00	0.00	0.00	0.05 (0.02)	0.31 (0.08)	0.33 (0.24)

Standard errors are given in parentheses (omitted when $\hat{\psi}$ = 0.0). Occupancy probability was derived from the models using Eq. (7.4) for 1981 onwards.

differ fundamentally from those dealing with covariate relationships discussed earlier in this chapter, focusing instead on constraints that could be made upon the dynamic probabilities themselves. Obviously the two approaches could be combined to assess not only the factors that may affect occupancy, for instance, but also whether changes in occupancy appear to be Markovian or random in nature. Our basic philosophy for investigating competing hypotheses, as mentioned throughout this book, is simple: articulate the different hypotheses through different model structures, fit all models to the observed data, and formally compare them to determine the strength of evidence for each model and hence each hypothesis.

TABLE 7.4 Estimated Probability of Colonization ($\hat{\gamma}$) by House Finch Within Each of 26 100-km Distance Bands (d) Every Five Years for the Period 1976–2001

	$\hat{\gamma}$				
Band (d)	1976	1981	1986	1991	1996
0	0.58 (0.07)	0.91 (0.04)	0.90 (0.04)	0.66 (0.11)	0.63 (0.17)
1	0.52 (0.06)	0.86 (0.05)	0.88 (0.04)	0.64 (0.10)	0.61 (0.16)
2	0.46 (0.05)	0.81 (0.06)	0.86 (0.05)	0.63 (0.10)	0.60 (0.15)
3	0.41 (0.05)	0.73 (0.06)	0.83 (0.05)	0.62 (0.09)	0.58 (0.14)
4	0.36 (0.04)	0.64 (0.06)	0.80 (0.05)	0.60 (0.09)	0.56 (0.14)
5	0.31 (0.04)	0.54 (0.06)	0.77 (0.06)	0.59 (0.08)	0.54 (0.13)
6	0.26 (0.04)	0.43 (0.07)	0.73 (0.06)	0.57 (0.08)	0.52 (0.12)
7	0.22 (0.04)	0.33 (0.07)	0.69 (0.06)	0.56 (0.07)	0.51 (0.11)
8	0.18 (0.04)	0.24 (0.07)	0.64 (0.05)	0.54 (0.07)	0.49 (0.10)
9	0.15 (0.04)	0.17 (0.07)	0.59 (0.05)	0.52 (0.06)	0.47 (0.09)
10	0.13 (0.04)	0.12 (0.06)	0.54 (0.05)	0.51 (0.06)	0.45 (0.08)
11	0.10 (0.04)	0.08 (0.06)	0.49 (0.05)	0.49 (0.05)	0.43 (0.07)
12	0.08 (0.04)	0.06 (0.05)	0.44 (0.04)	0.48 (0.05)	0.41 (0.07)
13	0.07 (0.03)	0.04 (0.04)	0.39 (0.04)	0.46 (0.04)	0.40 (0.06)
14	0.06 (0.03)	0.02 (0.03)	0.34 (0.04)	0.45 (0.04)	0.38 (0.06)
15	0.04 (0.03)	0.02 (0.02)	0.30 (0.04)	0.43 (0.04)	0.36 (0.05)
16	0.04 (0.02)	0.01 (0.02)	0.26 (0.04)	0.42 (0.04)	0.35 (0.05)
17	0.03 (0.02)	0.01 (0.01)	0.22 (0.04)	0.40 (0.04)	0.33 (0.05)
18	0.02 (0.02)	0.00	0.19 (0.04)	0.39 (0.04)	0.31 (0.05)
19	0.02 (0.02)	0.00	0.16 (0.03)	0.37 (0.05)	0.30 (0.06)
20	0.01 (0.01)	0.00	0.13 (0.03)	0.36 (0.05)	0.28 (0.06)
21	0.01 (0.01)	0.00	0.11 (0.03)	0.34 (0.05)	0.27 (0.06)
22	0.01 (0.01)	0.00	0.09 (0.03)	0.33 (0.06)	0.25 (0.07)
23	0.01 (0.01)	0.00	0.08 (0.03)	0.32 (0.06)	0.24 (0.07)
24	0.01 (0.01)	0.00	0.06 (0.02)	0.30 (0.06)	0.23 (0.08)
25	0.00	0.00	0.05 (0.02)	0.29 (0.07)	0.21 (0.08)

Standard errors are given in parentheses (omitted when $\hat{\gamma} = 0.0$).

MARKOVIAN, RANDOM AND NO CHANGES IN OCCUPANCY

The colonization and local extinction probabilities defined in the unconditional explicit dynamics model allow Markovian changes in occupancy; in other words, the probability that a site is occupied this season depends upon the state of occupancy last season. If a site was occupied last season, then the probability that the site is occupied this season is $(1 - \varepsilon_t)$, whereas if a site was unoccupied last season, the probability of occupancy this season is γ_t. In many applications it may be reasonable to expect that the changes in occupancy could be Markovian in nature, particularly when a "site" is defined in terms of the behavioral nature of the species at certain times of the year. For instance in a study of northern spotted owls (*Strix occidentalis caurina*), Olson *et al.* (2005) defined a "site" as a potential breeding territory. As many bird species will return to the same breeding or nesting area each year, one would expect that potential breeding territories that were occupied in year 1 would be more likely to be occupied in year 2 than those unoccupied in year 1. When the sampling season is defined to coincide with the owl's breeding season and each season is separated by a single year, modeling changes in occupancy as a Markov process seems reasonable. However, there are two other types of changes in occupancy that may be of biological interest: (1) random changes in occupancy; and (2) no changes in occupancy (or closure of the system).

If changes in the occupancy state of sites over time is random, then the probability of occupancy in season t does not depend upon the occupancy state of the site in season $t - 1$. Random changes in occupancy may suggest that the species as a whole has a low degree of site fidelity (not to be confused with [although likely to be a function of] the site fidelity of individuals), as the species tends to occur randomly at sites within the area over time, rather than being more likely to persist at the same sites. It may also be that changes in occupancy may appear to be random when the length of time between seasons is sufficiently large that several Markovian changes can occur (e.g., if the potential spotted owl territories were only surveyed every five years rather than annually).

Here, we develop the implicit dynamics model of Section 7.2 in a mechanistic manner in terms of probabilities of local extinction and colonization. We begin with the unconditional explicit dynamics model and impose the constraint $\gamma_t = (1 - \varepsilon_t)$ (or equivalently $\varepsilon_t = (1 - \gamma_t)$) for $t = 1, 2, \ldots, T - 1$; that is, the probability of a site becoming occupied is the same as the probability a site stays occupied. This now means that it is no longer necessary to estimate (and model) both colonization and extinction probabilities, as one is simply the complement of the other. This means fewer parameters are required in this model than in the full explicit dynamics model, which may result in a more parsimonious model for the data. Note the effect that this constraint has on Eq. (7.4) for recursively calculating occupancy probabilities:

$$\psi_{t+1} = \psi_t \gamma_t + (1 - \psi_t)\gamma_t$$
$$= \gamma_t = (1 - \varepsilon_t),$$

which corresponds to our intuitive interpretation of random changes in occupancy, that the probability of occupancy in one season does not depend upon whether the site was occupied in the previous season. Also note the effect this constraint has on the probability statement for detection histories derived under the unconditional explicit dynamics model. Consider the example given in Eq. (7.2):

$$\Pr(h_i = 110\,000\,010) = \psi_1 p_{1,1} p_{1,2}(1-p_{1,3})$$
$$\times \left\{ (1-\varepsilon_1)\prod_{j=1}^{3}(1-p_{2,j})(1-\varepsilon_2) + \varepsilon_1\gamma_2 \right\}$$
$$\times (1-p_{3,1})p_{3,2}(1-p_{3,3})$$
$$= \psi_1 p_{1,1} p_{1,2}(1-p_{1,3})$$
$$\times \left\{ \psi_2 \prod_{j=1}^{3}(1-p_{2,j})\psi_3 + (1-\psi_2)\psi_3 \right\}$$
$$\times (1-p_{3,1})p_{3,2}(1-p_{3,3})$$
$$= \psi_1 p_{1,1} p_{1,2}(1-p_{1,3})$$
$$\times \left\{ \psi_2 \prod_{j=1}^{3}(1-p_{2,j}) + (1-\psi_2) \right\}$$
$$\times \psi_3 (1-p_{3,1})p_{3,2}(1-p_{3,3})$$

This now has the same structure as the probability statement derived for this history using the implicit dynamics model of Section 7.2 [Eq. (7.1)]. This demonstrates that imposition of random changes in occupancy on the explicit models of Section 7.3 leads to the implicit dynamics model of Section 7.2. That is, even though the implicit dynamics model could be used to describe the resulting patterns in occupancy caused by any underlying mechanistic process (random, first-order Markovian or otherwise), the model makes the de facto assumption that any changes are random in nature. One would conclude changes to be random if the strength of evidence for such a model (or group of models) was sufficiently greater than the evidence for similar models in which changes are Markovian.

Rather than Markovian or random changes in occupancy, in some situations it may be reasonable to consider that the system is static or closed to changes in occupancy over the duration of the study. That is, the occupancy status of sites does not change between seasons. While perhaps a biological unreality over a longer time frame, when data have been collected for only a few seasons this assumption of no changes in occupancy may be reasonable, depending upon the target species and study design. Such a model could be considered by enforcing the constraint $\gamma_t = \varepsilon_t = 0$ for $t = 1, 2, \ldots, T-1$ (i.e., there is zero probability that sites change occupancy status). In effect, the unconditional explicit dynamics model reduces to the MacKenzie *et al.* (2002) single-season model detailed in Chapter 4 when these constraints are imposed, for example:

$$\Pr(h_i = 110\ 000\ 010) = \psi_1 p_{1,1} p_{1,2}(1 - p_{1,3})$$
$$\times \left\{ (1 - \varepsilon_1) \prod_{j=1}^{3} (1 - p_{2,j})(1 - \varepsilon_2) + \varepsilon_1 \gamma_2 \right\}$$
$$\times (1 - p_{3,1}) p_{3,2}(1 - p_{3,3})$$
$$= \psi_1 p_{1,1} p_{1,2}(1 - p_{1,3}) \left\{ \prod_{j=1}^{3} (1 - p_{2,j}) \right\} (1 - p_{3,1}) p_{3,2}(1 - p_{3,3}).$$

Given the assumption that the occupancy state of sites is not changing between seasons, when the species was not detected during the second season it must have been present but undetected because the species was detected at the site during the other seasons.

EQUILIBRIUM

An assumption required by many of the published techniques for analyzing single-season occupancy data is that the metapopulation is stationary or in a state of equilibrium (e.g., Hanski 1994a)—in other words, that the net number of sites that are colonized each season is equal, on average, to the net number of sites where the species goes locally extinct. Given the changeable nature of most ecosystems around the world and the outside pressures they are frequently subjected to, the validity of this assumption must be questioned. Yet it is an assumption that goes untested in many metapopulation studies. However, using the unconditional, explicit dynamics model of MacKenzie et al. (2003), it is a biological assumption that can be assessed.

The equilibrium assumption could be defined in two ways. First, it may be expressed in terms of occupancy, where the probability of occupancy is constant for the duration of the study, that is, $\psi_t = \psi$ for $t = 1, 2, \ldots, T$ (or equivalently $\lambda_t = \lambda'_t = 1$). Second, it could also be defined in terms of colonization and local extinction probabilities being constant across time (i.e., $\gamma_t = \gamma$ and $\varepsilon_t = \varepsilon$ for $t = 1, 2, \ldots, T - 1$). A process characterized by these equalities is typically referred to as a stationary Markov process. When these probabilities are constant, the population may be at (or heading towards) an equilibrium level of occupancy, which can be calculated as $\psi_{Eq} = \gamma/(\gamma + \varepsilon)$. These definitions are not equivalent, as in the former the overall level of occupancy may be constant even though the colonization and local extinction probabilities vary over time (but balance each other each season), and in the latter situation occupancy may not be constant, either because the population is still heading towards the new equilibrium level or as an artifact of how sites were sampled from the population (see Section 7.7). Regardless of exactly how "equilibrium" is defined, it is a biological hypothesis that can be represented by applying appropriate constraints to the parameters of the unconditional, explicit dynamics model.

It should also be noted that the sites could be in a state of equilibrium with respect to occupancy, while either Markovian, random, or no changes in occupancy are occurring between seasons. Each combination represents a different set of biological mechanisms that may be of interest in different circumstances.

EXAMPLE: NORTHERN SPOTTED OWL

To illustrate the above approach for investigating occupancy dynamics using the unconditional model, we now revisit the example considered by MacKenzie et al. (2003) of the northern spotted owl (*Strix occidentalis caurina*) in California. As part of the example, we also illustrate the idea of formally comparing assumptions commonly used in investigations of metapopulations, specifically, that the population is at a point of equilibrium in terms of occupancy dynamics. We also illustrate how one could compare models to make inferences about whether changes in occupancy are best described as Markovian or random, or whether the occupancy is static. As with the single-season example, different hypotheses about the occupancy state of the biological system and the associated vital rates are expressed by a suite of candidate models. The candidate models are then ranked according to AIC model selection procedures.

In northern California, monitoring of potential spotted owl habitat commenced even before the species was listed as threatened by the United States Fish and Wildlife Service in 1990 (U.S. Fish and Wildlife Service 1990). The data considered by MacKenzie et al. (2003) consisted of 55 potential breeding territories ($s = 55$ sites) surveyed each year between 1997 and 2000 ($T = 5$ breeding seasons). Each site was surveyed up to eight times per season to determine whether the site was occupied by a pair of breeding owls. Surveys were not conducted simultaneously across sites, resulting in some missing observations ($K_{max} = 8$ surveys, $K_{average} = 5.3$ surveys); however, when conducted, surveys followed a well-established protocol that was consistent across years (Franklin et al. 1996). For simplicity, we assume that detection probabilities were constant for all surveys within seasons, but the modeling could be extended to allow detection probabilities to vary in time or in accordance with measured covariates. In addition, MacKenzie et al. (2003) found strong evidence that detection probabilities were year specific, $p(year)$, so we adopt the same detection probability structure for this analysis.

Our primary goal is to illustrate how practitioners could examine changes in occupancy and equilibrium issues with models presented in this chapter.

Specifically, we explore whether our set of potential breeding sites was in a state of equilibrium in terms of occupancy dynamics for breeding pairs over the five-year period, and whether the changes were better represented by random or Markovian processes.

We represent these hypotheses with six competing models (Table 7.5). First, we consider a model in which the occupancy status of sites does not change, (i.e., by constraining both colonization and extinction probabilities equal to zero; denoted as $\psi(\cdot)p(year)$). This model suggests that owl territories were well established and the breeding pair occupancy state was static for all five breeding seasons. Next, we explore possible random changes in occupancy, where the probability that a breeding pair occupies a site does not depend on whether the site was occupied in the previous breeding season. We would expect this type of occupancy change if owl pairs had low site fidelity or if breeding pairs choose sites randomly each year. This is probably not the case for spotted owls, but we are able to formally assess this expectation by setting $\varepsilon = (1 - \gamma)$. Further, the population may be at a point of equilibrium or not. The model $\psi(1997)\gamma(year)\{\varepsilon = 1 - \gamma\}p(year)$ suggests that changes in occupancy are random, and the probability of occupancy may vary among years (i.e., population may not be in equilibrium). As noted previously, equilibrium could be defined in two different ways, either in terms of occupancy or in terms of the dynamic parameters. Here we use the latter definition, where equilibrium suggests colonization, and local extinction probabilities are constant over time. We denote the model representing the hypotheses of a population at equilibrium with random changes in occupancy as $\psi(1997)\gamma(\cdot)\{\varepsilon = 1 - \gamma\}p(year)$. Previous capture-recapture studies have shown spotted owls to be territorial, with high site fidelity and little migration among adult breeders; thus *a priori* we would expect Markovian changes in occupancy (Franklin *et al.* 1996). To allow Markovian changes in occupancy, we remove the constraint $\varepsilon = (1 - \gamma)$, such that both colonization and local extinction probabilities are separately estimated. The models $\psi(1997)\gamma(\cdot)\varepsilon(\cdot)p(year)$ and $\psi(1997)\lambda(year)\varepsilon(year)p(year)$ represent the equilibrium and nonequilibrium hypotheses with Markovian changes in occupancy, respectively. Finally, we consider one other model: $\psi(\cdot)\gamma(\cdot)p(year)$. This model suggests that occupancy is constant for all five years, unlike the previous models with random or Markovian changes where occupancy was estimated for the first year (1997) and occupancy in subsequent years is determined through the dynamic processes. Further, colonization probability is also constant each year. Hence, the local extinction probability must also be constant and furthermore is defined completely by the occupancy and colonization probability [e.g., $\varepsilon = \gamma(1 - \psi)/\psi$, from Eq. (7.4)]. As the local extinction probability can be derived exactly from the other parameters in the model, we drop the parameter ε from the model notation to represent that it is not directly estimated. This

TABLE 7.5 Summary of Model Selection Procedure Results for Northern Spotted Owl Example

Model	ΔAIC	w	No. Pars.	$-2l$	$\hat{\psi}_{1997}$	$\hat{\gamma}_{1997}$	$\hat{\gamma}_{1998}$	$\hat{\gamma}_{1999}$	$\hat{\gamma}_{2000}$	$\hat{\varepsilon}_{1997}$	$\hat{\varepsilon}_{1998}$	$\hat{\varepsilon}_{1999}$	$\hat{\varepsilon}_{2000}$
$\psi(\cdot)\gamma(\cdot)p(year)$	0.00	0.62	7	1337.95	0.60	0.19	0.19	0.19	0.19	0.13*	0.13*	0.13*	0.13*
$\psi(1997)\gamma(\cdot)\varepsilon(\cdot)p(year)$	1.57	0.28	8	1337.52	0.62	0.18	0.18	0.18	0.18	0.14	0.14	0.14	0.14
$\psi(1997)\gamma(year)\varepsilon(year)p(year)$	3.69	0.10	14	1327.64	0.63	0.11	0.07	0.39	0.12	0.09	0.13	0.24	0.12
$\psi(1997)\gamma(\cdot)|\varepsilon = 1 - \gamma)p(year)$	91.58	0.00	7	1429.53	0.62	0.59	0.59	0.59	0.59	0.41*	0.41*	0.41*	0.41*
$\psi(1997)\gamma(year)|\varepsilon = 1 - \gamma)p(year)$	97.37	0.00	10	1429.32	0.62	0.60	0.57	0.60	0.57	0.40*	0.43*	0.40*	0.43*
$\psi(\cdot)p(year)$	202.61	0.00	6	1542.56	0.84	0.00	0.00	0.00	0.00	0.00	0.00	0.00	0.00
Model Averaged					0.61	0.18	0.18	0.21	0.18	0.13	0.13	0.15	0.13

ΔAIC is the relative difference in AIC values compared with the top-ranked model; w is the AIC model weight; No. Pars. is the number of parameters in the model; $-2l$ is twice the negative log-likelihood value; $\hat{\psi}_{1997}$ is the estimated occupancy probability in 1997; $\hat{\gamma}$ and $\hat{\varepsilon}$ represent the estimated colonization and local extinction probabilities, respectively.

*indicates derived estimates.

211

final model represents a stationary Markov process, which is sometimes assumed in metapopulation studies that use so-called incidence functions.

Table 7.5 presents the six candidate models ranked according to AIC. Consistent with our *a priori* expectation, changes in occupancy are best represented by a Markov process. Models with random or no changes in occupancy have essentially no support (ΔAIC values are much greater than 10). Additionally, the hypothesis that these 55 sites are in some form of equilibrium state appears to be well supported. The two top-ranked models both represent some form of equilibrium situation and have a combined model weight of 90%. The model-averaged estimates (Chapter 3) suggest the occupancy probability in 1997 was 0.61; colonization probabilities from 1997 to 2000 were relatively constant at ≈ 0.20; and local extinction probabilities over the same period at ≈ 0.14. The equilibrium occupancy defined by these constant rates of extinction and colonization, $\approx 0.20/(0.20 + 0.14) = 0.59$, is very close to the estimated occupancy. This result is not surprising, as the model with greatest weight reflects a stationary Markov process that has reached equilibrium.

7.5. VIOLATIONS OF MODEL ASSUMPTIONS

The assumption of no unmodeled heterogeneity in any of the rate parameters (occupancy, colonization, extinction, or detection probabilities) is one of several assumptions for the multiple-season models presented in the chapter. Additional assumptions include: (1) occupancy state at each site does not change over surveys within a season; that is, consistent with Pollock's robust design, sites are "closed" to changes in occupancy within seasons or primary periods; (2) detection of species and detection histories at each location are independent; and (3) the target species are never falsely detected (i.e., species are identified correctly).

If these assumptions are not met, some or all estimators may be biased, and inferences about factors that influence both occupancy and occupancy dynamics may be erroneous. Even within the capture-recapture arena, there has been little investigation of the effects of heterogeneity on robust design estimators; rather, it is believed that these estimators behave in a manner similar to those for separate closed and open capture-recapture models (Williams *et al.* 2002). Assumption violations for single-season (closed) models were presented in Chapter 4. In this section we briefly review anticipated impacts of occupancy closure violations and possible solutions and then focus primarily on the impacts of assumption violation on rate parameters estimated between seasons (i.e., during the open periods). We caution readers that investigations of assumption violations within the occupancy context are just beginning and that information in this section is based mostly on the analogy with capture-recapture population models that may not always have parallels to occupancy studies.

As mentioned in previous chapters, the closed occupancy state assumption within seasons can be relaxed, provided changes in occupancy are random (Kendall et al. 1997; Kendall 1999). The species of interest is viewed as occupying some area larger than that of a site and having some non-negligible probability of being present in the site at the time of any survey. In these situations, the occupancy estimator is approximately unbiased but interpreted as the proportion of sites "used" by the target species, and detection probability is the probability the species is present at the time of the survey and detected at occupied or used sites.

Nonrandom movement of a species in and out of sample units likely causes bias in occupancy estimators; nevertheless, if movement is always either only in or only out of the site(s) (i.e., immigration or emigration only), then Kendall (1999) describes ways in which surveys can be combined to likely eliminate bias in occupancy estimators. Kendall's (1999) recommendations involve pooling survey data into two surveys per season and then using models with survey-specific detection probabilities. Specifically, for the case of only emigration, the first survey is retained for each site, and the last $K - 1$ surveys are combined into a second "survey." In the case of only immigration, the first $K - 1$ surveys are combined and treated as the initial survey, and survey K becomes the second survey. Under this approach, approximately unbiased estimates can likely be obtained for either occupancy at the beginning of each season for emigration-only situations, or occupancy at the end of each sampling season for immigration-only situations (see Chapter 4 and Kendall 1999 for details). Kendall (1999) also mentions that this pooling approach is valid within the robust design context, yielding unbiased estimates of survival rate between primary periods. In the context of multiple-season occupancy models, we would anticipate that similar pooling to accommodate emigration- or immigration-only movement within seasons would yield approximately unbiased estimates of extinction and colonization probabilities. Bias in occupancy estimates will likely remain if analyses are conducted using more than two surveys per season or models with constant detection probability (Chapter 4; Kendall 1999). Kendall's (1999) work suggests that unmodeled heterogeneity or permanent trap response (see below) in detection probabilities will cause bias in occupancy and vital rate estimators.

Another option for dealing with the closure assumption is to restrict the data to include surveys between times when the availability of the species is uninterrupted (i.e., during periods of closure), as demonstrated by MacKenzie et al. (2003) with tiger salamanders in Minnesota. Here, detection/nondetection information was only included during a time period when the life history of the species dictated that individuals would be confined to the pond (eggs, larvae, and early metamorphs). Time periods when adults may be migrating to ponds or when metamorphs may be transitioning to a terrestrial life phase were not included in the analysis. Again, investigators should use their knowledge about the phenology of the target species and design their studies to try to minimize violations in the closure assumption.

The impact of unmodeled variation in occupancy, colonization, and extinction probability among sites is virtually unexplored, and more thorough simulation studies are still needed. Effects of heterogeneous survival rates have been investigated for open population capture-recapture estimators (Nichols et al. 1982; Pollock and Raveling 1982; Pollock et al. 1990). However, the analogy between extinction and the complement of survival is not sufficiently close that we are comfortable in drawing inferences about effects on extinction estimators based on inferences about survival estimators. Recall, for example, that in order to use capture-recapture models to estimate parameters of occupancy dynamics, Barbraud et al. (2003) equated temporary emigration parameters with colonization and extinction. We are aware of no investigation of the effects of heterogeneity on the temporary emigration estimators presented by Kendall et al. (1997), so we conclude that the effects of heterogeneous rates of extinction and colonization are a topic of future investigation.

7.6. MODELING HETEROGENEOUS DETECTION PROBABILITIES

There has been virtually no attempt to extend the models of Chapter 5 that deal with heterogeneous detection probabilities to the situation with multiple seasons. Of course, the covariate modeling described in Section 7.3 should prove very useful when the primary covariates associated with heterogeneity can be identified. However, it is possible to consider at least three reasonable approaches to modeling in the face of heterogeneous detection probabilities that cannot be associated with measured covariates. One approach involves extension of the finite mixture approach of Norris and Pollock (1996) and Pledger (2000) to the robust design. This extension has been incorporated into program MARK for analysis of capture-recapture data (White and Burnham 1999). Basically, the mixtures used for abundance estimation in closed models are viewed as defining groups of animals that may have group-specific rates of survival. There may or may not be a covariance structure relating group-specific survival and capture probabilities. Such modeling could be easily adapted to the type of multi-season modeling described above. Alternatively, occupancy data could be used with currently available software (MARK) to compute estimates under a conditional modeling approach. The modeling would follow the approach of Barbraud et al. (2003) outlined above, in which temporary emigration parameters for animals that were and were not temporary emigrants the previous period are equated with local extinction rate and the complement of colonization rate. The advantage of this approach is that it can be implemented now with existing software.

Another approach to the modeling of detection heterogeneity in multiple-season data would be to extend the abundance distribution modeling of Royle and Nichols (2003; see Chapter 5). Under this model, the probability of detecting the species at a site (p_{i,N_i} is the probability of detecting at least 1 individual at the site if there are N_i individuals at the site) is modeled as a function of the detection probability of individual animals (r): $p_{i,N_i} = 1 - (1 - r)^{N_i}$. We could similarly model the local extinction probability as the product of the probabilities of all individuals dying or dispersing from the site. Define ϕ as apparent survival probability, the probability that an individual present at a site in one primary period is present in the next primary period. Then we can define site-level extinction probability for a site with N_i individuals at period $\pm$ as: $\varepsilon_{i,N_i} = (1 - \phi)^{N_i}$. We might even consider modeling colonization probability as a function of either the overall abundance distribution for all sites or abundances of neighboring sites. The main point is that we could extend the more mechanistic abundance modeling of Chapter 5 to the situation of multiple seasons in a manner that hopefully would deal with heterogeneous detection, extinction, and possibly colonization probabilities associated with site-specific variation in abundance.

A third approach would be a "random effects" model in which each parameter for a specific site is a random value from some defined distribution, similar to the approach outlined in Chapter 5. In some circumstances it may be reasonable to consider that some random values for a site may be correlated. For instance, sites with a high detection probability (due to abundance, say) may have a low extinction probability. As such, the "random effects" could be modeled as a random draw from a multivariate distribution with some covariance or correlation structure. Such a model could be difficult to implement using a maximum likelihood approach, but could be very easily implemented using MCMC algorithms, including software such as WinBUGS.

7.7. STUDY DESIGN

In Chapter 6 we examined some of the issues related to the design of a single-season occupancy study. In particular we emphasized the importance of considering how the biology of the species corresponds to the model assumptions and timing of the surveys, and also presented results on the "optimal" allocation of resources. We believe the issues considered in Chapter 6 (e.g., site selection, defining a "season," and number of repeat surveys) are equally pertinent to the design of multiple-season occupancy studies, as a multiple-season study is simply a sequence of single-season studies.

Very little has been published on design issues related to multiseason occupancy studies, the exceptions being Field *et al.* (2005) and MacKenzie (2005b). Field *et al.* (2005) considered a situation in which trend in occupancy (specifically a decline) over a three-season time period is of interest, subject to budgetary constraints. Based upon a hypothesis testing objective, they assessed the effect the number of repeated surveys has on the power of the test to detect a decline (assuming that increasing the number of repeat surveys decreases the number of units that can be sampled, given a fixed budget). Field *et al.* (2005: 476) recommended that "2 to 3 visits to each site would perform adequately for most species [and] fewer than the optimal number of visits resulted in a harsher penalty than making more [visits]." MacKenzie (2005b) noted that Field *et al.* (2005) assumed detection probability was equal in each of the three seasons, an assumption which the analysis of empirical data sets has shown to be unlikely (e.g., MacKenzie *et al.* 2003; Bailey *et al.* 2004; MacKenzie *et al.* 2005; Olson *et al.* 2005). As such, a greater number of repeat surveys (visits) per season should be used, a suggestion more in line with the recommendations of MacKenzie and Royle (2005) (as reported in Chapter 6). Field *et al.* (2005) found that a general strategy of sampling more sites, less intensively, for rare (i.e., low occupancy) species and fewer sites, more intensively, for common species was optimal for detecting a decline in occupancy. Given the similarities of this result (and others) to those of MacKenzie and Royle (2005) in the single-season case, we believe that many of the recommendations given in Chapter 6 are equally applicable to multi-season occupancy studies. In this section we consider many of the additional issues that must be considered while designing a multiple-season occupancy study.

TIME INTERVAL BETWEEN "SEASONS"

It is important to contemplate the length of time that should separate two "seasons." While the type of practical situations we have had experience with are generally those in which "seasons" are clearly defined by the species biology (i.e., breeding seasons), an arbitrary length of time could be used if that is consistent with the objectives of the study. The main considerations are the timescale at which it is believed the processes of colonization and local extinctions operate and how that relates to the objectives of the study. It is important to note that the estimated parameters relate to the length of time between two seasons and that attempts to "rescale" these estimated quantities onto other timescales may be misleading. For example, suppose a population is surveyed during a single week each year. Colonization and local extinction probabilities therefore relate to potential changes in the occupancy status of sites over a 12-month period. Now suppose that managers decide they want 6-monthly estimates (e.g., summer and winter estimates) without the collection of 6-monthly data. It is possible to compute average 6-month estimates of colonization and extinction probabilities that assume constant instantaneous rates over the 12-month interval. For example, under this assumption of constant rates, the probability of the species not going locally extinct over a 12-month period can be written as the product of the probabilities of not going locally extinct in the two 6-month periods: $(1 - \varepsilon_{12}) = (1 - \varepsilon_6)^2$, where $\varepsilon_{\Delta t}$ denotes the probability of going extinct during an interval of duration Δt. Thus, an estimator for average 6-month extinction probability can be obtained from a 12-month extinction probability estimate as:

$$\hat{\varepsilon}_6 = 1 - \sqrt{(1 - \hat{\varepsilon}_{12})}.$$

However, because of the assumption of constancy, the above estimator cannot be used to draw inferences about potential seasonal variation in local extinction probability. Such inferences depend on the collection of additional data about when the colonization or local extinction event occurred (or indeed whether multiple such events may have occurred). In the metapopulation literature, some authors have postulated a "rescue effect" (e.g., Brown and Kodric-Brown 1977; Hanski 1991, 1994a, 1997; Moilanen 1999, 2002), where between two survey periods the species may go locally extinct at a site or patch, and the patch is then recolonized by members of the species from a nearby patch. While we do not argue whether a rescue effect may be a biological reality, we hold the view that without additional information it cannot be reliably estimated. Present attempts to estimate parameters associated with the rescue effect rely on restrictive assumptions about the number of extinction and colonization events that can occur between surveying periods (e.g., Clark and Rosenzweig 1994). Our primary recommendation is to tailor the timing of sampling to the level of temporal resolution at which questions are being addressed. In addition, it may be possible in some situations to use ancillary data (e.g., on marked individuals) to address very specific mechanistic questions (e.g., regarding rescue effects) about dynamic processes (Chapter 10; MacKenzie and Nichols 2004).

Similarly, we suggest that a consistent time interval between seasons be used. While we can think of a number of ad hoc approaches that could be used to allow for inconsistent time intervals in collected data (e.g., by reparameterizing the dynamic parameters as above, including a "time interval" covariate, or inserting missing observations to the detection histories to represent missing "seasons"), parameter estimates may be misleading and difficult to interpret. Again, the nature of the questions being asked should dictate the appropriate time intervals between periods of data collection.

SAME VS. DIFFERENT SITES EACH SEASON

The explicit dynamics models detailed in Section 7.3 rely on information collected from the same sites each season to estimate colonization and local extinction probabilities, and thus rates of change in occupancy. Using the unconditional model, not all sites would have to be surveyed each season because of the ability to incorporate missing values, but generally the expectation would be that a reasonable fraction of the sites would be surveyed every season. However, the implicit dynamics model of Section 7.2 could be used equally well in situations in which the same or different sites are surveyed each year, because there is no necessary linkage of specific sites at multiple seasons, and therefore no necessary modeling of changes in the occupancy status of sites between seasons.

In many of the studies with which we are most familiar, most sites are surveyed over multiple seasons; However, this is not the only general design that could be used. Another type of design that is occasionally used in ecological studies is the rotating panel design (Urquhart and Kincaid 1999). In a rotating panel design, rather than attempting to survey sites randomly selected over the entire area of interest each season, sites are selected from within a smaller subarea, and each season a different subarea is selected (e.g., surveying different watersheds within a forest or national park each year). Clearly, an advantage of such a design is that each season the logistical costs may be lower, as less time will probably be required to travel among sites. Another justification that is sometimes given for using such a design is that the spatial coverage of a study can be increased. In other words, if in any given season there are only enough resources to survey 20 sites, by using a three-year rotating panel design, the total sample size (viewed over the entire three years) can be increased to 60.

We recently conducted a simulation study to compare the relative benefits of using a rotating panel–type design versus a design in which all sites are surveyed over multiple seasons (which was briefly reported on in MacKenzie 2005b). For any of the scenarios considered, the simulation study compared designs that had the same number of sites surveyed per season but differed in terms of which sites were surveyed in consecutive seasons (e.g., the same 20 sites surveyed each season; 20 sites surveyed per season with a five-year rotation period, hence 100 sites surveyed in all; or 10 sites surveyed each season and an additional 10 sites surveyed per season on a five-year rotation period, hence 60 sites surveyed in all). One of the main results from our simulation study is that in terms of estimating the trend in occupancy (using either an implicit or explicit dynamics model), the overall design of the study is relatively unimportant. The precision of the trend estimate is largely determined by the number of sites surveyed each season (provided sufficient repeat surveys have been conducted within each season). This means that using a rotating panel design with a total of 100 sites surveyed over five seasons (i.e., 20 different sites per season) does not provide any more information with regard to a *seasonal* trend in occupancy than a design in which the same 20 sites are surveyed over the same period. That is, using a rotating panel design does not increase the effective sample size for estimating a seasonal trend. Furthermore, this result assumes that all sites that could be surveyed in a particular season have the same occupancy, colonization, and local extinction probabilities, in other words, sites that are selected to be surveyed in two consecutive seasons do not come from two different subpopulations with different occupancy-related parameters. With a rotating panel design in which sites are selected from different subareas each season, this assumption potentially could be violated. Resulting inferences from a rotating panel design could be inaccurate, and possibly misleading; the reasons for an observed change in occupancy may be due to: (1) a change in occupancy across the landscape over time; (2) the fact that different sites within the landscape have been sampled; or (3) a combination of both. It will not be possible to make strong statements about a trend in occupancy because it could be reasonable to argue that the observed change was a result of surveying different locations characterized by a different average level of occupancy, rather than temporal changes in occupancy within locations (trend). While additional structure could be built into a model to reflect any potential effect of the subareas, the inclusion of additional parameters to be estimated will increase the level of uncertainty in the trend parameter (i.e., increase its standard error), further hampering one's ability to make strong inference about trend. We believe that for the purpose of studying the processes or trends in occupancy, a far more practical approach is to survey the same sites each season, thereby removing any confounding spatial variation.

Furthermore, a rotating panel design will not provide the necessary information required to make seasonal estimates of colonization and local extinction probabilities. In some situations, the appropriate management action to a downward trend in occupancy, for example, could be different if colonization probability was high compared to if it was zero. Hence, in order to understand the mechanisms of change in the population and make appropriate decisions when necessary, rotating panel designs may be of limited usefulness.

MORE SITES VS. MORE SEASONS

In some situations it may be possible to consider a tradeoff between number of sites surveyed per season and number of seasons for which the study is conducted. For example, given a fixed total budget, is it better to monitor a population with 100 sites over 10 years, or 200 sites over 5 years? Of course, the 10-year study provides information about a period of time for which no formal inference is possible based on the 5-year study. Thus, treatment of this decision as a design tradeoff is predicated on an assumption of similar dynamics over the entire 10-year period of interest. Nevertheless, "trend" is frequently viewed as a parameter characterizing a relatively long time period, such that such design considerations are reasonable. MacKenzie (2005b) gave evidence that such a tradeoff is possible by considering the coefficient of variation (CV) of the estimated trend in occupancy on the logistic scale (and using an implicit dynamics model for simplicity). Figure 7.3 portrays the approximate CV for an increasing trend of 0.2 unit on the logit scale per season (i.e., the odds of occupancy increase by $e^{0.2} = 1.22$ each season). Solid lines indicate the approximate CV (obtained via simulation) for the trend estimate, assuming designs in which 50, 100, or 200 sites are surveyed each season. Clearly a tradeoff exists, for note that a similar level of precision can be achieved by surveying more sites over fewer seasons versus surveying fewer sites over a longer time period (e.g., 200 sites for four seasons vs. 50 sites for eight seasons). An implication of this result is that if precise information about a trend in occupancy is required within a short time frame, more sampling effort will be needed each season than if a longer time frame is available. An alternative interpretation of this result is that if available funding only permits a small number of sites to be surveyed each season, then management and stakeholders must understand

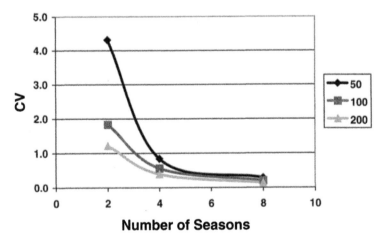

Number of Seasons

FIGURE 7.3   Coefficient of variation (CV) for a 0.2 seasonal trend in occupancy on the logit scale (approximated via simulation) from an implicit dynamics model, given that 50, 100, or 200 sites are surveyed for two, four, or eight seasons. Solid lines are for ease of interpretation only. Within seasons, three surveys were conducted at all sites with a detection probability of 0.5. Source: MacKenzie (2005b).

that a longer time frame will be required to provide decisive information about the system. Hence managers should be prepared to make a long-term commitment to the program. Finally, we caution readers that the results presented in Fig. 7.3 are illustrative only and pertain to a specific situation. While we believe this tradeoff to exist generally, the magnitude of the tradeoff should be assessed on a case-by-case basis.

MORE ON SITE SELECTION

The final issue we address here with respect to the design of multiple-season occupancy studies is site selection. In order to generalize to the entire area of interest, we have so far assumed that sites are selected in some probabilistic manner (i.e., the probability of a site being selected from the population of sites can be defined). If sites are selected in a non-probabilistic manner, then any generalization of the occupancy-related parameters to beyond the sites that were surveyed must be made with caution. This point is particularly relevant if sites are included in the sample due to prior knowledge about their expected occupancy state, which may create bias in parameter estimates. For example, we have seen some studies in which sites are selected because the species has been known to historically inhabit those sites. Depending upon the time frame between that historic knowledge and the present study, it may be reasonable to expect that a sample of historic sites will exhibit a higher proportion of sites occupied than a random sample of sites from the area of interest. While estimates of colonization and local extinction probabilities should remain unbiased, as they are conditional upon the occupancy state of sites in the previous season, any estimate of a change or trend in occupancy could be biased because of the recursive relationship between occupancy, colonization and local extinction. For example, consider Fig. 7.4. Suppose the system was in a state of equilibrium; thus the probability of a site being occupied was constant each season. In other words, there is no trend in occupancy. If the sites are selected in such a manner that the proportion of sites occupied in the sample is much higher than the level in the overall population, then an apparent downward trend in occupancy could be observed at the sampled sites even though no such trend actually exists. A similar result can occur when sites are added to those being surveyed based upon knowledge of the likely occupancy state (e.g., monitoring begins at a new site because of casual reports of the species being present there).

While the surveying of "historic" sites alone could create apparent trends in occupancy, it is one approach for targeting the surveying to provide better estimates of local extinction probabilities (as these are conditional upon sites being occupied, and a sample of "historic" sites may have a higher fraction of

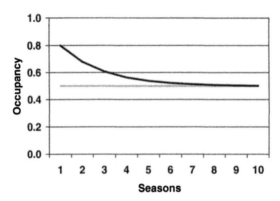

FIGURE 7.4 Apparent trend in occupancy caused by having an "unrepresentative" sample of sites with a higher occupancy probability than the population of interest. The solid grey line represents the true level of occupancy in the population over time, and the solid black line represents the level of occupancy in the sample of sites that have the same colonization and local extinction probabilities as the population in general.

occupied sites). To provide accurate trend estimates about the population in general while utilizing the "historic" site information, one could balance this information with data from a second sample of sites of unknown historic status. The population is now effectively stratified into sites of known and unknown historic occupancy status; hence, a form of stratified random sample could be used to select sites. Of course an assessment should be made of whether a more complicated stratified random sample is likely to achieve the objective of the study more efficiently than a simple random sample. As with other design issues, there is no single approach that is likely to be optimal in all situations. Thus, we are simply recommending that sites should be selected for surveys in a manner that is consistent with study objectives.

7.8. DISCUSSION

We believe that this chapter may be the most important in the book. Although most previous occupancy studies have focused on single-season patterns (the topic of Chapters 4–6), the objectives of most of these previous investigations involved dynamic processes (see review of Chapter 2). Because of the difficulties inherent in attempts to infer process from observation of pattern (see Chapters 1 and 2), we believe that studies of sites tending over multiple seasons are likely to provide the strongest inferences about occupancy dynamics and the processes that produce these dynamics. We thus believe that the models of this chapter should see a great deal of use and that future work should focus on extensions and elaborations of these approaches.

The implicit dynamics modeling approach of Section 7.2 essentially involved multiple applications of the single-season models of Chapter 4 to species detection data from a sequence of seasons. However, if occupancy dynamics are best viewed as a Markov process, as will be reasonable in many situations, then the explicit dynamics models of Section 7.3 should provide better descriptions of the data. In cases for which they are appropriate, the implicit dynamics models of Section 7.2 yield time-specific estimates of occupancy, as well as estimates of rate of change or "trend" in occupancy. Trend estimates are the focus of many current animal monitoring programs, and are justified as providing a basis for prioritization of conservation efforts. Species and areas in which rapid reductions in occupancy are occurring make prime targets for conservation efforts. However, estimation of trends does not provide much information about the causes of observed declines or, more importantly, about the kinds of management actions that are likely to reverse them. It is possible to model rate of change in occupancy as functions of environmental or management covariates, and such modeling can prove useful. However, we believe that direct modeling of the processes governing change is likely to be even more useful.

The models of Section 7.3 explicitly incorporate parameters for the vital rates responsible for changes in occupancy, rates of local extinction, and colonization. Covariate modeling can be used to investigate effects of environmental variables and management actions on these rate parameters. In addition, it is possible to investigate fundamental properties of the stochastic processes governing changes in occupancy. For example, we described competing models for random versus Markov processes and discussed investigation of such topics as equilibria and process stationarity (Section 7.4).

The remainder of the chapter dealt with model assumptions, suggested approaches for dealing with heterogeneity, and made recommendations for study design. We believe that the models of this chapter deserve much more attention than they have thus far received, as they permit direct investigation of such topics as metapopulation dynamics, range dynamics, and the relationship between occupancy dynamics and habitat change. These models also provide alternative means of investigating population dynamics that do not require detailed studies of marked individuals. For example, Barbraud et al. (2003) were able to draw inferences about bird movement (shifting of colony sites) by modeling the vital rates in one location as a function of vital rates in a neighboring location. These inferences were indirect, and thus not as strong as those based on observed movements of marked animals (e.g., Nichols 1996; Kendall and Nichols 2004). However, such indirect inferences can be obtained for areas too large to permit comprehensive capture-recapture studies, thus providing a useful complement to more detailed intensive investigations. The models for multiple species over multiple seasons outlined in Chapter 8 should prove very useful, for example, for dealing with competition between native species of conservation concern and related species experiencing range expansions (see Olson *et al.* 2005). Finally, we believe that similar models of occupancy dynamics hold great promise for investigations of such topics as multi-species community dynamics (Chapter 9) and joint habitat-occupancy dynamics (Chapter 10).

CHAPTER 8

Occupancy Data for Multiple Species: Species Interactions

Chapters 4–7 have focused on the issues related to estimating and modeling occupancy patterns and dynamics for a single species. However, often more than one species may be of primary interest, in terms of either studying relationships between multiple species, or investigating patterns and dynamics at the community level. As discussed in Chapter 2, there is a wide range of applications where occupancy-type thinking has been used to make inference about multiple species, with few attempts to formally incorporate detection probabilities, despite acknowledgments that they are a reality of many sampling situations. In this and the following chapter we outline how the occupancy modeling approaches that have been developed to date (which explicitly incorporate the estimation of detection probability) could be applied to, or extended for, multi-species or community-level studies. This chapter deals with a small number of species and focuses on hypotheses about species interactions, whereas Chapter 9 focuses on entire communities with possibly large numbers of species but does not explicitly deal with interspecific interactions. Many of the concepts and models discussed in the following chapters are the result of very recent or current research; hence, the contents of this chapter and the next are not as well developed as those of the previous chapters. The models

225

are conceptually sound, but we have not had the opportunity to develop associated software or conduct example analyses for all of the suggested methods.

In this chapter we turn our attention to the problem of studying relationships among the occurrence patterns of multiple species from detection/nondetection data. Such approaches have a long history of use in ecology, particularly to identify nonrandom patterns in the species co-occurrence matrix (Chapter 2). A large body of literature has focused on appropriate formulation of the "null model," associated test statistics, deriving the null distribution of the test statistics, and the resulting debate between proponents of the various concepts and ideas. However, other factors (e.g., habitat preferences and physiological tolerances) often may introduce nonrandom patterns to species incidence matrices that are unrelated to interspecific interactions. These could be incorporated into the "null model" by identifying them *a priori*, although we prefer to think of such factors as hypotheses that one may wish to investigate (e.g., one may wish to estimate the magnitude of any such effect). A second facet of species co-occurrence data relevant to inferences about interspecific interactions is detectability. Very little attention has been paid to the practicality of the sampling in which species are unlikely to be detected with certainty (exceptions are provided by Cam *et al.* 2000 and MacKenzie *et al.* 2004b). By accounting for detection probability explicitly, absolute measures of co-occurrence (such as the probability that two species co-occur at a site) can be estimated. When the intent is to compare the level of co-occurrence at multiple points, stronger inference about changes or differences can be made if detection probabilities have been incorporated into the inferential procedure. The modeling procedure of MacKenzie *et al.* (2004b) detailed below can be viewed as an approach that unifies these two concepts of incidence or occurrence of different species and detection probability within a single modeling framework for investigating species co-occurrence patterns at a single point in time. However, in tune with our arguments throughout this book, the processes of co-occurrence cannot be reliably inferred from observed patterns. We end this chapter by developing a model that examines how co-occurrence patterns change over time when data are available from multiple sampling seasons, thus allowing stronger inference to be made about the underlying processes.

8.1. DETECTION PROBABILITY AND INFERENCES ABOUT SPECIES CO-OCCURRENCE

Imperfectly detecting species will result in misleading inferences about species co-occurrence patterns. Traditionally, the nondetection of a species at a location is interpreted as an absence in the species co-occurrence matrix. However,

TABLE 8.1　Probability Structure for the Possible Co-occurrence Patterns of Two Species

		Species B	
		Present	Absent
Species A	Present	ψ^{AB}	$\psi^{A} - \psi^{AB}$
	Absent	$\psi^{B} - \psi^{AB}$	$1 - \psi^{A} - \psi^{B} + \psi^{AB}$

ψ^{AB} is the probability both species are present at a site, ψ^{A} is the probability species A is present at a site regardless of the presence or absence of species B, and ψ^{B} is the probability species B is present at a site regardless of the presence or absence of species A.

if the nondetection is a "false absence", both the value and the null distribution of the test statistic are incorrect (in comparison to the "true" situation with respect to the presence of the species). That is, the null distribution of the test statistic for the observed data (detection/nondetection) is different from the null distribution for the true situation (presence/absence). What effect, then, would these differences due to imperfect detection of the species have on our resulting inferences?

To address this issue, consider the simple case in which the potential for interaction between two species is investigated using a simple 2×2 contingency table (e.g., Forbes 1907; Dice 1945; Cole 1949; Pielou 1977; Hayek 1994). In Table 8.1 we give the probability for each of the four possible outcomes for whether the two species are present at a site, where ψ^{AB} is the probability that both species A and B are present at a site (i.e., the probability of co-occurrence); ψ^{A} is the probability that species A occupies a site regardless of whether species B is also present (hence $\psi^{A} - \psi^{AB}$ is the probability of species A occurring without species B); ψ^{B} is the probability that species B occupies a site regardless of whether species A is also present (hence $\psi^{B} - \psi^{AB}$ is the probability of species A occurring without species B); and $1 - \psi^{A} - \psi^{B} + \psi^{AB}$ is the probability that both species are absent. These are also graphically presented in Figure 8.1. Note that the probabilities given here are general in that they do not assume the species occur independently of one another. From this contingency table, a χ^2 test could be used to determine whether the rows and columns of the table (i.e., the presence or absence of each species) are independent. Alternatively, an odds ratio could be calculated to estimate the magnitude of any dependence between the rows and columns, which is the approach we take here (recall the odds ratio was discussed in Chapter 3 with respect to the logit link).

The odds that species B is present at a site given that species A is also present is $\psi^{AB}/(\psi^{A} - \psi^{AB})$, whereas the odds that species B is present given that species

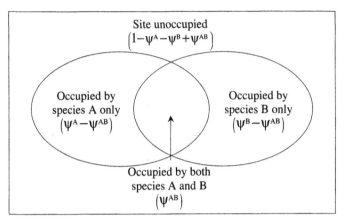

FIGURE 8.1 Venn diagram representing the probability for each occupancy state. Left-hand ellipse represents the probability that the site is occupied by species A, ψ^A, and the right-hand ellipse represents the probability that the site is occupied by species B, ψ^B. The intersection of the two ellipses is the probability that the site is occupied by both species, ψ^{AB}.

A is absent is $(\psi^B - \psi^{AB})/(1 - \psi^A - \psi^B + \psi^{AB})$. The odds ratio ($OR$) will therefore be:

$$OR = \frac{\psi^{AB}/(\psi^A - \psi^{AB})}{(\psi^B - \psi^{AB})/(1 - \psi^A - \psi^B + \psi^{AB})}$$
$$= \frac{\psi^{AB}(1 - \psi^A - \psi^B + \psi^{AB})}{(\psi^A - \psi^{AB})(\psi^B - \psi^{AB})} \tag{8.1}$$

OR will in fact have the same form regardless of whether the odds are defined in terms of species B being present (as above) or alternatively in terms of species A. If the two species occur independently, then $OR = 1$. Furthermore, under independence, the probability of both species occurring at a site will be the product of the marginal occurrence probabilities, that is, $\psi^{AB} = \psi^A \times \psi^B$. OR values less than 1 indicate that the species have a smaller probability of co-occurring than expected under a hypothesis of independence, and values greater than 1 indicate a higher probability of co-occurring.

Now consider Table 8.2, which is the probability structure for a 2×2 contingency table of whether each species was *detected* or not (i.e., detection/nondetection data, not presence/absence). Here the odds ratio is:

$$OR' = \frac{\psi^{AB} p^A p^B (1 - \psi^A p^A - \psi^B p^B + \psi^{AB} p^A p^B)}{(\psi^A p^A - \psi^{AB} p^A p^B)(\psi^B p^B - \psi^{AB} p^A p^B)}, \tag{8.2}$$

TABLE 8.2 The Probability of *Detecting* Species A and/or Species B Given Both May Be Detected Imperfectly

		Species B	
		Detected	Not Detected
Species A	Detected	$\psi^{AB}p^{A}p^{B}$	$\psi^{A}p^{A} - \psi^{AB}p^{A}p^{B}$
	Not Detected	$\psi^{B}p^{B} - \psi^{AB}p^{A}p^{B}$	$1 - \psi^{A}p^{A} - \psi^{B}p^{B} + \psi^{AB}p^{A}p^{B}$

ψ^{AB} is the probability both species are present at a site, ψ^{A} is the probability species A is present at a site regardless of the presence or absence of species B, and ψ^{B} is the probability species B is present at a site regardless of the presence or absence of species A. The probability of detecting species A in a survey is p^{A}, and the probability of detecting species B in a survey is p^{B}.

where p^{A} and p^{B} are the probabilities of detecting each species in the survey. Clearly, Eqs. (8.1) and (8.2) are not equivalent. While we do not offer any definitive proof here, when species are detected imperfectly, the odds ratio indicates the correct direction of the relationship (i.e., less than or greater than 1), but the magnitude of the relationship is underestimated (i.e., always estimated closer to 1 than it should be). That is, if at least one of the species is detected imperfectly, when using a contingency table approach that does not account for detection probabilities, the estimated level of co-occurrence will underestimate the level of the interaction.

For example, suppose that within a region the probability of species A being present at a site (ψ^{A}) is 0.6 and the probability of species B being present (ψ^{B}) is 0.3. Further, suppose that the two species are competitors, hence do not tend to co-occur very frequently, and the probability of both species being present at a site (ψ^{AB}) is 0.1 (note that if they occupied sites independently, then $\psi^{AB} = 0.6 \times 0.3 = 0.18$). From Eq. (8.1) the true odds ratio would be:

$$OR = \frac{0.1(1 - 0.6 - 0.3 + 0.1)}{(0.6 - 0.1)(0.3 - 0.1)}$$

$$= \frac{0.1 \times 0.2}{0.5 \times 0.2}$$

$$= \frac{0.02}{0.10} = 0.2.$$

Now suppose that in a survey for the species at a site, both are detected imperfectly, and the probability of detecting species A (P^{p}) is 0.4 and the probability of detecting species B (P^{p}) is 0.7. The odds ratio for detecting the species from Eq. (8.2) is therefore:

TABLE 8.3 The Probability of *Detecting* Species A and/or Species B Given Both May Be Detected Imperfectly, Where the Detection of One Species depends upon Whether Only One or Both Species Are Present at a Site

		Species B	
		Detected	Not Detected
Species A	Detected	$\psi^{AB}r^A r^B$	$\psi^{AB}r^A(1-r^B) + (\psi^A - \psi^{AB})p^A$
	Not Detected	$\psi^{AB}(1-r^A)r^B + (\psi^B - \psi^{AB})p^B$	$1 - (\psi^A - \psi^{AB})p^A - (\psi^B - \psi^{AB})p^B$ $- \psi^{AB}(r^A + r^B - r^A r^B)$

ψ^{AB} is the probability both species are present at a site, ψ^A is the probability species A is present at a site regardless of the presence or absence of species B, and ψ^B is the probability species B is present at a site regardless of the presence or absence of species A. The probability of detecting species A in a survey, given only species A is present, is p^A; the probability of detecting species A in a survey, given both species A and B are present, is r^A; the probability of detecting species B in a survey, given only species B is present, is p^B; and the probability of detecting species B in a survey, given both species A and B are present, is r^B.

$$OR' = \frac{0.1 \times 0.4 \times 0.7(1 - 0.6 \times 0.4 - 0.3 \times 0.7 + 0.1 \times 0.4 \times 0.7)}{(0.6 \times 0.4 - 0.1 \times 0.4 \times 0.7)(0.3 \times 0.7 - 0.1 \times 0.4 \times 0.7)}$$

$$= \frac{0.028(1 - 0.24 - 0.21 + 0.028)}{(0.24 - 0.028)(0.21 - 0.028)}$$

$$= \frac{0.028 \times 0.578}{0.212 \times 0.182} = 0.419.$$

Hence, because the species are detected imperfectly, we might conclude that the level of competition or exclusion between the two species is less severe than it actually is.

Now consider the more general and realistic case in which the probability of detecting each species depends upon whether the other species is also present at the same site (Table 8.3). The odds ratio in this situation is:

$$OR'' = \frac{\psi^{AB}r^A r^B[1 - (\psi^A - \psi^{AB})p^A - (\psi^B - \psi^{AB})p^B - \psi^{AB}(r^A + r^B - r^A r^B)]}{[\psi^{AB}r^A(1-r^B) + (\psi^A - \psi^{AB})p^A][\psi^{AB}(1-r^A)r^B + (\psi^B - \psi^{AB})p^B]} \quad (8.3)$$

where r^A and r^B are the probabilities of detecting each species given that both species are present at a site, and p^A and p^B are the detection probabilities given that only one species is present. It can be shown (by considering simple examples) that the odds ratio calculated from detection/nondetection data can provide completely misleading information with regards to the actual co-occurrence of species. That is, the odds ratio may not indicate the correct direction of any relationship or may even indicate a spurious relationship (in terms of occupancy and co-occurrence). We can only speculate that similar results hold when investigating co-occurrence patterns for greater than two species, but feel confident in our assertion that this would be so.

8.2. A SINGLE-SEASON MODEL

MacKenzie et al. (2004b) extended the single-species single-season model of MacKenzie et al. (2002) as a means of incorporating detection probability into inferences about co-occurrence patterns of multiple species. They noted that the model could be applied to any number of species in theory, but due to the large number of parameters required, it would be most practical to model only a small number of species (≤ 4), and they presented their model for the two-species case. For a larger number of species, the majority of the parameters would be required to model high-order interactions (in the statistical sense) among a number of species (e.g., interactions between four or more species). In many real-life situations we imagine there would be insufficient data to estimate such high-order interactions reliably, and even if there were, such interactions could be very difficult to interpret. Therefore, one practical means of reducing the number of parameters is to only model interactions between two or three species (i.e., only model the lower-order interactions). Due to the newness of this modeling technique, we have not had the opportunity to explore the merits of such a procedure, but it offers one approach for applying the general procedure to a large number of species in a reasonable manner. However, the suggestion of modeling only a small number of species does have its advantages. It encourages researchers to carefully consider *a priori* the species on which they wish to focus, rather than taking a "shotgun" approach and collecting data on a large number of species and looking for "significant" relationships that may be spurious. We would advocate the former approach as much closer to our view of science. For simplicity, here we only consider the two-species model detailed by MacKenzie et al. (2004b), but suggest that it could be extended in an obvious manner.

GENERAL SAMPLING SITUATION

The general sampling situation envisaged here is very similar to the single-species case, although now data are being collected on more than one species at each site. Multiple (not necessarily an equal number of) surveys are conducted at s sites to detect the target species, with the intent of providing a snapshot of the system with respect to species co-occurrence patterns. For the duration of the surveying, sites are closed to changes in the occupancy state with respect to each species (i.e., a species is either always present or always absent from the site over the surveying period), or changes in occupancy are completely at random for each species.

A detection history for each species at a site can be recorded to denote the sequence of detections and nondetections at that site. For example, the detection history $h_i^A = 101$ represents that site i was surveyed on three occasions, with species A being detected only in the first and third surveys. The detection history $h_i^B = 000$ represents that species B was never detected at site i.

STATISTICAL MODEL

In Table 8.4 the model parameters defined by MacKenzie *et al.* (2004b) are presented (recall that Figure 8.1 illustrates the relationship between ψ^A, ψ^B, and ψ^{AB}). Note also that their use of different detection probability parameters for the cases in which one species and two species are present at a site results in an occupancy model that is very general and permits the possibility that detection probability of one species depends on whether the site is occupied by the other species (e.g., as is hypothesized by U.S. biologists for northern spotted owls and barred owls in the Pacific Northwest; see Chapter 2).

The basic procedure for constructing a two-species model is the same as that used previously; namely, create a verbal description of the processes that

TABLE 8.4   Notation for the Parameters Used in the Single-season, Two-species Model

Parameter	Description
ψ^{AB}	Probability of both species being present
ψ^A	Probability of occupancy for species A, regardless of occupancy status of species B
ψ^B	Probability of occupancy for species B, regardless of occupancy status of species A
p_j^A	Probability of detecting species A during the jth survey, given only species A is present
p_j^B	Probability of detecting species B during the jth survey, given only species B is present
r_j^{AB}	Probability of detecting both species during the jth survey, given both species are present
r_j^{Ab}	Probability of detecting species A, but not B, during the jth survey, given both species are present
r_j^{aB}	Probability of detecting species B, but not A, during the jth survey, given both species are present
r_j^{ab}	Probability of detecting neither species during the jth survey, given both species are present; $= 1 - r_j^{AB} - r_j^{Ab} - r_j^{aB}$

Source: MacKenzie *et al.* 2004b.

may have resulted in a particular combination of detection histories being observed, and then translate the description into a mathematical equation involving the defined model parameters. The resultant mathematical equation is the probability of observing the combination of detection histories for the two species at that site. For example, suppose the following pair of detection histories was observed at site i, $h_i^A = 110$ and $h_i^B = 000$. A verbal description of the events giving rise to these detection histories would be:

both species are present at the site, with species A, but not species B, being detected in the first and second surveys, and neither species detected in the third survey

or

only species A is present at the site, with it being detected in surveys 1 and 2, but not in survey 3.

Translating this description into a mathematical equation using the model parameters gives the probability of observing these detection histories as:

$$Pr(h_i^A = 110, h_i^B = 000) = \psi^{AB} r_1^{Ab} r_2^{Ab} r_3^{ab} + (\psi^A - \psi^{AB}) p_1^A p_2^A (1 - p_3^A).$$

To aid the formulation of their model, MacKenzie *et al.* (2004b) suggest that sites could be considered in one of four mutually exclusive states of occupancy when modeling two species (more generally, there are 2^k possible states for k species): (1) occupied by both species A and B; (2) occupied by species A only; (3) occupied by species B only; or (4) occupied by neither species. Therefore a row vector could be defined that denotes the probability of a site being in each of the four respective states immediately before the site is surveyed during the sampling season, for example:

$$\phi_0 = [\psi^{AB} \quad \psi^A - \psi^{AB} \quad \psi^B - \psi^{AB} \quad 1 - \psi^A - \psi^B + \psi^{AB}].$$

Next, define a column vector $p^{\{h_i^A\}\{h_i^B\}}$ that contains the probability of observing the detection histories observed at site i, conditional upon each occupancy state. For example, if both species were present at the site, then $Pr(h_i^A = 110, h_i^B = 000|$both species present$) = r_1^{Ab} r_2^{Ab} r_3^{ab}$, while if the site was only occupied by species A, then $Pr(h_i^A = 110, h_i^B = 000|$only species A present$) = p_1^A p_2^A (1 - p_3^A)$. Hence, the column vector $p^{\{110\},\{000\}}$ would be defined as:

$$p^{\{110\},\{000\}} = \begin{bmatrix} r_1^{Ab} r_2^{Ab} r_3^{ab} \\ p_1^A p_2^A (1 - p_3^A) \\ 0 \\ 0 \end{bmatrix}$$

Note that the bottom two conditional detection probabilities are zero. Respectively, these elements relate to the states where only species B is present or neither species is present. As species A was detected during the surveys, which would be precluded if the site was in either of these two states, the conditional detection probabilities for this pair of histories must be zero.

The unconditional probability of observing any given pair of detection histories can then be calculated as:

$$\Pr(h_i^A, h_i^B) = \phi_0 p^{\{h_i^A\}\{h_i^B\}},$$

for example:

$$\Pr(h_i^A = 110, h_i^B = 000)$$

$$= [\psi^{AB} \quad \psi^A - \psi^{AB} \quad \psi^B - \psi^{AB} \quad 1 - \psi^A - \psi^B + \psi^{AB}] \begin{bmatrix} r_1^{Ab} r_2^{Ab} r_3^{ab} \\ p_1^A p_2^A (1 - p_3^A) \\ 0 \\ 0 \end{bmatrix}$$

$$= \psi^{AB} r_1^{Ab} r_2^{Ab} r_3^{ab} + (\psi^A - \psi^{AB}) p_1^A p_2^A (1 - p_3^A)$$

Some further examples of detection histories and the unconditional probabilities of observing them are given in Table 8.5.

Assuming that the detection histories collected at the s sites are independent, the model likelihood is simply defined as:

$$L(\psi^{AB}, \psi^A, \psi^B, r, p | h_1^A, h_1^B, \ldots, h_s^A, h_s^B) = \prod_{i=1}^{s} \Pr(h_i^A, h_i^B).$$

To obtain parameter estimates, the likelihood could be maximized to give maximum likelihood estimates, or prior distributions could be assigned for each parameter and, by regarding the likelihood function as the probability of observing the data, their posterior distributions could be obtained.

REPARAMETERIZING THE MODEL

The parameters as defined in Table 8.4 may not always be the most meaningful or natural forms with which to investigate species co-occurrence patterns. In this section we briefly discuss alternative parameterizations as suggested by MacKenzie *et al.* (2004b). We believe that in most applications, interpretation of the reparameterized version of the model will be more natural. In Section 8.4 we describe how these parameterizations can be used to address interesting biological hypotheses about the system.

If two species co-occur at sites independently, then the following relationship should hold: $\psi^{AB} = \psi^A \times \psi^B$. The level of co-occurrence between the two species could therefore be quantified by the expression:

TABLE 8.5   Example Detection Histories (h^A, h^B) and the Probabilities of Observing Them $(\Pr(h^A, h^B))$ Using the Single-season, Two-species Model

h^A	h^B	$\Pr(h^A, h^B)$
011	010	$$= \begin{bmatrix} \psi^{AB} \\ \psi^A - \psi^{AB} \\ \psi^B - \psi^{AB} \\ 1 - \psi^A - \psi^B + \psi^{AB} \end{bmatrix}^T \begin{bmatrix} (1 - r_1^{AB} - r_1^{Ab} - r_1^{aB})r_2^{AB}r_3^{Ab} \\ 0 \\ 0 \\ 0 \end{bmatrix}$$ $$= \psi^{AB}(1 - r_1^{AB} - r_1^{Ab} - r_1^{aB})r_2^{AB}r_3^{Ab}$$
000	101	$$= \begin{bmatrix} \psi^{AB} \\ \psi^A - \psi^{AB} \\ \psi^B - \psi^{AB} \\ 1 - \psi^A - \psi^B + \psi^{AB} \end{bmatrix}^T \begin{bmatrix} r_1^{aB}(1 - r_2^{AB} - r_2^{Ab} - r_2^{aB})r_3^{aB} \\ 0 \\ p_1^B(1 - p_2^B)p_3^B \\ 0 \end{bmatrix}$$ $$= \psi^{AB}r_1^{aB}(1 - r_2^{AB} - r_2^{Ab} - r_2^{aB})r_3^{aB} + (\psi^B - \psi^{AB})p_1^B(1 - p_2^B)p_3^B$$
000	000	$$= \begin{bmatrix} \psi^{AB} \\ \psi^A - \psi^{AB} \\ \psi^B - \psi^{AB} \\ 1 - \psi^A - \psi^B + \psi^{AB} \end{bmatrix}^T \begin{bmatrix} \prod_{j=1}^{3}(1 - r_j^{AB} - r_j^{Ab} - r_j^{aB}) \\ \prod_{j=1}^{3}(1 - p_j^A) \\ \prod_{j=1}^{3}(1 - p_j^B) \\ 1 \end{bmatrix}$$ $$= \psi^{AB}\prod_{j=1}^{3}(1 - r_j^{AB} - r_j^{Ab} - r_j^{aB}) + (\psi^A - \psi^{AB})\prod_{j=1}^{3}(1 - p_j^A)$$ $$+ (\psi^B - \psi^{AB})\prod_{j=1}^{3}(1 - p_j^B) + (1 - \psi^A - \psi^B + \psi^{AB})$$

$$\varphi = \frac{\psi^{AB}}{\psi^A \psi^B}.$$

That is, φ is the ratio of how much more or less likely the species are to co-occur at a site compared to what would be expected if they co-occurred independently. MacKenzie *et al.* (2004b) termed the quantity φ a *species interaction factor* (SIF). Values of φ less than 1 would suggest species co-occur less frequently than if they were distributed independently (e.g., possibly exclusion or avoidance), while values greater than 1 would suggest a tendency for species to co-occur more frequently than expected under independence. If the species occupy sites independently, then $\varphi = 1$. Hence, rather than deriving $\hat{\varphi}$ from estimates of ψ^A, ψ^B, and ψ^{AB}, often it may be advantageous to reparameterize the model as:

$$\psi^{AB} = \psi^A \psi^B \varphi,$$

so that φ may be estimated (and modeled) directly. There is, however, a natural relationship among the occupancy probabilities that restricts the values that φ can possibly take, reflecting limits to the degree of overlap that is possible between the two species; in other words,

$$\max(\psi^A + \psi^B - 1, 0) \le \psi^{AB} \le \min(\psi^A, \psi^B).$$

For example, if ψ^A and $\psi^B = 0.6$, then the two species must co-occur at a minimum of 20% of the locations, while if they always co-occur, then it can only be at 60% of sites at most. This implies that φ can only take values between $\max((\psi^A + \psi^B - 1)/(\psi^A \psi^B), 0)$ and $\min((1/\psi^B), (1/\psi^A))$. This restriction must be enforced when estimating the SIF, which may cause numerical problems in some instances.

Another parameterization that has been recently suggested to us is to define the SIF in terms of the odds ratio of occupancy for the two species (φ_{OR}), as in Eq. (8.1) (David Fletcher, Proteus Wildlife Research Consultants; Richard Barker, University of Otago, personal communications). The potential advantage of such a parameterization is that φ_{OR} may take any value between 0 and infinity; hence, there is less restriction on the allowable values of the SIF, which may help avoid some of the convergence problems experienced by MacKenzie et al. (2004b) when fitting a model with covariates (see the example below). We have not yet explored how well this reparameterization may work in practice.

Similar reparameterizations can also be used on the r parameters. Rather than estimating the probability for detecting/not detecting both species in a single survey (given both species are present at a site), it may be more natural to estimate the marginal detection probabilities for both species (i.e., the probability of detecting species A regardless of whether species B was also detected in the same survey, given both species were present) and a second SIF. Let r_j^{AB} be redefined as:

$$r_j^{AB} = r_j^A \times r_j^B \times \delta,$$

where r_j^A is the marginal probability of detecting species A during survey j, r_j^B is the marginal probability of detecting species B during survey j, and δ is the SIF for the detection probabilities. δ has a similar interpretation to φ above; that is, values less than 1 suggest it is less likely to detect both species in a survey than if the species were detected independently, while values greater than 1 suggest that it is more likely to detect both species. The other r parameters can then be derived as follows:

$$r_j^{Ab} = r_j^A - r_j^{AB}$$
$$= r_j^A (1 - r_j^B \delta)$$
$$r_j^{aB} = r_j^B - r_j^{AB}$$
$$= (1 - r_j^A \delta) r_j^B$$
$$r_j^{ab} = 1 - r_j^{Ab} - r_j^{aB} - r_j^{AB}$$
$$= 1 - r_j^A - r_j^B + r_j^{AB}$$
$$= 1 - r_j^A - r_j^B + r_j^A r_j^B \delta$$

Note that once again there are constraints upon the allowable values for δ when it is defined in this manner, so there may be some advantage to also defining this SIF in terms of an odds ratio.

INCORPORATING COVARIATE INFORMATION

The probability that a species occupies a location potentially may be affected by specific characteristics of the site. For example, some species may prefer particular habitat types over other available habitats (e.g., have a higher occupancy probability at locations near permanent water sources); require a minimum patch size for a sustainable population; or show reduced probability of occurrence in isolated patches (e.g., Verner et al. 1986; Hanski 1999; Scott et al. 2002). Different species preferring different habitats may well be responsible for the nonrandom patterns observed in the species-occurrence matrix when using analytic methods that do not account for such factors. Similarly, the probability of detecting species at a site may be affected by site-specific covariates (e.g., open old-growth forest vs. dense rejuvenating forest). Detection probabilities may also be affected by the conditions at the time of the survey, such as air temperature, cloud cover, or time since a rain event.

Using the basic modeling approach outlined above, allowing these probabilities to be functions of measured covariates is simply achieved in a manner similar to that used in Chapters 4 and 7. MacKenzie et al. (2004b) stated that the choice of link function used depends upon the parameterization used for the model. Generally, modeling the data in terms of the marginal probabilities (ψ^A, ψ^B, r^A, r^B, p^A and p^B) and the species interaction factors (φ and δ) will provide a more natural interpretation than modeling the possible discrete outcomes of the observed data (e.g., modeling the probabilities related to the four possible outcomes for observing both species during a survey, r^{AB}, r^{Ab}, r^{aB}, r^{ab}). The logit link function can be used to incorporate covariate information for the marginal probabilities, and the log link function [Eq. (8.4); where θ_i is the value of the respective SIF being modeled at site i] can be used for the species interaction factors, as one would want them to be nonnegative (otherwise one will obtain negative probabilities) and have a value of 1 when there is no effect (i.e., $\ln(0) = 1$):

$$\ln(\theta_i) = \beta_0 + \beta_1 x_{1i} + \ldots + \beta_u x_{ui} \tag{8.4}$$

MacKenzie $et\ al.$ (2004b) suggest that the multinomial logistic model (a generalization of the logit link to situations in which there are more than two possible outcomes) could be used to incorporate covariate information with the original parameterization. However, the reparameterized version offers a more natural interpretation of covariate effects, and hence we do not present such an approach here. Interested readers are directed to MacKenzie $et\ al.$ (2004b) for further details on this alternative method.

MISSING OBSERVATIONS

As in single-species studies, "missing observations" in the detection histories may occur due to logistical constraints (it is simply not possible to survey all locations virtually simultaneously), study design, or unforeseen circumstances such as a vehicle breakdown en route. These can be easily accommodated in the multi-species modeling in the same manner as in Chapters 4 and 7. For survey occasions when no data were collected, the respective detection probabilities do not appear within the mathematical equation denoting the probability of observing that detection history. Hence, no detection probability parameters appear in the model likelihood for that site at that survey occasion.

We generally imagine situations in which the detection/nondetection of both species arise from the same information source (i.e., two frog species are detected/not detected during a single five-minute calling survey), and a "missing observation" at a particular site at a certain survey occasion thus applies to both species. However, in some circumstances missing observations may occur for one species but not for the other. For example, if different field methods are required to detect each species (e.g., camera traps to detect carnivores and track plates to detect their prey), and one method functions correctly but the other method fails, this would create a survey with a missing observation for only one (or, more generally, a subset) of the species. One view of this problem is that the second species may or may not have been detected had the data been collected. To incorporate this idea, we can use the same technique we have used elsewhere when faced with the possibility of multiple explanations for the same set of data: include both possibilities within the probability statement for the detection history. Consider the detection histories $h_i^A = 101$ and $h_i^B = 0\text{-}0$, where there is a missing observation for species B in the second survey. A verbal description of the detection histories would be:

both species are present at the site, with species A detected in the first and third surveys but not species B, and in the second survey species A was not detected and species B may *or* may not have been detected
or
only species A is present at the site, with it being detected in surveys 1 and 2, but not in survey 3.

Translating this description into a mathematical equation gives:

$$\Pr(h_i^A = 101, h_i^B = 0\text{-}0) = \psi^{AB} r_1^{Ab} (r_2^{aB} + r_2^{ab}) r_3^{Ab} + (\psi^A - \psi^{AB}) p_1^A p_2^A (1 - p_3^A).$$

The term $r_2^{aB} + r_2^{ab}$ combines the possibilities that species A was not detected and species B may or may not have been detected if the observation had been recorded. Stated differently, the sum, $r_2^{aB} + r_2^{ab}$, reflects the probability that species A was not detected, but conveys no information about whether or not species B might have been detected.

8.3. ADDRESSING BIOLOGICAL HYPOTHESES

MacKenzie *et al.* (2004b) showed that when using this modeling approach, it is possible to formulate three interesting biological hypotheses about the system by constraining various parameter values: (1) level of co-occurrence between species; (2) independence of detecting the species; and (3) whether detection of each species depends upon the presence of the other species. By fitting a set of candidate models that represent the various biological hypotheses, the degree of support for each can be evaluated using formal statistical methods.

In many applications, the first hypothesis will be of primary interest and is equivalent to searching for nonrandom patterns in the species co-occurrence matrix. Recall from the above discussion that if the two species co-occur independently, then $\psi^{AB} = \psi^A \times \psi^B$ or, expressed in terms of the SIF, $\varphi = 1$. The degree of support for whether species occur at sites independently can therefore be evaluated by comparing two models: (1) a full model in which each of the parameters ψ^A, ψ^B, and φ are estimated; and (2) a reduced model in which only ψ^A and ψ^B are estimated and the value for φ is set equal to 1. A formal comparison of these two models allows an informed decision to be made about the level of support for each hypothesis.

The second question relates to the detection of the species: At sites where both species exist, is the probability of detecting species A during a survey of the site affected by whether species B is also detected, or vice versa? That is, are the species detected independently during the survey of a site? For example, suppose the two species of interest are predator and prey species (e.g., a hawk and rabbits). At a site where both species are present, if a hawk is seen during a survey, does that decrease the chances of seeing rabbits during the same survey? The level of support for this hypothesis can be assessed in the same manner as above, using the detection probability SIF (δ) by fitting two models, one where δ is estimated and one where δ is set equal to 1.

The third biological hypothesis is whether the probability of detecting species A in a survey is affected by whether species B is also present at the site; in other words, does $r_j^A = p_j^A$? Similarly for species B, does $r_j^B = p_j^B$? Note that this issue is distinct from the question of whether detections of the two species occur independently during a survey given that both species are present (i.e., does $\delta = 1$?). For example, in the previously mentioned case of northern spotted owls and barred owls in the Pacific Northwest, it is thought that detection probability of northern spotted owls (NSO) may be lower when barred owls are also present at a site, that is, $r_j^{NSO} < p_j^{NSO}$. There are a number of models that could be considered here, as there are a number of sub-hypotheses one may be interested in. For example, it may be of interest whether the presence of the other species affects the detection probability for just one or for both of the species. That is, possible constraints that could be applied are (1) none (i.e., r_j^A, r_j^B, p_j^A, p_j^B are all estimated separately); (2) $r_j^A = p_j^A$; (3) $r_j^B = p_j^B$; or (4) $r_j^A = p_j^A$ and $r_j^B = p_j^B$. The level of support for each hypothesis can therefore be determined, although we advise that one should always attempt to consider *a priori* which hypotheses are of prime importance rather than considering all possible alternatives.

8.4. EXAMPLE: TERRESTRIAL SALAMANDERS IN GREAT SMOKY MOUNTAINS NATIONAL PARK

MacKenzie *et al.* (2004b) illustrated the use of their modeling approach with monitoring data collected on terrestrial salamanders in Great Smoky Mountains National Park (GSMNP), located on the Tennessee and North Carolina borders in the United States of America. Data were collected from 88 sites within the Roaring Fork Watershed (GSMNP, Mt. LeConte USGS Quadrangle) that were located adjacent to trails and spaced approximately 250 m apart. Two parallel transects were sampled at each site: a natural cover transect (50 m long × 3 m wide) and a coverboard transect consisting of five stations placed 10 m apart. MacKenzie *et al.* (2004b) pooled the detection data from the two transects for each site in their analysis. Sites were sampled five times between April 4, 1999, and June 27, 1999, with approximately two weeks between successive sampling occasions. Co-occurrence patterns were considered for two terrestrial salamander species: Jordan's salamander (*Plethodon jordani*; PJ) and members of the *Plethodon glutinosus* (PG) complex including *Plethodon glutinosus* and *Plethodon oconaluftee*. In their cursory analysis, MacKenzie *et al.* (2004b) were interested in determining whether there is any evidence that the two species exhibit strong co-occurrence patterns, after allowing for any elevational gradient in occupancy probabilities. They utilized the reparameterized version of the model (i.e., in terms of species interaction factors) and assumed (based on biology and the practicalities of the sampling) that throughout the study the species were detected independently when both were present at a site (i.e., $\delta = 1$).

MacKenzie *et al.* (2004b) conducted their analysis in two parts to illustrate how the non-inclusion of covariates that may be related to habitat preferences (for example) of the individual species could influence one's inference about species co-occurrence. First, they considered a simple set of models that assumed the occupancy and detection probabilities for both species were unaffected by elevation. The only model with a substantial level of support (based upon AIC) indicated that the two species avoided each other, $\hat{\varphi}(\widehat{SE}) = 0.67$ (0.11), and that detection probabilities were different when both species were present rather than just one of the species (for PG: $\hat{p} = 0.54$ and $\hat{r} = 0.48$; for PJ: $\hat{p} = 0.91$ and $\hat{r} = 0.55$). In the second part of their analysis they considered models that utilized the elevation covariate. Allowing occupancy probabilities to vary with elevation resulted in numerical problems (technically, a lack of convergence) when the SIF, φ, was also included in the model (a point we return to below); hence it does not appear in their model selection summary. However, once site elevation was incorporated as a covariate for the occupancy and detection probabilities, the top-ranked model from the first stage of the analysis had essentially no support in comparison. That is, including site elevation as a covariate substantially improved the models. The summaries of the model selection procedure for both parts of the analysis have been amalgamated into Table 8.6. Because of the convergence problems, MacKenzie *et al.* (2004b) could not make definitive statements about the level of co-occurrence between the species once elevation was included as a covariate for occupancy, but comparison of some of the lower-ranked models in Table 8.6 certainly indicates that the apparent avoidance of the two species detected in the initial modeling may have been due to differences in elevational preferences. For example, the models $\psi(S)\varphi(\cdot)r(S \times E)p(S \times E)$ (where S and E denote species and elevation, respectively) and $\psi(S)r(S \times E)p(S \times E)$ (which assumes $\varphi = 1$) have a similar level of support, although one would expect model $\psi(S)\varphi(\cdot)r(S \times E)p(S \times E)$ to have much greater support given the results of the initial analysis if much of the apparent co-occurrence pattern could *not* be explained by elevational preferences.

TABLE 8.6 Summary of Model Fit and Selection Statistics for the Salamander Data Considered by MacKenzie *et al.* (2004b)

Model	ΔAIC	w	NPar	$-2l$
$\psi(S \times E)r(S \times E)p(S \times E)$	0.0	0.76	12	617.3
$\psi(S \times E)r(S \times E)p(S)$	2.3	0.24	10	623.6
$\psi(S \times E)r(S)p(S \times E)$	38.8	0.00	10	660.1
$\psi(S \times E)p(S \times E)$	50.2	0.00	8	675.6
$\psi(S \times E)r(S)p(S)$	50.8	0.00	8	676.1
$\psi(S)r(S \times E)p(S \times E)$	51.8	0.00	10	673.2
$\psi(S)\varphi(\cdot)r(S \times E)p(S \times E)$	52.5	0.00	11	671.8
$\psi(S)\varphi(\cdot)r(S)p(S)$	109.3	0.00	7	736.6
$\psi(S)r(S)p(S)$	117.6	0.00	6	747.0
$\psi(S)\varphi(\cdot)p(S)$	130.0	0.00	5	761.4
$\psi(S)p(S)$	142.7	0.00	4	776.0

ΔAIC is the absolute difference in AIC values relative to the model with the smallest AIC, w is the AIC model weight, NPar is the number of estimated parameters in the model, and $-2l$ is twice the negative log-likelihood value. The terms in parentheses represent the factors in the model for the respective parameter, with "S" denoting that species has been used as a factor, "E" indicating use of elevation as a factor, and "$\cdot$" indicating a parameter set equal across species and elevation. Absence of the φ parameter in the model notation implies $\varphi(\cdot) = 1$, and absence of $r(S)$ implies $r(S) = p(S)$. (Modified from Tables 4 and 5 of MacKenzie *et al.* 2004b).

Returning to the convergence problems experienced by MacKenzie *et al.* (2004b), consider Figure 8.2. This plot indicates the allowable lower and upper bounds for φ on the natural log scale (i.e., $\ln(\varphi)$) based upon parameter estimates from the top-ranked model in Table 8.6. The only straight line that could be drawn on the plot that stays fully between the indicated limits is $\ln(\varphi) = 0$, implying that the only allowable value for φ across the range of observed elevations was 1. Subsequent to the MacKenzie *et al.* (2004b) paper, we have investigated other models in which φ is a curved or stepwise function with respect to elevation, but have found none (biologically reasonable or otherwise) that have a similar level of support to the current top-ranked model. Redefining the SIF in terms of the odds ratio may resolve this issue to some degree, as there is no limit on the range of allowable values (other than it must be a value greater than 0).

8.5. STUDY DESIGN ISSUES

Due to the newness of this approach, we have not yet had the opportunity to give in-depth consideration to study design issues as they pertain to investigating relationships between species, nor do we know of any relevant work

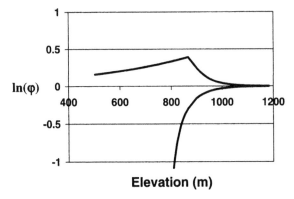

FIGURE 8.2 Allowable limits for the species interaction factor ϕ (on the natural log scale) as a function of elevation based upon the top-ranked model of the terrestrial salamander example, which incorporated elevation as a covariate for occupancy probabilities.

from other research groups. Many of the issues raised in Chapter 6 for single-species studies are also pertinent to the multi-species case, such as defining and selecting sites, allocating effort, and so forth.

One aspect of study design that deserves special attention in the multi-species case is the appropriate definition of a "season." In the single-species case the main considerations were: (1) the time period over which it is reasonable to consider the occupancy state of sites as unchanged, or over which changes occur completely at random; and (2) whether "occupancy" or "use" provides the most relevant information with respect to the study objectives. Recall that a season represents a snapshot of the population at a particular point in time, and inference about the species is based upon this snapshot. The equivalent of season length in photography is shutter speed. If the shutter speed is too fast, then photographs tend to be dark and are not an accurate representation of the subject, while if the shutter speed is too slow, the subject may move, blurring the image and again resulting in a poor image. The same applies in occupancy studies; too short a season may result in insufficient information about the system, while too long a season may result in a distorted view of the system. In photography, different shutter speeds can be used for different situations depending upon the desired "shot," and the same applies here; different season lengths may be appropriate to achieve some objectives but not others. These points hold true for multi-species studies, with the added consideration that interpretation of "co-occurrence" may also be influenced by how a "season" is defined. For example, suppose that over a shorter time frame (e.g., a week) two species tend to co-occur very rarely, as they tend to be competitors for the same resources, or perhaps they are predator and prey species.

However, over a longer time frame (e.g., three months) they tend to co-occur more frequently, as they prefer similar habitats or are highly mobile species. Inferences about how the species co-occur may be very different depending upon whether a short or long exposure time was used to capture a snapshot of the population. When determining the appropriate length of time for a "season" (i.e., the period over which the repeat surveys will be conducted), it is very important to pay careful attention to the known biology of the species and the nature of the co-occurrence patterns that are of interest according to the study objective.

A similar argument also applies when defining a "site." Here the issue is the spatial scale of co-occurrence rather than a temporal one. Too small and the resolution may be inappropriate; too large and the resolution blurs and many species will appear to co-occur. Again, as part of the study's objective, careful consideration needs to be paid to the spatial scale of the co-occurrence questions that are of interest.

8.7. EXTENSION TO MULTIPLE SEASONS

The modeling approach of MacKenzie *et al.* (2004b) is useful for examining patterns of how species co-occur within a general region at a specific point in time. As in the single-species case, it provides no insight on the interactive processes between the species that may have taken place and produced the observed pattern. Only with the system being observed at systematic points in time can some understanding of the underlying processes be gained. In this section, we suggest how the MacKenzie *et al.* (2004b) single-season model can be extended to multiple seasons. This is a recent development that we have not had the opportunity to apply to a specific situation as yet, but we know of many applications in which such modeling would be highly relevant, particularly for assessing the impact of invasive species (e.g., the previously mentioned northern spotted and barred owls; black ducks and mallards in the northeastern United States).

Here we assume that data have been collected from the same sites for multiple (T) seasons, and within each season multiple surveys are conducted to detect a number of species. We develop the modeling approach for only two species, but again suggest it can be extended to a greater number of species if desired. By considering that each site may be in one of the four mutually exclusive states described above (occupied by both species, by species A only, by species B only, or occupied by neither species), the dynamic processes of change in the occupancy state of a site between seasons can be simply represented by defining a transition probability matrix (ϕ_t), much like the single-species unconditional explicit dynamics model (MacKenzie *et al.* 2003,

TABLE 8.7 Notation for the Dynamic Parameters Associated with Changes in the Occupancy Status of Sites for Multiple Species

Parameter	Description
ε_t^{AB}	probability both species A and B go locally extinct between seasons t and $t + 1$
ε_t^A	probability species A goes locally extinct between seasons t and $t + 1$, given both species present in season t
ε_t^B	probability species B goes locally extinct between seasons t and $t + 1$, given both species present in season t
υ_t^A	probability species A goes locally extinct between seasons t and $t + 1$, given species B was absent in seasons t and $t + 1$
υ_t^B	probability species B goes locally extinct between seasons t and $t + 1$, given species A was absent in seasons t and $t + 1$
γ_t^{AB}	probability both species A and B colonize a site between seasons t and $t + 1$
γ_t^A	probability species A colonizes a site between seasons t and $t + 1$, given both species absent in season t
γ_t^B	probability species B colonizes a site between seasons t and $t + 1$, given both species absent in season t
η_t^A	probability species A colonizes a site between seasons t and $t + 1$, given species B was present in seasons t and $t + 1$
η_t^B	probability species B colonizes a site between seasons t and $t + 1$, given species A was present in seasons t and $t + 1$
ω_t^{AB}	species A is replaced by B between seasons t and $t + 1$
ω_t^{BA}	species B is replaced by A between seasons t and $t + 1$

Chapter 7). Rows of ϕ_t denote the states of sites in season t, and columns represent the states of sites in season $t + 1$. Each element within ϕ_t therefore represents the probability of a site changing from one occupancy state to another between seasons. A general representation of ϕ_t would be:

$$\phi_t = \begin{bmatrix} AB \to AB & AB \to A & AB \to B & AB \to U \\ A \to AB & A \to A & A \to B & A \to U \\ B \to AB & B \to A & B \to B & B \to U \\ U \to AB & U \to A & U \to B & U \to U \end{bmatrix}$$

where $X \to Y$ denotes the probability of transitioning from occupancy state X in season t to state Y in season $t + 1$. The state U indicates that the site is unoccupied by both species. Importantly, the elements of each row must sum to 1, as a site must transition to one of the four states.

In the single-species case, the dynamic processes of site occupancy are colonization and local extinction. In the multiple-species situation, the dynamic processes are not only colonization and local extinction of either species, but also the process of one species displacing the other. More generally, the vital rates (colonization and extinction) for one species may be influenced by the current occupancy status of the other species. In Table 8.7 we define parameters that could be used to capture these dynamic processes of multiple-species site occupancy. Using these parameters, the transition probability matrix would be:

$$\phi_t = \begin{bmatrix} 1-\varepsilon_t^A-\varepsilon_t^B+\varepsilon_t^{AB} & \varepsilon_t^B-\varepsilon_t^{AB} & \varepsilon_t^A-\varepsilon_t^{AB} & \varepsilon_t^{AB} \\ \eta_t^B & 1-\omega_t^{AB}-\eta_t^B-\upsilon_t^A & \omega_t^{AB} & \upsilon_t^A \\ \eta_t^A & \omega_t^{BA} & 1-\omega_t^{BA}-\eta_t^A-\upsilon_t^B & \upsilon_t^B \\ \gamma_t^{AB} & \gamma_t^A-\gamma_t^{AB} & \gamma_t^B-\gamma_t^{AB} & 1-\gamma_t^A-\gamma_t^B+\gamma_t^{AB} \end{bmatrix}.$$

Note the flexible nature of this transition probability matrix. It allows, for example, the probability of a site being colonized by a species to depend upon whether the other species is already present at the site (i.e., the probability an unoccupied site is colonized only by species B is $\gamma_t^B-\gamma_t^{AB}$, while the probability of species B colonizing a site already occupied by species A, without displacing species A, is η_t^A). In many situations, the parameters ω_t^{AB} and ω_t^{BA} may be of particular interest, as they represent the probability that one species is displaced by the other between successive seasons (e.g., where an endemic and an invasive species may compete for resources).

To incorporate detection probabilities into the model, for each season a detection probability vector can be defined that denotes the probability of observing a given sequence of detections and nondetections for both species within that season, conditional upon the occupancy state of the site—in other words, the same approach that was used for the above single-season model.

The probability of observing any pair of detection histories for two species could be generally expressed as:

$$\Pr(h_i^A, h_i^B) = \phi_0 \prod_{t=1}^{T-1} D\left(p_t^{\{h_{i,t}^A\},\{h_{i,t}^B\}}\right) \phi_t p_T^{\{h_{i,T}^A\},\{h_{i,T}^B\}},$$

which has the same form as the equation to calculate the probability of observing a given detection history for multiple seasons in the single-species case (Chapter 7), except the number of possible occupancy states has increased from two to four. Note that ϕ_0 denotes initial occupancy states, as defined for the single-season model of Section 8.2. The model likelihood would be calculated in the same manner as for the other models considered in this book, that is:

$$L(\phi, p|h_1^A, h_1^B, \ldots, h_s^A, h_s^B) = \prod_{i=1}^{s} \Pr(h_i^A, h_i^B).$$

This basic framework could be easily extended to incorporate covariate information and missing observations, as was done previously in this chapter and in Chapters 4 and 7. Also, various biological hypotheses could be formulated and translated into the model by applying various constraints to the model parameters, that is, $\psi^{AB} = \psi^A \times \psi^B$, or $\omega_t^{AB} = \varepsilon_t^A \times \gamma_t^B$. As with the single-season modeling, we are considering different parameterizations that basically reflect the difference in colonization or extinction probabilities of one species depending on occupancy status of the other species.

8.7. DISCUSSION

In this chapter we have presented techniques that permit investigation of relationships between multiple species while accounting for the imperfect detection of the species. These models are very flexible and should be suitable for a wide range of applications. Not only can these models be used to determine the level of support for various biological hypotheses, but by accounting for detection probability, the absolute levels of co-occurrence can be estimated or modeled directly as well. We believe such approaches will provide more reliable inferences about the patterns and processes of species co-occurrence than other methods that have been used to date, particularly when used in combination with a sound experimental design (Chapter 1).

The multiple-season model of Section 8.7 should be particularly relevant to researchers working with species of high conservation importance that are under threat from invasive species that prefer similar habitats or occupy a similar niche. The η and ω parameters represent the probability that a second species colonizes a site already occupied by the first and the probability of the first species being replaced by the second, respectively. We suggest that these quantities will often be of prime interest when investigating relationships between native and invasive species. These models also provide the inferential basis for theoretical patch-based models that use multi-species occupancy patterns as state variables. For example, such models have been developed to investigate predator-mediated coexistence of competing species (Caswell 1978) and the interaction of competition, disturbance, and heterogeneity in determining structure and dynamics of multi-species communities (Caswell and Cohen 1991a,b). The modeling of Section 8.7 is ideal for estimation of the various transition probabilities required for such Markov process modeling.

CHAPTER 9

Occupancy in Community-level Studies

In the previous chapter we considered modeling approaches that could be used to investigate interspecific relationships among multiple species. A different kind of multi-species study focuses on the state variable of species richness, defined as the number of species (either in total or within a predetermined group) within a predefined area of interest (Chapter 1). Investigations focus on changes in this state variable as a function of local rates of species colonization and extinction, without regard to how species directly interrelate. Traditionally, species accumulation curves have been used to estimate species richness, where the number of species encountered is plotted as a function of search effort within the sites (see reviews in Soberon and Llorente 1993; Colwell and Coddington 1994; Flather 1996). The intent is that the accumulation curve will flatten off or asymptote to the total number of species present at the site with increasing search effort. Another approach that has seen recent application is to use closed-population capture-recapture models to estimate the number of species not encountered at a site, based upon the species that were encountered at least once (e.g., Burnham and Overton 1979; Bunge and Fitzpatrick 1993; Williams *et al.* 2002).

Here we suggest that many of the methods described in Chapters 4–7 could be applied to community-level studies, as often the data collected in such studies are the presence/absence (or, more correctly, the detection/nondetection) of multiple species. In some instances, the basic sampling situation is different from that considered earlier and requires model parameters to be interpreted differently, but the practical application of the modeling is very similar (much like the application of capture-recapture methods to community-level studies). In other instances, we simply view the multi-species studies as the combination of a number of single-species studies, all conducted at the same set of sites. We structure our thoughts by considering two different situations. The first is when sampling of the community takes place at only a single site (or a small number of sites) within a relatively small area of interest. The second situation is when the community is considered at a larger scale with sampling taking place at many sites, as in the single-species situations considered earlier. We have structured the chapter in this manner because the same basic models are used, but are applied differently in each situation.

9.1. INVESTIGATING THE COMMUNITY AT A SINGLE SITE

In this section we consider a basic sampling situation in which surveys are conducted for multiple species at a single site. As in the single-species case, species are not detected perfectly; hence, repeated surveys are required to estimate detection probability. Each species may be detected or not detected during a survey; hence, a detection history can be constructed denoting the sequence of detections and nondetections for each species. The period of repeated surveying is again referred to as a season, and the assumption is made that the occupancy status of the site for each species does not change during the season, or that changes occur completely at random (i.e., the members of the local species pool present at the site are constant during a season). Repeated surveys for the species may be conducted either temporally (e.g., daily) or spatially (e.g., on small plots within the "site"). Once again, the exact nature of the repeat surveys and definition of a "season" influence how model parameters should be interpreted. The site may be surveyed for multiple seasons (e.g., years), where changes in the membership of the local species pool present at the site may occur between seasons.

When data are collected in this form, Williams *et al.* (2002: Chapter 20) provide a concise summary of methods that could be used to estimate community-related parameters using capture-recapture techniques. They utilize an analogy between the above sampling situation and capture-recapture studies in which the detection/nondetection of species is analogous to the detection/nondetection of individual animals. The number of species present at the site (species richness) can then be estimated using capture-recapture methods that are appropriate for estimating the number of individual animals in a study area. That is, the total number of species that may be present at a site is estimated from data collected on the species that were detected at least once. Williams *et al.* (2002) give some guidance on the types of estimation methods they consider to be most appropriate depending upon whether repeated surveys have been conducted temporally or spatially; however, they note that generally they would expect that heterogeneity in the detection probabilities of different species should be accounted for. If data are collected over multiple seasons, then estimates of community parameters such as rates of turnover and extinction can be obtained (Nichols *et al.* 1998a), and these estimates can be used to investigate hypotheses about community dynamics (e.g., Boulinier *et al.* 1998b, 2001; Doherty *et al.* 2003a,b).

Species richness at a particular location is likely to be determined not only by local ecological conditions but also by the regional species pool (e.g., see Cornell and Lawton 1992; Cornell 1993; Karlson and Cornell 1998; Cam *et al.* 2000a). In order to focus on local ecological determinants of richness, Cam *et al.* (2000a) defined *relative species richness* as the ratio of richness at a site to the number of species in the regional pool. This quantity can be viewed as a measure of community completeness and should be greater at sites with favorable ecological conditions than at sites with unfavorable conditions (e.g., as might be caused by urbanization). Cam *et al.* (2000a) estimated relative richness at a site as the ratio of estimated richness at the site to the known or estimated size of the local species pool.

Occupancy modeling provides a means of modeling and estimating relative richness directly, rather than via a two-step approach, as in Cam *et al.* (2000a). Note that the techniques used by Cam *et al.* (2000a) to estimate species richness at a site (see Williams *et al.* 2002; and references therein) place no constraint upon the total number of species that may reside in the community. As noted above, however, in many situations it is reasonable to suggest that such a limit is known. This may arise, for example, when researchers have precompiled a list of species in which they are interested. In locations that have been studied for a long period of time, investigators will frequently have a list of the species corresponding to the species pool (e.g., Cam *et al.* 2000a). Similar to Williams *et al.* (2002), an analogy can be drawn between estimating the proportion of sites occupied by a single species and estimating the proportion of species on a list that occupy a single site (i.e., each species is considered as a "site" in the context of Chapters 4–7). When the list is the regional species pool, then the proportional occupancy of a site (probability that a member of the pool is present at the site) is exactly the relative richness parameter of Cam *et al.* (2000a). The occupancy approach and methods detailed in Chapters 4–7 permit the direct estimation and modeling of this parameter, ψ, as well as of the rate parameters that cause the relative species richness and species composition to change over time. Importantly, as covariate information can be easily incorporated into the above methods, information related to the individual species (i.e., characteristics of different species such as body size or degree of specialization) can be used to model the dynamic community parameters, even for those species on the list that are never detected at the site. This cannot be done using capture-recapture techniques, as without specification of a species list, covariate information about species that are never detected is unknown.

FRACTION OF SPECIES PRESENT IN A SINGLE SEASON

Suppose a list of s species is composed, and interest lies in what fraction of the species on the list is present at a site. Repeated surveys are conducted for the s species, and a detection history for each of the species can be constructed (e.g., Table 9.1). Because species are detected imperfectly, some species that were not detected at the site may have in fact been present (i.e., a false absence), while others could be genuinely absent from the site (i.e., not part of the local community during that season). The analogy between the estimation problem here and that of Chapters 4–5 is obvious, and as such the same estimation techniques can be used. The methods of Chapter 5 may be particularly useful to account for the heterogeneity in detection probabilities among species, but the ability to incorporate covariate information for all species on

TABLE 9.1 Example of a Partial Species List for 10 Bird Species with the Associated Detection History (h_i) for Each Species from Four Surveys Conducted in the Mid-Atlantic Region of North America. Examples of Species-specific Covariates Are Also Indicated

Species	h_i	Songbird	Body Size
Northern parula, *Parula Americana*	1001	yes	small
American robin, *Turdus migratorius*	0111	yes	large
Indigo bunting, *Passerina cyanea*	0100	yes	medium
Song sparrow, *Melospiza melodia*	0011	yes	medium
Northern mockingbird, *Mimus polyglottos*	0010	yes	large
Blue-winged warbler, *Vermivora pinus*	0000	yes	small
Northern cardinal, *Cardinalis cardinalis*	0000	yes	large
Ruby-throated hummingbird, *Archilochus colubris*	1000	no	small
Mourning dove, *Zenaida macroura*	0101	no	large
Hairy woodpecker, *Picoides villosus*	0000	no	large

the list may mitigate the usual problems caused by heterogeneity (i.e., much of the heterogeneity among species may be modeled in terms of covariates). As inference is to be made only about those species that appear on the list, the species list represents the entire population of interest. To accurately represent the uncertainty in the estimated fraction or proportion of species present, the finite population methods of Section 4.5 are particularly applicable. However, if interest is in the underlying probability of a species on the list being present at the location, then it is not necessary to correct for the finite population and the methods of Section 4.4 can be used.

CHANGES IN THE FRACTION OF SPECIES PRESENT OVER TIME

The analogy with the single-species case can be continued to investigate changes in the members of the species pool present at a site over time using the unconditional explicit dynamics models of MacKenzie et al. (2003: Chapter 7). That is, changes to the species number and composition through the processes of colonization and local extinction can occur between seasons, but not within seasons. Now the probability of colonization (γ_t) relates to the probability that a species absent from the site in season t occupies the site in season $t + 1$. Similarly, local extinction probability (ε_t) is the probability that a species present at the site in season t is absent in season $t + 1$. Reparameterized versions of this model may also be used when analogous quantities are of interest (e.g., the rate of change in the fraction of species present at the site). This kind of dynamic model, in which members of the species pool join and leave the community at a local site, is well known to theoretical ecology (e.g., MacArthur and Wilson 1963, 1967; MacArthur 1972; Ricklefs and Schluter 1993; Boulinier et al. 2001). Methods such as those suggested here should permit formal inference about such topics of theoretical interest (e.g., MacArthur and Wilson 1963, 1967; Boulinier et al. 2001) as the existence of dynamic equilibria (e.g., using the methods of Chapter 7).

Incorporating covariate information about the species on the list, even for those species never detected, may be very useful for making reliable inference about the processes of change within the local species pool. For example, native species may have lower colonization and higher local extinction probabilities than invasive exotic species, or species with large body size may have higher local extinction probabilities than small species. Individual species characteristics cannot be incorporated into most of the estimation and modeling methods described by Williams et al. (2002), and ability to use species covariates represents an important advantage of the occupancy estimation approach. As suggested in Chapter 7, heterogeneity in the model parameters not accounted for by covariates can be accommodated by the use of "random effects" models.

9.2. INVESTIGATING THE COMMUNITY AT MULTIPLE SITES

Species richness often may be of interest at more than a single site. The methods outlined above could clearly be extended to more than one site, where each site may have the same or different species lists (pools). An alternative to estimating the fraction of species present for each of a large number of sites is to again estimate the proportion of sites occupied by each species (i.e., the initial context considered in Chapters 4–7), but to model all species simultaneously. Inference about the community can then be made based upon the joint modeling of the species. "Species richness" could then be defined as the number of species present at a single site, or as the number of species present in the community in the larger area from which the sampled sites were selected. An advantage of modeling all species simultaneously is that a common parameter may be shared by different species. This kind of multi-species modeling has several advantages. It should yield gains in precision, such that sharing parameters may allow more robust inferences to be made about some species than when each species is considered individually (e.g., MacKenzie et al. 2005). In other situations, interesting biological questions involve possible similarities in species responses to environmental and habitat characteristics. For example, site occupancy by species within the same guild (a group of species believed to exhibit similar characteristics, such as foraging habits, nesting habitat, etc.) would be expected to exhibit similar relationships to habitat covariates. Guild membership can be hypothesized a priori, and hypotheses can be tested about similar relationships among species within, but not between, groups. Another advantage of this multi-species modeling approach is that now species-specific detection probabilities can be directly estimated, unlike when using the approach suggested in the previous section on single-site analyses.

In this section we outline our ideas on how the joint modeling of multiple species at a large number of sites could be used to make inferences at the community level. First, we discuss single-season studies with two different kinds of objectives: (1) investigations directed at modeling probabilities of occupancy and/or detection as functions of species- and site-specific covariates; and (2) the estimation of species richness, both at sampled locations and for larger areas from which these locations are selected. Then, we briefly discuss occupancy studies for multiple species over multiple seasons.

SINGLE-SEASON STUDIES: MODELING OCCUPANCY AND DETECTION

The joint modeling of the data from each site may be used to address interesting biological hypotheses, such as whether groups of similar species at different sites have a similar occupancy probability. Discussion of covariate modeling for the single-site investigations described in Section 9.1 emphasized species-specific covariates such as body size, specialization, and the exotic versus native dichotomy. In this section, we can consider analysis of multiple sites with the same species pool. Such analyses would be effectively controlling for influences of the regional species pool, and covariate modeling would investigate the role of local site characteristics (e.g., habitat, disturbance) in determining local species richness and dynamics. Because species identities are retained in these analyses, models can incorporate covariates associated with individual species as well as those associated with local sites.

We consider the same basic sampling framework as that assumed in Chapters 4 and 5; repeated detection/nondetection surveys are conducted at s sites within a single season. However, data now are collected on M species rather than just a single species. This type of data could be used to assess interspecific relationships between species (i.e., Chapter 8), but here we assume species co-occur independently. Choice of which M species to model will depend upon study objectives, but it may be: (1) a small number of indicator species that were detected at least at one of the sites; (2) all species that were detected at least once; or (3) a list of species defined *a priori* that includes some species never detected at any of the sites (note that in this case the detection probability for species never detected must be assumed equal to that of other species that were detected, or related by some form of detection function). Modeling of the M species can be conducted as in Chapters 4 and 5, including investigation of the effect of different covariates on different species.

Generalizing the notation of Chapters 4 and 5, let ψ_{mi} be the probability that species m is present at site i, and p_{mij} be the probability of detecting species m in the jth survey of site i (given presence of species m). The most general models will contain different detection and occupancy parameters for all M species. The effect of different covariates may also be different for different species. However, reduced-parameter models could be considered where different species share common parameters. For example, many frog and toad species become more or less detectable with changes in temperature due to behavioral changes. Rather than estimating a different "temperature effect" for each species (i.e., logit(p_{mij}) = $\alpha_m + \beta_m \times T_{ij}$, where T_{ij} is the temperature recorded during survey j of site i, and β_m is the effect of temperature on detection probability for species m, which could be denoted as model p (*Species*

$\times$ *Temperature*)), it may be reasonable to consider a model in which the effect of temperature is the same for all, or a subset, of species (i.e., logit(p_{mij}) = $\alpha_m + \beta \times T_{ij}$, which could be denoted as model p (*Species* + *Temperature*)). Under this model, each species may have a different average detection probability, but changes in species-specific detection with temperature occur in parallel, reflecting the similar effect of temperature. Sharing parameters among species in this manner is effectively a form of aggregating or pooling data, and in this situation, model selection techniques can be used to indicate the level of aggregation among species that is best supported by the data (MacKenzie et al. 2005).

The effect of site-specific covariates (e.g., habitat type, patch size, etc.) on occupancy probabilities can also be investigated for each species, and it may be reasonable to hypothesize that the effect of some covariates is similar across a range of species. For example, a single "habitat effect" could be estimated that applies to the occupancy of all species in a group. Thus, we might investigate an additive model for occupancy (e.g., ψ(*Species* + *Habitat*)), with occupancy varying across species, but with all species (at least within a group of interest) otherwise showing the same relationship to a habitat covariate (i.e., the same slope parameter for the occupancy by habitat relationship). Note that incorporating site-specific covariates in the analysis of multi-species data in this manner is one approach for examining factors that may affect local species richness (e.g., Boulinier et al. 1998b, 2001). Across the landscape, different species may have different habitat preferences. By modeling these preferences for multiple species within that landscape, areas that are preferred by a greater or lesser number of species could be identified. That is, the expected number of species present at site i would be $\sum_{m=1}^{M} \psi_{mi}$ or, using the results from Chapter 4, the estimated number given the nondetection of some species at site i would be $n_i + \sum_{m=n+1}^{M} \hat{\psi}_{condl,mi}$ where n_i is the number of species detected at site i and $\hat{\psi}_{condl.mi}$ is the estimated probability of occupancy for species m at site i, conditional upon its nondetection.

SINGLE-SEASON STUDIES: SPECIES RICHNESS ESTIMATION

Dorazio and Royle (2005) used this type of joint modeling approach when they developed a model for estimating species richness and related community parameters. In this section, we briefly review that model. Their model is based on the type of design considered previously, which is a modification of conventional designs for estimating species richness (e.g., Bunge and

Fitzpatrick 1993; Boulinier et $al.$ 1998a; Williams et $al.$ 2002), where s sites are sampled within a region containing N distinct species. Here we detail their model, assuming each site is surveyed the same number of times (K), but this is not necessary.

A common objective of community modeling efforts is estimation of the parameter N (i.e., total number of species in the region). However, alternative summaries of the community under study might also be of interest. For example, we might wish to estimate the number of species at a single site or at a group of sites, perhaps even the collection of sampled sites. This last objective might be of interest when the samples are not randomly selected or representative. In this case, a purported estimate of N may be biased, as some species within the region may have zero probability of occurring at one of the s sites that were sampled within the region. Hence, because of the nonrandom sampling, an estimator of N will be inconsistent with our definition of N above. However, the number of species present at the collection of sampled sites (M; which will be less than or equal to N) is consistent and can be reasonably estimated regardless of the design imposed on sample site selection. More complex summaries of community structure might also be of interest, such as metrics describing the similarity of the communities among sites, or a species-area curve. Finally, there may be some interest in estimating these quantities in a manner that takes into account site-specific habitat differences and their effects on species occurrence. While Dorazio and Royle (2005) did not consider this problem explicitly, this extension can be considered within the modeling framework they proposed, using the approaches described in the previous section.

The data arising under the design described above consist of the site- and species-specific detection histories. By assuming the probability of detecting each species was constant at each site, the detection histories can be summarized by the detection frequencies y_{mi}, the number of times that species $m = 1, 2, \ldots, N$ was detected in K visits to site $i = 1, 2, \ldots, s$ (although note the methods of Dorazio and Royle (2005) could be extended to allow nonconstant detection probabilities). In general, not all species in the region will be detected in the surveys. Let x (which will be less than or equal to N) be the number of distinct species detected during the K visits to all s sites. Let $\mathbf{y}_m = (y_{m1}, y_{m2}, \ldots, y_{ms})$ denote the vector of the s site-specific detection frequencies of species m. For our purposes, it is convenient to order the observation vectors such that $m = 1, 2, \ldots, x$ correspond to the data for the x observed species, and $m = x + 1, x + 2, \ldots, N$ correspond to the unobserved (i.e., "all zero") species. It is also useful to introduce an $N \times s$ matrix of binary indicator variables, $\mathbf{Z}$, the elements of which denote the presence ($z_{mi} = 1$) or absence ($z_{mi} = 0$) of species m at site i. These are simply the latent (unobserved) occupancy state variables (as discussed in Chapter 4) for the m species. As in the single-species case, Z is only partially observed in the sense that if species m is detected at site i (i.e., $y_{mi} > 0$), then the species must be present and $z_{mi} = 1$, but if $y_{mi} = 0$ (i.e., species m is not detected at site i), two mutually exclusive possibilities exist for the value of z_{mi}: (1) species m is present at site i ($z_{mi} = 1$) but went undetected, or (2) species m is absent from site i ($z_{mi} = 0$).

As in the previous section, let ψ_m denote the probability of occurrence of species m. The z_{mi}'s are assumed to be independent Bernoulli random variables with density function $f(z_{mi}|\psi_m) = \psi_m^{z_{mi}}(1 - \psi_m)^{1-z_{mi}}$. Conditional on species m being present ($z_{mi} = 1$), the number of detections of species m at any site (y_{mi}) is assumed to be a random value from a binomial distribution with index K and parameter p_m. Conversely, if species m is absent from site i, then y_{mi} is assumed to equal zero with probability 1. These considerations define the joint distribution of the data y_{mi} and the occupancy state variables z_{mi}. Because the occupancy states, z_{mi}, are only partially observed, it is convenient in some cases to remove them from the likelihood by integration, yielding the familiar zero-inflated binomial density seen previously (Chapters 4 and 5), which we will denote by $g(y_{mi}|\psi_m,p_m)$.

Dorazio and Royle (2005) extended the single-species occupancy model to a community of species, allowing for species-specific differences in rates of occurrence and detection, by specifying normal distributions on the logits of the two parameters. That is:

$$\text{logit}(\psi_m) = \mu_\psi + u_m,$$

and:

$$\text{logit}(p_m) = \mu_p + v_m,$$

where u_m and v_m are species-specific random effects assumed to be normally distributed with mean 0 and variances σ_u^2 and σ_v^2, respectively. This is a form of aggregation (as discussed in the previous section) in which, rather than estimating independent species-specific occurrence and detection probabilities, a distributional form has been assumed. Note that Dorazio and Royle (2005) provided an argument that u_m and v_m should be positively correlated, and their model allowed for an additional parameter $\sigma_{uv} = Cov(u_m,v_m)$. Note also that the detection model here is that used in the logistic-normal model of heterogeneous detection of species (Coull and Agresti 1999), described in Chapter 5, except here it is applied to allow for heterogeneity among species rather than among sites for a single species.

As done for the heterogeneous detection probability models described in Chapter 5, estimators of model parameters based on the marginal density of the observed data can be developed. Assuming that the s observations of each species are independent, the marginal probability of the observation vector $\mathbf{y}_m$, say, $g(y_i|\mu_\psi,\mu_p,\sigma_u^2,\sigma_v^2,\sigma_{uv})$, may be obtained by integrating the zero-inflated likelihood for species m, $g(y_{mi}|\psi_m, p_m)$, over the bivariate joint distribution of (u_m, v_m). The integration may be done numerically, although, as noted by Dorazio and Royle (2005), this can be computationally intensive to implement. More importantly, estimates of the u_m parameters (related to the occupancy probability of species m) and their uncertainties are necessary for estimating summaries of community structure [e.g., Eq. (9.2)], and so it is disadvantageous to remove them from the likelihood.

To resolve such estimation problems, Dorazio and Royle (2005) developed a Bayesian framework for analysis of the model based on the likelihood conditioned on the species that were detected at least once. They note that the likelihood may be factored into two components (Sanathanan 1972), one for the detections of the observed species, conditional on x, and a second for the binomial distribution of x given the unknown community size N. Bayesian analysis based on the conditional likelihood avoids the tedious integration needed to estimate N directly by maximum likelihood, retains the species-specific parameters in the model, and yields direct estimates of them. In addition, an estimate of species richness can be computed as a function of model parameters. That is, given the estimates $\hat{\mu}_\psi, \hat{\mu}_p, \hat{\sigma}_u^2, \hat{\sigma}_v^2$, and $\hat{\sigma}_{uv}$:

$$\hat{N} = \frac{x}{1 - g(0|\hat{\mu}_\psi, \hat{\mu}_p, \hat{\sigma}_u^2, \hat{\sigma}_v^2, \hat{\sigma}_{uv})}, \tag{9.1}$$

where the term in the denominator is the probability of detecting a species in the community. Within the MCMC algorithm based on the conditional likelihood, $\hat{N}$ is just a function of model parameters, and MCMC samples of $\hat{N}$ can be obtained at each iteration of the algorithm by substituting the current values of each model parameter in Eq. (9.1).

The motivation for introducing the latent (i.e., unobserved) occupancy state variables (the z_{mi}'s) is that estimators of many ecologically important quantities are naturally expressed as functions of them. Note that we adopted a similar construction for estimating the number of occupied sites in a finite population of sites (Chapter 4, Section 4.5). If the z_{mi} were fully observed (i.e., if $p = 1$ for all species and sites), then the presence or absence of each species would be known exactly and certain quantities of interest could be computed directly. For example, the number of species occurring at site i (a quantity that is conceptually similar to the number of occupied sample sites considered in Chapter 4) is $N_i = \sum_{m=1}^{N} z_{mi}$. As another example, the number of species in common at two sites, say i and l, is $N_{il} = \sum_{m=1}^{N} z_{mi} z_{ml}$. Neither N nor all of the z_{mi}'s are known. However, estimators of such quantities may be obtained naturally from these expressions. For example:

$$\hat{N}_i = \sum_{m=1}^{x} [z_{mi} \times I(y_{mi} > 0) + \hat{z}_{mi} \times I(y_{mi} = 0)] + \sum_{m=x+1}^{\hat{N}} \hat{z}_{mj}, \qquad (9.2)$$

which is the number of species actually detected at site i, plus the expected occurrence at site i of species that were detected in the study but not at site i, plus the expected number of occurring species that were not detected at any of the s sites in the sample (recall the indicator variables $I(E) = 1$ if the expression E is true, or 0 if the expression E is false). Reasonable estimators of $\hat{z}_{mi}$ are their expectations conditional on the data, and these involve the species-specific occurrence probabilities, ψ_m, and detection probabilities, p_m. The Bayesian implementation directly yields estimates of the z_{mi}'s and species-specific occurrence probabilities. Thus, a sample from the posterior distribution of N_i can be obtained by substituting current values of N and the unknown elements of z_{mi}. Similar reasoning can be used to derive an estimator of the total number of species present among all s sample locations (M) and also indices of similarity in species composition among sites, such as Dice's Index (Dice 1945). See Dorazio and Royle (2005) for details.

As mentioned above, the general approach of Dorazio and Royle (2005) could be extended to accommodate more complex models with respect to ψ_m and p_m. For example, models could include the effect of local site characteristics on one or both of the probabilities, or detection probabilities could be modeled as a function of covariates that may change with each survey of each site (e.g., temperature). The synthetic treatment of occupancy and species richness estimation by Dorazio and Royle (2005) provides the framework for addressing a wide variety of ecologically relevant questions. Dorazio et al. (in review) have further elaborated on this framework, employing a data augmentation procedure described by Royle et al. (in review), to yield a very general and efficient implementation of the multispecies site occupancy model using the software package WinBUGS.

EXAMPLE: AVIAN POINT COUNT DATA

We refrain from providing a detailed example illustrating application of the community model, and instead briefly summarize the case study provided by Dorazio and Royle (2005). They made use of avian point count data collected along a North American Breeding Bird Survey (BBS) route comprised of 50 sample locations (roadside "stops") that were sampled 11 times during the breeding season in 1991. For these data, 75 unique species were observed.

The estimated species richness (posterior mean) for these data was $\hat{N} = 93.3$, with a posterior standard deviation of 9.0. The result was similar to that which they obtained by maximum likelihood, because the sample size was large and the prior distributions were relatively noninformative. The estimated number of species occurring on the 50 samples was $\hat{M} = 82.3$ (again, this is a posterior mean), with a 95% Bayesian confidence (or "credible") interval reported as (79.3, 86.1). Note that the interval bounds are not integers because M and N are derived parameters under the conditional formulation of the model [Eq. (9.1)]. The estimated (posterior) mean and standard deviation of the species occurrence parameters were $\hat{\mu}_\psi = -1.49$ and $\hat{\sigma}_\psi = 2.22$, respectively. The estimated posterior mean and standard deviation of the species detection parameters were $\hat{\mu}_p = -1.85$ and $\hat{\sigma}_p = 1.14$, respectively. These results indicate that substantially more heterogeneity is present in occurrence probability than in detection probability, a result illustrated graphically in Figure 9.1 (taken from Dorazio and Royle 2005), which shows the estimated distributions for both ψ_m and p_m. Thus, "detection failures in many bird species are attributed to low rates of occurrence, as opposed to simply low rates of detection" (Dorazio and Royle 2005). This is important because ecologically important summaries of community structure are functions of occurrence probability, whereas the detection parameters are merely nuisance parameters. Failing to properly decompose variation in "net detection probability" (i.e., the product

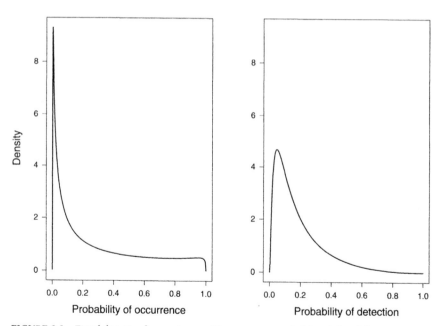

FIGURE 9.1 Fitted densities for species-specific occurrence probability (ψ) and detection probability (p) for avian point count data collected from a North American Breeding Bird Survey route with 50 sampled locations that was surveyed 11 times [Source: Dorazio and Royle (2005)].

of detection probability and occurrence probability) will have the effect of increasing estimates of uncertainty in these quantities (and possibly leading to bias as well). Dorazio and Royle (2005) also provided estimates of the number of species occurring at each site and the similarity in species richness among sites. Finally, they also estimated the correlation between u and v to be $\rho_{uv} = 0.74$, and argued this to be a consequence of the relationship of both occupancy and detection probability to abundance.

MULTIPLE-SEASON STUDIES

The approach described earlier in this section of analyzing the data for multiple species simultaneously can obviously be used to investigate changes in the community over time. As in the single-species case, only by examining the community at systematic points in time can one reliably make inference about the processes of change occurring within that community. Applying the single-species unconditional explicit dynamics model (Chapter 7) to multiple species simultaneously is one approach for doing so.

Once again, the joint modeling of multiple species allows questions to be asked concerning whether groups of species display similar processes of change within the community. For example, rather than estimating a separate extinction probability between all successive seasons for each species [i.e., a $\varepsilon(Species \times Time)$ model; Figure 9.2(a)], it may be reasonable to consider a model in which extinction probabilities vary over time in a parallel fashion for each species [i.e., a $\varepsilon(Species + Time)$ model; Figure 9.2(b)]. This may be particularly advantageous for making inference about a species that is rarely encountered but has a life history strategy similar to that of one or more common species. On its own, there may be insufficient data to make any reliable inferences about the former species, but by sharing parameters with one or more species that are detected with greater frequency, such inference may be possible (provided it is biologically reasonable to model those species jointly, of course). Parallel temporal variation in local extinction or colonization probabilities may also provide a means of assessing guild membership, based on the thinking that members of a guild are likely to respond to environmental variation in similar ways.

Site- (and season-) specific covariates can also be used to model local rates of extinction and colonization across species. Community analyses based on aggregations of forest bird species have provided inferences about the influence of habitat covariates (e.g., fragmentation statistics) on aggregated rates of species extinction and turnover (Boulinier et al. 1998b, 2001). Analyses using additive models of species and habitat (or habitat change) would provide a more detailed assessment of this type of hypothesis. For example, the work of

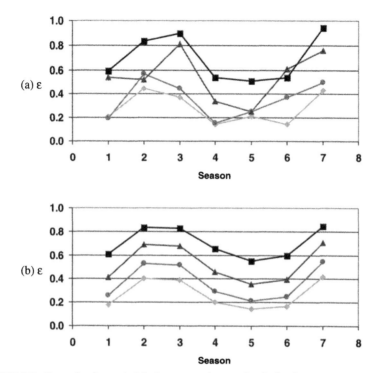

FIGURE 9.2 Example of two models that one might consider for local extinction probabilities (ε) when modeling multiple species simultaneously; (a) ε is species and season specific, model $\varepsilon(Species \times Time)$; and (b) ε for the group of species varies in parallel (on the logistic scale), model $\varepsilon(Species + Time)$.

Boulinier et al. (1998b, 2001) relied on a priori groupings of forest bird species as "area sensitive" or not. Investigation of additive species-plus-fragmentation models (e.g., $\varepsilon(Species + Fragmentation)$) with common slope parameters relating species-specific extinction probabilities to fragmentation statistics would provide an objective means of assessing species-specific group membership.

Finally, we note that the models of Dorazio and Royle (2005) can be extended to multiple seasons as well, permitting direct estimation of changes in species richness and related parameters over time. In particular, extension of this approach will permit inference about changes in species richness at multiple scales, ranging from the local site to aggregations of sites to the total area from which sampled sites are selected. Such investigations should be capable of providing inferences about the relative contributions to community change of processes acting at different scales, a topic of substantial interest to community ecologists (e.g., see Cornell and Lawton 1992; Cornell 1993; Karlson and Cornell 1998; Cam et al. 2000a).

9.3. DISCUSSION

In this chapter, we have outlined two approaches for application of the occupancy models described in Chapters 4–7 to community-level studies when interspecific relationships among species are not of interest. One approach uses a species list at a single or small number of sites, and the other approach is based on joint modeling of multiple species at a larger number of sites. Both approaches could be used to address similar questions, and the question of which approach may be more appropriate largely depends upon the quality of the data and the objective(s) of the study. We believe the second approach will work best with data from at least a moderate number of sites (>30). Data collected at fewer sites are unlikely to contain enough resolution (in terms of the occupancy state of a site for each species) to provide meaningful results. For example, with only 20 occupied sites, an observed 10% decrease in occupancy corresponds to 2 sites becoming unoccupied, which may easily lie within the bounds of random chance. However, with 100 sites, an apparent 10% decrease corresponds to 10 sites becoming unoccupied, which may be much less likely if the population really is stable.

There is the potential for having a very large number of parameters in the model when using the second approach, especially when data have been collected over multiple seasons. As such there may be a seemingly infinite number of possible models that could be considered. It is very important in such cases that *a priori* reasoning, biological knowledge of the system, and hypotheses of interest are used to limit the number of models that will be fit to the data. Provided the models are sufficiently complex to capture the main features of the data, robust inference can still be made about the system without requiring highly complex models. The fact that there may be a "better" model for the collected data outside of the model set should not necessarily be a cause of concern when sound, rational thought has been used to construct the set of plausible models. Indeed, by limiting the model set in such a manner, the likelihood is lessened of finding models that capture some random aspect of the data well but do not portray the underlying system accurately and hence cannot be used for prediction. That said, however, one should not restrict the model set so much that the models considered only represent a single hypothesis or view of the world. In our experience we also believe that the model set should usually include models that permit variation in detection probability over time and season.

We believe that both of the approaches presented in this chapter offer advantages over previous inferential approaches for dealing with ecological communities. We believe that the development of approaches that deal adequately with detection probabilities (e.g., Burnham and Overton 1979; Bunge and Fitzpatrick 1993; Williams *et al.* 2002) represented an important step forward. The new occupancy-based approach outlined in Section 9.1 can be viewed as similar to these previous methods, with a substantial advantage provided by the ability to develop species-specific covariate models. For example, these methods permit formal investigation of the relationships between species occupancy and rate of local extinction, and such species-specific characteristics as body size and degree of specialization.

The other new approach based on multi-species occupancy studies of multiple sites (Section 9.2) also offers opportunities for modeling of occupancy and related vital rates using both site-specific and species-specific covariates. Additive models including species effects can be used to investigate such topics as guild membership and common responses to environmental and habitat changes. Studies of the relationship between species characteristics and local vital rates (extinction and colonization) have previously been based on aggregations of species (e.g., the study of avian sexual dimorphism by Doherty *et al.* 2003a), whereas additive models permitting species effects would represent an inferential improvement. The modeling of Dorazio and Royle (2005) provides a useful approach to estimation of species richness at various geographic scales, ranging from local sample sites to aggregations of such sites to the entire area from which samples are selected. Their approach also permits inference about other community-level questions, including species-area relationships and similarities in species composition across sites.

In summary, we believe that the methods presented in this chapter should represent substantial improvements over previous approaches to inference in community ecology. The use of multi-species occupancy studies at many sites, with temporal replication within each season over multiple seasons, provides a rich database with which to address many of the important questions in community ecology.

CHAPTER 10

Future Directions

As stated in the Preface, we view this book not as the final word on occupancy modeling but as a starting point for what we believe to be a promising and useful area of methodological research. As we developed the methods presented here, we encountered problems and extensions that we believe to be interesting and useful, but that fall outside the initial list of topics that we decided to pursue in this book. In order to ensure that we completed this volume, we put those topics aside with the idea that we will pursue them at a later time. Here we discuss five of the more interesting and potentially important topics on that list. For each topic, we will provide some information on the motivation for considering it and then an outline approach to the subsequent modeling that is needed. Note that these ideas are in various stages of development, with some more complete than others. However, even in the case of the better-developed ideas, we have not considered model forms that will facilitate computation, but instead present the basic models in a manner designed to facilitate comprehension.

10.1. MULTIPLE OCCUPANCY STATES

The previous chapters of this book have treated occupancy as a dichotomous state; that is, a site is either occupied or it is not. In discussing these models with potential users, a question that has arisen many times involves how to treat the situation in which an occupied site might be characterized as belonging to one of two or three classes. The situation that is most frequently discussed involves breeding state and whether or not animals are breeding at an occupied site. The motivation for these considerations involves ideas about variation in the contributions of occupied sites to metapopulation dynamics (e.g., Pulliam 1988), with generally greater value and importance associated with sites at which reproduction does occur (Fig. 10.1). Here we present an approach to estimation of single-season occupancy with multiple states. We develop the model using breeding and nonbreeding as the two occupancy states, but the reader can see that the development can deal with other occupancy states and can be extended to deal with any number of states as well.

The sampling situation we consider is identical to that presented in Chapter 4, with the exception that when evidence of occupancy is found, an additional assessment is made of whether the evidence does or does not provide an indication that breeding activity is taking place. For example, for pond-breeding amphibians, finding egg masses, new metamorphs, or animals in amplexus might constitute evidence of breeding in the pond. In studies of the spotted owl, investigators routinely establish different sets of criteria that constitute evidence of occupancy and of reproduction or nonreproducton (e.g., Franklin *et al.* 2004). Evidence for occupancy typically involves simply hearing owls. Evidence of reproduction is provided by various kinds of information, including adult behavior when adults are fed mice by investigators. Reproducing adults typically take mice to feed them to young, so if adults eat multiple mice themselves and/or cache or refuse mice, this is taken as evidence of nonreproduction. Of course, finding nests and observing young being fed are also positive signs of reproduction.

A common characteristic of the above sampling situations, and of most other efforts to obtain evidence of breeding or successful reproduction, is uncertainty. The usual situation will be that some kinds of evidence provide unambiguous evidence (proof) of reproduction, whereas nonreproduction typically involves absence of evidence of reproduction and thus cannot provide certain state assignment. This sampling problem is similar to that involved in the estimation of occupancy itself, with evidence of occupancy (1 in the detection history) being unambiguous but evidence of nonoccupancy (0 in the detection history) being ambiguous (Chapter 4). This same problem arises in multi-state capture-recapture modeling, where misclassification can occur in

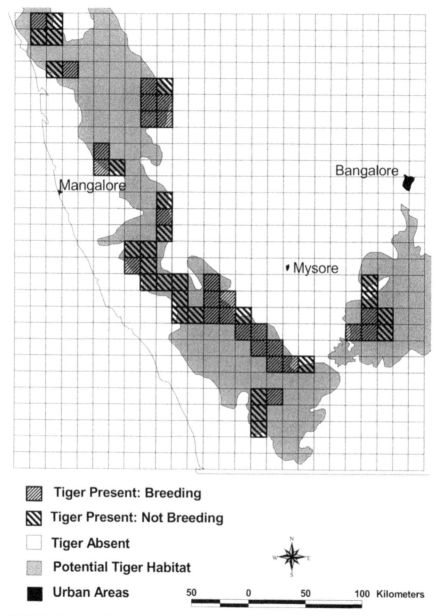

Tiger Present: Breeding

Tiger Present: Not Breeding

Tiger Absent

Potential Tiger Habitat

Urban Areas

50   0   50   100 Kilometers

FIGURE 10.1 Hypothetical survey grid cells overlaid on potential tiger habitat in an area of southern India. Cells are coded to show presence/absence of tigers and evidence of breeding. Such information can be viewed as a first step in identifying source and sink habitat for the purpose of making land management decisions. (Source: Stith and Kumar 2002)

the assignment of animal to state (Fujiwara and Caswell 2002; Lebreton and Pradel 2002; Kendall *et al.* 2003; Nichols *et al.* 2004). We thus propose an approach to estimation that admits uncertainty in occupancy state assignment. We assume that evidence of breeding is unambiguous and permits correct classification, whereas absence of such evidence admits two possibilities: reproduction occurred and was not detected, or reproduction did not occur.

We extend the detection history notation of Chapter 4 to deal with two occupancy states. Thus, "0" still denotes no detection of occupancy by the species, "1" denotes detection of occupancy but no definitive evidence of breeding, and "2" denotes detection with definitive evidence of reproduction. So history $h_i = 1021$ indicates a site at which the species was detected at sampling period 1 with no definitive evidence of breeding at that time, no detection at sampling period 2, detection at sampling period 3 with definitive evidence of breeding, and detection at period 4 with no definitive evidence of breeding. Thus, the three types of detection history observations carry different levels of uncertainty. Nondetection of the species at a sampling occasion (0) indicates either absence of the species or occupancy in one of the two possible reproductive states. Detection with no definitive evidence of reproduction (1) indicates occupancy in one of the two reproductive states. Detection with definitive evidence of reproduction (2) permits unambiguous assignment of the site to the group of reproductive sites.

In order to model this situation, we simply extend the notation and modeling of Chapter 4. We retain the two sets of model parameters reflecting probabilities of occupancy and detection, but we add superscripts to denote true occupancy state (1 = nonbreeding; 2 = breeding). We also add a new set of parameters to deal with the probability of correct classification of a site at which occupancy is detected. We define parameters for site i as follows:

$\psi_i^1 = \Pr$ (patch i is occupied by nonbreeders);
$\psi_i^2 = \Pr$ (patch i is occupied by breeders);
$p_{ij}^1 = \Pr$ (detection in period j | patch i occupied by nonbreeders);
$p_{ij}^2 = \Pr$ (detection in period j | patch i occupied by breeders);
$\delta_{ij} = \Pr$ (identified as breeders in period j | patch i occupied by breeders and animals detected).

In the absence of site-specific covariates, the subscript i is omitted and the modeling is based on a group of sites assumed to have similar characteristics (see Chapter 4).

Modeling then proceeds in a manner similar to that shown in Chapter 4. Consider the detection history $h_i = 1021$. The probability for that history can be written in terms of the above parameters as:

$$\Pr(h_i = 1021) = \psi_i^2 p_{i1}^2 (1 - \delta_{i1})(1 - p_{i2}^2) p_{i3}^2 \delta_{i3} p_{i4}^2 (1 - \delta_{i4}).$$

Because definitive evidence of reproduction was observed during at least one sampling occasion, this site is known to be in state 2, and this occurs with probability ψ_i^2. At sampling occasion 1, the species was detected (associated probability p_{i1}^2; note that the superscript 2 applies because that is the correct state of the site), but definitive evidence of reproduction was not observed $(1 - \delta_{i1})$. The species was not detected at sampling occasion 2 $(1 - p_{i2}^2)$. The species was detected at occasion 3 (p_{i3}^2), and definitive evidence of reproduction was observed (δ_{i3}) at that time. The species was again detected at sampling occasion 4 (p_{i4}^2), but no definitive evidence of reproduction was observed $(1 - \delta_{i4})$.

Now consider a history with more uncertainty because definitive evidence of reproduction is never observed, $h_i = 0101$. The associated probability can be written as:

$$\Pr(h_i = 0101) = \psi_i^1[(1 - p_{i1}^1)p_{i2}^1(1 - p_{i3}^1)p_{i4}^1]$$
$$+ \psi_i^2[(1 - p_{i1}^2)p_{i2}^2(1 - \delta_{i2})(1 - p_{i3}^2)p_{i4}^2(1 - \delta_{i4})]$$

The above probability is written as the sum of two terms. The first term is associated with the possibility that the site was occupied by the species but that no reproduction occurred there (ψ_i^1). The product in brackets following this term simply reflects either detection or nondetection, given that the site was one at which no reproduction occurred. The second term in brackets deals with the alternative possibility that the species indeed reproduced at the site (ψ_i^2) but that no definitive evidence of reproduction was ever observed. Thus, at every occasion at which the species was detected, the model requires the product of a detection probability (p_{ij}^2) and the complement of the probability of correct classification $(1 - \delta_{ij})$.

Finally, consider the history characterized by the largest degree of uncertainty, $h_i = 0000$, and its associated probability:

$$\Pr(h_i = 0000) = \psi_i^1 \prod_{j=1}^{4}(1 - p_{ij}^1) + \psi_i^2 \prod_{j=1}^{4}(1 - p_{ij}^2) + (1 - \psi_i^1 - \psi_i^2).$$

This history is written as the sum of three terms. The first term indicates that the site was occupied, that no reproduction occurred during the season (ψ_i^1), and that the species was not detected at any of the four sampling occasions $\left(\prod_{j=1}^{4}(1 - p_{ij}^1)\right)$. The second term of the sum indicates that the site was occupied, that reproduction did occur (ψ_i^2), and that the species was not detected $\left(\prod_{j=1}^{4}(1 - p_{ij}^2)\right)$. The final term of the sum simply specifies the probability that the site was not occupied $(1 - \psi_i^1 - \psi_i^2)$.

The likelihood for all of the detection histories can then be created in the same manner as in Chapter 4, by multiplying the probabilities for the detection histories at all s sites:

$$L(\psi^1, \psi^2, p^1, p^2, \delta | h_1, h_2, \ldots, h_s) = \prod_{i=1}^{s} \Pr(h_i).$$

Although we have not yet implemented this model, it appears that parameters should be identifiable. As with the models of Chapter 4, it should be possible to either maximize the likelihood directly or to use a Bayesian approach to obtain parameter estimates.

The assumptions underlying this model are the same as those of the simpler models of Chapter 4, with a few additions. We permit the detection probabilities to be different for occupied sites at which reproduction does and does not occur, although a natural model to consider is one with the following constraints: $p_{ij}^1 = p_{ij}^2$. We assume that the time frame of the sampling (the defined season) includes periods over which definitive evidence of reproduction is possible to detect. For example, the approach would not be sensible if all sampling was done early in the reproductive season, and if the only definitive evidence of reproduction involved late-season events (e.g., presence of new metamorphs). The subscripting of the correct classification parameters with time does permit the probability of correct classification to be greater at some times than at others.

Note that in the modeling presented above, we assume that the definitive evidence of reproduction can be obtained independently at each sampling occasion at which the species is detected. Thus, our first detection history, h_i = 1021, includes a "1" in sampling occasion 4 following a "2" in occasion 3. This is a reasonable assumption in many sampling situations, but we can also envision situations in which it is not possible to make independent assessments following the initial detection of evidence for reproduction. For example, once the location of an active bird nest at a site is determined, it will likely be checked at each subsequent visit to the site. In such a situation, both occupancy and occupancy state are essentially known following the initial sampling occasion at which the nest is found. We would thus modify the above detection history as h_i = 102−, where the final "−" simply indicates that there is no independent information on either detection probability or correct classification probability provided by the final visit to the known-location nest. We would model this modified detection history as:

$$\Pr(h_i = 102-) = \psi_i^2 p_{i1}^2 (1 - \delta_{i1})(1 - p_{i2}^2) p_{i3}^2 \delta_{i3}.$$

The only difference between the above expression and that provided previously involves whether or not the history includes information about detection or classification at period 4.

We will not provide a complete development of multi-season models for the case of multiple occupancy states, but we note that such modeling should be possible and also useful. The occupancy, detection, and correct classification parameters remain the same as in the single-season modeling developed above. In addition, we require site extinction probabilities that represent transitions from occupied to absent. The modeling of Chapter 7 must be modified to permit local extinction to be state-specific, so we will now have extinction parameters that potentially differ between sites at which reproduction does and does not occur. Similarly, probabilities of colonization (sites moving from unoccupied to occupied) will be state-specific with potentially different probabilities of colonizing with and without reproduction. Finally, the modeling will require transition probabilities of moving between the two occupied states, no reproduction to reproduction and vice versa.

The motivation for multi-season modeling again involves the relative values of occupied sites at which reproduction does and does not take place. For example, we can envision situations in which occupancy itself would be unchanged over time, but where the transition probabilities associated with moving from the reproductive state to the nonreproductive state were relatively high, such that the proportion of sites at which reproduction occurred was decreasing. This information would be important to the assessment and modeling of occupancy dynamics.

10.2. INTEGRATED MODELING OF HABITAT AND OCCUPANCY

One point made by users of occupancy modeling involves the interpretation of occupancy estimates in situations in which the amount of suitable habitat changes. For example, assume that the number of patches of suitable habitat is reduced by 50% between two times, t and $t + \Delta$. If occupancy is estimated for the set of suitable patches at each time step, and if the occupancy of suitable habitat patches remains the same for the two times, then the occupancy estimates would indicate no change, although in reality the metapopulation has shown a decline in *number* of occupied patches of 50%. One way to avoid this problem in interpretation in this simple situation is to base occupancy on the original set of suitable patches, which includes many patches that are then unsuitable at time $t + \Delta$. However, we believe that a more general and useful solution is to model not only occupancy but also suitable habitat in cases where this is likely to change from season to season (e.g., year to year). Statements about the status of the species of interest would then depend not only on the occupancy of suitable habitat but also on the suitability of all potential habitat. This sort of occupancy modeling has already appeared in the more theoretical literature of metapopulation dynamics and has led to the concept of extinction thresholds (see Chapter 2), which represent the fraction of potential habitat patches that are suitable or represent good habitat for the species (Lande 1987, 1988; Merila and Kotze 2003). Recently, Verheyen *et al.* (2004) modified Hanski's (1994a) incidence function approach to metapopulation modeling to include aspects of habitat turnover and dynamics. Dynamical modeling by Ellner and Fussman (2003) emphasized the importance of habitat succession to metapopulation persistence. In the approach described below we recommend a general approach to estimation that incorporates habitat dynamics and still deals with uncertainty in occupancy detection.

In Chapter 7, we described a robust sampling design that involved repeat sampling of sites within each season for a number of seasons. This design permits estimation of occupancy for each season, as well as parameters (e.g., local rates of colonization and extinction) responsible for changes in occupancy from season to season. Now consider that the set of potential habitat sites is identified at the beginning of such a study and that an assessment is made each year of the suitability of habitat at each sampled site. Although multiple states of suitability are possible, assume the simplest case, in which habitat is either suitable or not. For amphibians that breed in vernal pools each spring, suitability could be assessed as the site either having some water (suitable) or being completely dry (not suitable). Further assume that this characterization of habitat (suitable or not) applies to the entire season (i.e., a habitat cannot be suitable at one visit and not the next; this assumption can be relaxed if needed). We can then write a habitat history for each sampled site that specifies the suitability (1 = suitable, 0 = not suitable) for the site at each season or primary sampling period. Thus, a possible habitat history for a site for a five-season study would be $\mathbf{h}'_i = 1\ 1\ 0\ 1\ 1$, where the prime distinguishes the habitat history at site i ($\mathbf{h}'_i$) from the detection history for the species of interest ($\mathbf{h}_i$). This site, i, contained suitable habitat in seasons 1, 2, 4, and 5, but not in season 3. We emphasize that the "0" indicates the unsuitable habitat state and not nondetection. Indeed, we are assuming the ability to unambiguously assign the appropriate habitat state to a site, so there is no possibility of either nondetection or misclassification.

We now consider modeling of habitat histories using an approach similar to that used for detection histories. Define the following suitability and transition probabilities for describing habitat state and dynamics:

ξ_{it} = Pr (site i is suitable [state = 1] at season t for the species of interest);
η_{it}^{10} = Pr (site i is not suitable at season $t + 1$ | suitable at t);
η_{it}^{01} = Pr (site i is suitable at season $t + 1$ | not suitable at t).

Models of habitat dynamics will entail use of these parameters to model site habitat history data. For example, consider the following habitat history for site i, $\mathbf{h}'_i = 1\ 1\ 0\ 1\ 1$, and its associated probability:

$$Pr(h_i' = 1\,1\,0\,1\,1) = \xi_{i1}(1 - \eta_{i1}^{10})\eta_{i2}^{10}\eta_{i3}^{01}(1 - \eta_{i4}^{10}).$$

The patch is suitable in the first season, and the probability associated with this event is ξ_{i1}. The patch remained suitable at time 2, and the probability associated with the transition from suitable to suitable is $(1 - \eta_{i1}^{10})$, in other words, the patch did not become unsuitable. The habitat in the patch made the transition to unsuitable between periods 2 and 3 (η_{i2}^{10}) and then became suitable again in the subsequent time step (η_{i3}^{01}). The patch remained suitable in the final sampling period $(1 - \eta_{i4}^{10})$. Note that we include no detection probabilities in this model, as we assume that suitability can be assessed with certainty. This assumption also may be relaxed in situations for which it is not appropriate. In this situation the survey design would be extended to incorporate habitat assessments at each survey, as well as assessments of occupancy.

Using the above approach, we can directly estimate the parameters relevant to habitat suitability and dynamics using the following likelihood:

$$L(\xi_1, \eta \,|\, h_1', h_2', \ldots, h_s') = \prod_{i=1}^{s} Pr(h_i'). \tag{10.1}$$

Such a model would permit direct investigation of habitat dynamics. Covariate modeling could be used to identify site- or time-specific factors associated with these dynamics. We could then develop a separate model for occupancy dynamics conditional on the set of particular habitat histories. The joint likelihood for both habitat and occupancy dynamics would then be given by the product of these two components.

An alternative modeling approach is to incorporate information about habitat suitability directly into site detection histories, so that such histories include both habitat suitability and occupancy detection data. We will proceed with this combined approach. Assume that there are two surveys each season. For example, a possible detection history of a site visited twice each season for five seasons would be $h_i = 01\ 11\ [00]\ 10\ 00$. The species of interest was detected at this site in seasons 1, 2, and 4, but not in seasons 3 or 5 (see Chapter 7). Note that the detection history entries for season 3 are enclosed by brackets, [00]. We propose this notation for the situation in which the site is necessarily unoccupied because the habitat patch is not suitable. We acknowledge that in some situations it may be more reasonable to permit some occupancy of unsuitable habitat, but for the development here we consider the situation in which a site that is not suitable cannot be occupied. Because of this notation identifying when the habitat at a site is unsuitable, it is possible to include both habitat suitability and occupancy detection in the same model structure.

Our modeling of the occupancy process is then viewed as conditional on the habitat process in the sense that occupancy dynamics depend on the suitability of the patch over time. We use the same parameters as in Chapter 7, with the exception that we allow colonization of a previously unoccupied patch to be a function of whether or not the patch was suitable in the previous season. In addition to the parameters of the habitat process, we thus require the following, where we use superscripts to denote habitat suitability (1 = suitable; 0 = not suitable):

ψ_{it}^1 = Pr (site i is occupied at season t | habitat is suitable at t);

ψ_{it}^0 ≡ 0 = Pr (site i is occupied at season t | habitat is not suitable at t);

ε_{it}^{11} = Pr (site i is unoccupied at $t + 1$ | site is occupied at t and habitat is suitable at $t + 1$);

ε_{it}^{10} ≡ 1 = Pr (site i is unoccupied at $t + 1$ | site is occupied at t and habitat is not suitable at $t + 1$);

γ_{it}^{11} = Pr (site i is occupied at $t + 1$ | site is suitable at $t + 1$, and suitable, but not occupied, at t);

γ_{it}^{01} = Pr (site i is occupied at $t + 1$ | site is suitable at $t + 1$, but not suitable at t);

$p_{it,j}$ = Pr (species is detected at site i during survey j of season t | species present at site i during season t).

As in Chapter 7, the survey (j) denotes the repeat sample within season t.

All of the parameters associated with occupancy dynamics are conditional on suitable habitat at either the beginning (extinction) or end (colonization) of each interval. Thus, the modeling of detection histories includes models of both habitat and occupancy dynamics. Consider the probability model for the above detection history as:

$$\Pr(h_i = 01\,11\,[00]\,10\,00)$$
$$= \xi_{i1}\psi_{i1}^1(1 - p_{i1,1})p_{i1,2}(1 - \eta_{i1}^{10})(1 - \varepsilon_{i1}^{11})p_{i2,1}p_{i2,2}\eta_{i2}^{10}\eta_{i3}^{01}\gamma_{i3}^{01}p_{i4,1}$$
$$(1 - p_{i4,2})(1 - \eta_{i4}^{10})(\varepsilon_{i4}^{11} + (1 - \varepsilon_{i4}^{11})(1 - p_{i5,1})(1 - p_{i5,2}))$$

The site was occupied in season 1, and the probability associated with this event is the product of the probability of habitat being suitable (ξ_{i1}) and the probability of suitable habitat being occupied (ψ_{i1}^1). In season 1, the species was detected in survey 2 ($p_{i1,2}$) but not in survey 1 ($1 - p_{i1,1}$). The habitat remained suitable in season 2 (there was no [00] in the detection history for season 2), and the associated probability is given by ($1 - \eta_{i1}^{10}$). The site was occupied again in season 2 ($1 - \varepsilon_{i1}^{11}$) and was detected during both surveys ($p_{i2,1}p_{i2,2}$). The habitat at the site became unsuitable the next season (η_{i2}^{10}), so the species necessarily was absent from the site (no detection or extinction parameters needed; $\varepsilon_{it}^{10} \equiv 1$). The site made the transition to suitable habitat the next season (η_{i3}^{01}) and was colonized (γ_{i3}^{01}), with the species being detected in survey 1 ($p_{i4,1}$) but not in survey 2 ($1 - p_{i4,2}$). The final season is again characterized by suitable habitat ($1 - \eta_{i4}^{10}$) but is ambiguous with respect to occupancy status. One possibility is that the species went locally extinct (ε_{i4}^{11}), and the other is that the species did not go extinct $(1 - \varepsilon_{i4}^{11})$ but was present and undetected $(1 - p_{i5,1})(1 - p_{i5,2})$. Thus, the parameterization of this history differs from that used in Chapter 7 in that there are parameters reflecting transitions in both habitat and occupancy state. With respect to occupancy dynamics, there are two different probabilities of local extinction, one for sites that remain suitable (ε_{it}^{11}) and another for those that make the transition to unsuitable ($\varepsilon_{it}^{10} \equiv 1$). Similarly, the colonization parameter differs for sites that are (γ_{it}^{11}) and are not (γ_{it}^{01}) suitable at season t. Probability of colonization over the period t to $t + 1$ is 0 for sites that are not suitable at $t + 1$, regardless of whether or not the habitat was suitable in season t.

The habitat and occupancy data, with their associated models, produce a likelihood of the following general form:

$$L(\xi_1, \psi_1, \varepsilon, \gamma, \eta, p \mid h_1, h_2, \ldots, h_s) = \prod_{i=1}^{s} \Pr(h_i),$$

where the probabilities associated with the detection histories are modeled as in the above example. Although we have not yet coded this model, we believe that we should be able to obtain estimates using maximum likelihood. An equivalent approach would be to model habitat dynamics with one part of the likelihood (Eq. 10.1), and then model occupancy dynamics, conditional on habitat suitability, as another part. The joint likelihood obtained as the product of these two components would be equivalent to, and yield the same estimates as, the combined approach described above.

As was the case with the multi-season models of Chapter 7, we can write time-specific suitability of sites and occupancy of suitable sites recursively. The probability that a site is suitable in season $t + 1$ can be written as:

$$\xi_{it+1} = \xi_{it}(1 - \eta_{it}^{10}) + (1 - \xi_{it})\eta_{it}^{01}.$$

This expression simply adds the fraction of suitable sites at season t that remain suitable to the fraction of sites at t that are unsuitable but make the transition from unsuitable to suitable. The analogous recursive expression for probability of occupancy of suitable habitat in season $t + 1$ is:

$$\psi_{it+1}^1 = \psi_{it}^1(1 - \eta_{it}^{10})(1 - \varepsilon_{it}^{11}) + (1 - \psi_{it}^1)(1 - \eta_{it}^{10})\gamma_{it}^{11} + (1 - \xi_{it})\eta_{it}^{01}\gamma_{it}^{01}.$$

This expression is more complicated than its counterpart in Chapter 7 (Eq. 7.4). The first of the three terms in the summation is the probability that an occupied site in season t remains suitable and does not exhibit a local extinction. The second term indicates the probability that a site is suitable, yet unoccupied, in season t, that the site remains suitable in season $t + 1$, and that it is colonized between seasons t and $t + 1$. The final term indicates the probability that a site is unsuitable in season t, but that it becomes suitable at season $t + 1$ and is colonized by $t + 1$.

It is also possible to compute parameters other than those defined above. For example, suppose that instead of defining extinction probability, ε_{it}^{11}, as applying to sites that contain suitable habitat at seasons t and $t + 1$ (denoted by the superscript "11"), we are interested in the probability that any randomly selected site occupied at primary period t is not occupied in season $t + 1$. We can compute this as $1 - (1 - \varepsilon_{it}^{11})(1 - \eta_{it}^{10})$, which is simply the complement of the probability that the patch remains suitable *and* the species does not go locally extinct, given suitability. Similarly, we can compute the probability that a randomly selected site that is unoccupied at season t is occupied at season $t + 1$ as: $\xi_{it}(1 - \psi_{it})(1 + \eta_{it}^{10})\gamma_{it}^{11} + (1 - \xi_{it})\eta_{it}^{01}\gamma_{it}^{01}$. The first term in this sum represents the probability that a site was suitable but unoccupied, remained suitable, and was colonized. The second term in the sum represents the probability that a site did not contain suitable habitat, became suitable, and was colonized. The sum is thus a weighted average of colonization rates for sites that were (γ_{it}^{11}), and were not (γ_{it}^{01}) suitable during season t, weighted by the probabilities of a site being in each habitat state (suitable, not suitable) at season t. Finally, the unconditional probability of occupancy for any randomly selected site i is given by $\xi_{it}\psi_{it}$, the product of the probabilities of being suitable and occupied given suitability.

We speculate that this sort of joint habitat and occupancy modeling may be very useful for situations in which habitat does not remain stable and suitable from season to season. First, we explicitly recognize that the state of the system with respect to occupancy involves both the state of potential habitat (ξ_{it}) and the occupancy of suitable habitat (ψ_{it}). Neither state variable by itself is adequate as a measure of system state. Second, because both state variables are relevant, we consider the simultaneous modeling of dynamics for both of them. Covariate modeling can be used to investigate functional relationships between potential factors of influence and the rate parameters governing change in both habitat and occupancy. We could even consider extinction thresholds (Lande 1987, 1988; Merila and Kotze 2003) explicitly by modeling extinction probabilities (ε_{it}^{11}) and colonization rates ($\gamma_{it}^{11},\gamma_{it}^{01}$) for suitable sites as a function of the fraction of habitat that is suitable (ξ_{it},ξ_{it+1}), based on the ideas that both rescue effect (Brown and Kodric-Brown 1977) and colonization are reduced by reductions in the fraction of potential sources of dispersers and colonists (Hanski 1999). In addition to investigation of extinction thresholds, such a modeling approach would address the more general question of relative vulnerability of species to land use change (Periera *et al.* 2004). Management interventions designed to increase habitat suitability could also be modeled using this framework. Although our example modeling has involved two habitat states, we could include several states reflecting different degrees of habitat suitability and estimate the transition probabilities associated with ecological succession. Such modeling would provide the estimation tools needed to investigate theoretical models of metapopulation persistence in the face of succession (Ellner and Fussman 2003). Finally, we note that combining dynamic modeling of habitat (this section) with the dynamic multi-species modeling of Chapter 8 would permit direct inference about the transition probabilities needed to parameterize the theoretical models of Caswell and Cohen (1991a,b) that relate community dynamics and equilibria to interspecific competition, habitat heterogeneity, and disturbance.

10.3. INCORPORATING INFORMATION ON MARKED ANIMALS

The definition of local extinction rate in the multi-season modeling of Chapter 7 and above is simply $\varepsilon_{it} = $ Pr(site unoccupied at season $t + 1$ | occupied at season t). We believe that this definition is adequate for a large number of questions. However, in some cases there is interest in a more mechanistic definition that better describes the processes occurring between the two times of observation. For example, the complement of extinction $(1 - \varepsilon_{it})$ can arise by any one of a number of possible sequences of events. One possibility is that at least some of the individuals present at t are still present at $t + 1$. Another possibility is that all individual animals present at t either die or move away, and new individuals move in to occupy the site at $t + 1$ (i.e., one manifestation of the rescue effect). If we knew that this latter sequence of events took place, then we might want to model the sequence as the product of a local extinction probability followed by colonization, all within the single interval. In some cases we might like to distinguish between these two possibilities of continuous occupation by at least some of the same individuals versus complete turnover of individuals.

Clark and Rosenzweig (1994) considered the estimation of local rates of extinction and colonization based on site occupancy data using models similar to those of Chapter 7, with the exception that they assumed process stationarity and detection probabilities of 1. Their modeling parameterization differed from ours in that they defined colonization, λ, as Pr(new arrival per unit time) and extinction, μ, as Pr(existing population becomes extinct, per time period). They went on to define extinction probability as we have, Pr(site not occupied at $t + 1$ | occupied at t) and parameterize it as $\delta = \mu(1 - \lambda)$, reflecting the product of probabilities of extinction of the existing population and the complement of the probability of colonization. This parameterization appears to be slightly more mechanistic than our parameterization in Chapter 7, but we note that this comes at the cost of assumptions about the number of extinction and colonization events that can occur within a time interval. For example, the observation, occupied at $t + 1$ and not occupied at t, can only reflect a single colonization event under the Clark and Rosenzweig (1994) model, rather than colonization, followed by local extinction, followed by another colonization. We do not view this as a criticism of the Clark and Rosenzweig (1994) approach. Instead, we simply note that rather than make such assumptions about the number of possible events per time interval, we have chosen to select parameters that are defined directly by observations at t and $t + 1$ (see similar discussion in Chapter 7 and Nichols et al. 1998a). We conclude that, using simple detection/nondetection data without auxiliary information, it is difficult to distinguish between a site that has been continuously occupied between two seasons (or periods of surveying) and a site where the species has gone locally extinct and then has been recolonized.

MacKenzie and Nichols (2004) suggest that one source of potentially useful auxiliary information is obtained from having marked individuals in the study population. They conceptualized how the unconditional explicit dynamics model described in Chapter 7 could be extended to include this type of information. They propose that including information from marked individuals in the manner they suggest would be most useful for species where a site is only occupied by a single individual or where small groups effectively exist as a single unit (e.g., breeding pairs). The marked individuals effectively provide information about the movement of individuals between sites, hence making it possible to distinguish sites that are continuously occupied by the same individuals from those occupied by different individuals through time. We do not provide details here, because they have already been provided for one approach (MacKenzie and Nichols 2004), and we are considering alternative approaches as well. We believe that the ability to model such processes more mechanistically permits access to investigation of a set of biologically relevant questions that cannot be addressed using our less mechanistic approach.

10.4. INCORPORATING COUNT AND OTHER DATA

The abundance models of Chapter 5 focused on heterogeneity of site-specific detection probabilities induced by variation in abundance across sites. In addition to providing an approach for dealing with heterogeneous detection, this modeling provides an estimate of the distribution of abundances over sites. As described in Chapter 5, this approach is based on an assumed distributional form for abundance. The additional information on site-specific abundance then comes from the relationship between site-level detection probability and abundance.

Royle (2004) extended the approach of Royle and Nichols (2003; see also Chapter 5) to the use of count data. Instead of simply recording presence or absence of a species, the investigator counts individuals detected. These counts are then used with prior distributions (e.g., Poisson, negative binomial) in N-mixture models to estimate the distribution of abundance across sample units. Of course, this distribution includes the proportion of sites with abundance equal 0, the quantity associated with the complement of occupancy (i.e., probability of absence).

In some survey situations, investigators may collect the information needed to directly estimate abundance at each visited site, or at least at a subset of such sites. For example, there are numerous approaches to estimation of abundance from observation-based counts: distance sampling (e.g., Buckland et al. 2001), multiple observers (Cook and Jacobson 1979; Pollock and Kendall 1987; Nichols et al. 2000), time at first sighting models (Farnsworth et al. 2002), sighting probability models (e.g., Caughley et al. 1976; Samuel et al. 1987), and marked subsamples (e.g., Arnason et al. 1991; White 1993). In addition, a number of approaches use marked animals, including closed and open capture-recapture models (Otis et al. 1978; Seber 1982; Pollock et al. 1990; Williams et al. 2002), catch per unit effort models (Seber 1982; Williams et al. 2002), trapping webs (Anderson et al. 1983; Buckland et al. 2001), and other methods based on the locations of captures with respect to trap grid configuration (Link and Barker 1994; Efford 2004). If any of the above methods is applied at each surveyed site (or at some sample of sites), then the resulting data can be used to inform a prior distribution of abundance across sites. Royle et al. (2004; also see Hedley et al. 1999) used this approach to estimate the abundance distribution based on avian point counts at which distance data were collected. Dorazio and Royle (2005) applied a similar model to removal counts of fish made at a number of sample locations. Although this approach extends beyond the scope of this book, we believe that it holds promise and deserves use and further development.

10.5. RELATIONSHIP BETWEEN OCCUPANCY AND ABUNDANCE

Numerous ecologists have recognized that knowledge about the relationship between occupancy and abundance can be used to estimate abundance from occupancy data. For example, if individuals are distributed randomly over space such that they follow a Poisson abundance distribution, then the proportion of sites at which at least one individual is detected (occupancy) can be used to estimate mean density (e.g., as $-\log(1 - \hat{\psi})$; see Chapter 5, also

Gerrard and Chiang 1970). This approach has a long history in plant and animal ecology (e.g., Gleason 1920; Fisher 1922), with many practitioners recognizing that individuals are not always randomly distributed, and that other distributions (e.g., negative binomial) may be needed (Dice 1948; Gerrard and Chiang 1970; Seber 1982). These approaches are very similar to the abundance models of Chapter 5 (also see Royle and Nichols 2003), with the exception that they do not deal with detection probability and the likelihood that some 0's will not represent true absences. We thus prefer the models of Chapter 5 to these approaches.

More recently, He and Gaston (2000, 2003; see application by Tosh et al. 2004) have revisited the issue of abundance estimation from occupancy data. Their most recent effort involves the combination of phenomenological models of occupancy-abundance and mean-variance relationships and yields the suggestion that the unified relationship permits estimation of abundance from occupancy data (He and Gaston 2003). Royle et al. (2005) recommended that a more reasonable approach to investigating these relationships was to note that the three quantities of interest to He and Gaston (2003)—abundance, occupancy, and spatial variance in abundance—are completely specified by, and thus emerge directly from, the species abundance distribution. In addition, He and Gaston (2003) acknowledge, yet do not deal with, the likelihood of detection probabilities less than 1. We again recommend the approaches of Chapter 5 as preferable to these other occupancy-based approaches to abundance estimation (also see Royle et al. 2005). We include this topic in this chapter not because it represents a methodological extension to methods presented previously, but because it represents an area of current research interest that should benefit greatly from adoption of our methods in future work.

10.6. DISCUSSION

The above list of topics for future work is clearly not exhaustive, constituting only a small subset of methodological work that should be useful. Some additional topics involve methodological extensions such as those of Sections 10.1–10.4. For example, consider situations in which uncertainty characterizes not only "absence" but also "presence." For example, in occupancy surveys based on animal sign, there may be situations in which tracks or scat of two species are sufficiently similar that they cannot be distinguished with certainty. If a subset of detections is known with certainty, or if relevant ancillary data with known detections can be collected, then we can include classification probabilities in the models of this book, using approaches similar to that of Section 10.1, for example. We can think of several other extensions of this sort, and we expect this list to grow rapidly as more investigators use these methods and encounter special cases that require tailoring of methods and models.

The amount of information included in this book on study design is relatively large, especially considering the very recent development of the models themselves, and should provide good guidance. Nevertheless, we see the need for additional work on various aspects of study design. We also see the need for more thought and conceptual development in some areas. For example, most of the development of these methods has been based on situations with discrete habitat patches (e.g., woodlots, ponds). In cases in which this has not been the case, we have usually thought in terms of sample units that are at least as large as the home range size of study organisms. In situations in which animal ranges are larger than sample units, we believe that methods in this book may still be applicable to questions about habitat use, although we admit that additional thought and attention should be devoted to these scenarios. For example, questions about lack of independence naturally arise when a single animal can be responsible for detections in multiple sample units. In cases in which this kind of dependence leads to problems in interpreting occupancy estimates, it may be necessary to develop models that incorporate dependencies. As should be the case for all methodological development, extensions and modifications will be driven by needs and sampling situations of users.

The most important part of the future of occupancy modeling will be the use of the models described in this book to provide inferences about interesting ecological questions and relevant conservation problems. As outlined in Chapter 2, a large amount of previous ecological work using occupancy data has not produced reliable inferences because of a failure to deal adequately with detection probability. We are currently aware of numerous projects in which investigators are adopting the methods presented in this book, and we look forward to learning results of these efforts. Many countries around the world have initiated large-scale ecological monitoring programs for a variety of reasons (Yoccoz et al. 2001). Although few such programs deal adequately with detectability, relatively minor changes in protocol (e.g., repeat surveys within each season or year) would permit use of the models described herein. We hope that such changes are implemented, as some of the earth's most pressing conservation problems result from changes in climate and large-scale land use. Such changes are likely to result in shifts in large-scale distribution patterns of plants and animals, with local extinctions in some areas and expansions (colonization) in others. Properly designed monitoring programs using the methods of this book would provide excellent opportunities for inference about responses of ecological systems to the important landscape and climate changes they experience in the near and distant future.

Some Important Mathematical Concepts

There are some important mathematical concepts that are widely used in this book that those with stronger backgrounds in mathematics or statistics often take for granted, but we have found that they can be stumbling blocks for many others. These are fundamental concepts, but have been relegated to an appendix to avoid distracting from the main focus of the book. The three topics briefly covered here are: (1) notation for summations and products, (2) vectors and matrices, and (3) differentiation and integration.

A.1. NOTATION FOR SUMMATIONS AND PRODUCTS

There is a very useful shorthand way of defining sums and products of, potentially, a large number of terms. For summations, the Σ (sigma) notation is used, which includes three main features: (1) the index variable and its first value, often placed below the Σ symbol; (2) the last value of the indexing variable, often placed above the Σ symbol; and (3) the actual term or expression that is to be summed, to the right of the Σ symbol. For example, in Eq. (A.1) "$i = 1$" indicates the indexing variable is i, starting from the value of 1, "5" indicates that terms up to $i = 5$ will be added together, and "y_i" is the actual term to be summed:

$$\sum_{i=1}^{5} y_i = y_1 + y_2 + y_3 + y_4 + y_5 \tag{A.1}$$

Some other examples are:

$$\sum_{j=1}^{10} j = 1 + 2 + 3 + 4 + 5 + 6 + 7 + 8 + 9 + 10,$$

$$\sum_{i=1}^{n} (y_i - \overline{y})^2 = (y_1 - \overline{y})^2 + (y_2 - \overline{y})^2 + \ldots + (y_n - \overline{y})^2,$$

$$\sum_{j=1}^{3} y_{ij} = y_{i1} + y_{i2} + y_{i3}.$$

Note that in this final example, there are two potential indexing variables, i and j, but the summation is only specified (and performed) over j.

A similar notation is used for defining products or multiples of variables represented by the Π (pi) symbol. The same three main features are also present here, although now the Π symbol indicates that terms are to be multiplied together rather than summed, for example,

$$\prod_{i=1}^{5} y_i = y_1 \times y_2 \times y_3 \times y_4 \times y_5.$$

Multiple Σ or Π symbols may also be used if there is more than one indexing variable over which the summation or multiplication is to be performed. In such instances, the operation is performed for the innermost (or rightmost) symbols first, then moving outward. For example:

$$\sum_{i=1}^{3} \sum_{j=1}^{4} y_{ij} = \sum_{i=1}^{3} (y_{i1} + y_{i2} + y_{i3} + y_{i4})$$

$$= (y_{11} + y_{12} + y_{13} + y_{14}) + (y_{21} + y_{22} + y_{23} + y_{24})$$

$$+ (y_{31} + y_{32} + y_{33} + y_{34})$$

A.2. VECTORS AND MATRICES

Vectors and matrices are simply arrays of numbers for which the numerical value and relative position within the array conveys some meaning. Each number of the array is called an *element*, which is indexed by the row and column of the array in which it appears. For example the element a_{ij} is the value in the ith row and jth column. In the context of this book, the position of the element in the array is often used to represent the occupancy state of a site in different seasons (e.g., Chapter 7). Below is a very brief introduction to vectors and matrices, illustrating some key concepts that are used in this book. For a more complete introduction suitable for biologists, we direct readers to Appendix B of Williams *et al.* (2002).

VECTORS

A vector is simply a one-dimensional array that consists of either a single row or single column; for example, a three-element row vector, v, would be: v = [5 1 3]. Note that in this book we use bold font to indicate a vector or matrix. An alternative notation is to underline the vector (or matrix) name, for example, $\underline{v}$.

MATRICES

Here we only consider two-dimensional matrices, but higher dimensional matrices are possible. A matrix can be considered as a series of row or column vectors. For example, a square matrix with three rows and three columns (i.e., a 3×3 matrix) would be:

$$\mathbf{A} = \begin{bmatrix} 4 & 2 & 1 \\ 3 & 5 & 0 \\ 1 & 4 & 2 \end{bmatrix}.$$

Matrices do not have to be square. For example, a matrix with three rows and two columns (i.e., a 3×2 matrix) would be:

$$\mathbf{B} = \begin{bmatrix} -1 & 6 \\ 0 & 2 \\ 1 & -3 \end{bmatrix}.$$

VECTOR AND MATRIX MANIPULATIONS

To "transpose" a vector or matrix, one simply swaps the rows and columns of the array. For example, the transpose of the vector v defined above (denoted here as $\mathbf{v}^T$) would be:

$$\mathbf{v}^T = \begin{bmatrix} 5 \\ 1 \\ 3 \end{bmatrix}.$$

For the matrices defined above, the respective transposes are:

$$\mathbf{A}^{\mathsf{T}} = \begin{bmatrix} 4 & 3 & 1 \\ 2 & 5 & 4 \\ 1 & 0 & 2 \end{bmatrix} \text{ and } \mathbf{B}^{\mathsf{T}} = \begin{bmatrix} -1 & 0 & 1 \\ 6 & 2 & -3 \end{bmatrix}.$$

Note that the elements of the first, second, and third rows of the original matrices now appear as the elements in the first, second, and third columns of the transposed matrices, respectively.

Matrix (and vector) multiplication is slightly more complicated. Suppose we wish to multiply the matrices $\mathbf{A}$ and $\mathbf{B}$ to give the resulting matrix $\mathbf{C}$ (i.e., $\mathbf{C} = \mathbf{AB}$). The element in the ith row and jth column of $\mathbf{C}$ is calculated by multiplying the elements in the ith row of $\mathbf{A}$ with the elements in the jth column of $\mathbf{B}$; that is, $c_{ij} = \sum_{k=1}^{n} a_{il}b_{kj}$. For example, in this case the element c_{11} would be:

$$
\begin{aligned}
c_{11} &= \sum_{k=1}^{3} a_{1k}b_{k1} \\
&= (4 \times -1) + (2 \times 0) + (1 \times 1) \\
&= -4 + 0 + 1 \\
&= -3
\end{aligned}
$$

Clearly, a requirement for matrix multiplication is that the number of columns in the first matrix ($\mathbf{A}$ in this case) must be equal to the number of rows in the second matrix (here, $\mathbf{B}$); otherwise the matrices are not conformable for multiplication. For example, $\mathbf{A}$ is a 3×3 matrix and $\mathbf{B}$ is a 3×2 matrix; hence, it is possible to calculate $\mathbf{C} = \mathbf{AB}$ as $\mathbf{A}$ has 3 columns and $\mathbf{B}$ has 3 rows. However, it would not be possible to calculate $\mathbf{D} = \mathbf{BA}$ as $\mathbf{B}$ has 2 columns while $\mathbf{A}$ has 3 rows. This leads to another important point: that the ordering of the matrix multiplication is important, and that for two square matrices ($\mathbf{E}$ and $\mathbf{F}$) with the same number of rows and columns, $\mathbf{EF} \neq \mathbf{FE}$, except where there is a special relationship between the two matrices. Below we provide some simple examples of matrix multiplication, but leave details of the working as an exercise.

$$
\begin{aligned}
\mathbf{C} &= \mathbf{AB} \\
&= \begin{bmatrix} 4 & 2 & 1 \\ 3 & 5 & 0 \\ 1 & 4 & 2 \end{bmatrix} \begin{bmatrix} -1 & 6 \\ 0 & 2 \\ 1 & -3 \end{bmatrix}. \\
&= \begin{bmatrix} -3 & 25 \\ -3 & 28 \\ 1 & 8 \end{bmatrix}
\end{aligned}
$$

$$d = vA$$

$$= [5 \quad 1 \quad 3] \begin{bmatrix} 4 & 2 & 1 \\ 3 & 5 & 0 \\ 1 & 4 & 2 \end{bmatrix}.$$

$$= [26 \quad 27 \quad 11]$$

$$e = vAv^T$$

$$= [5 \quad 1 \quad 3] \begin{bmatrix} 4 & 2 & 1 \\ 3 & 5 & 0 \\ 1 & 4 & 2 \end{bmatrix} \begin{bmatrix} 5 \\ 1 \\ 3 \end{bmatrix}$$

$$= 190$$

In practice, one would often perform matrix multiplication using computer software, either specialized mathematical software, or a spreadsheet package, as many of these include matrix manipulation functions.

A.3. DIFFERENTIATION AND INTEGRATION

Differentiation and integration are very extensive topics that are presented in detail in many books on calculus (e.g., Anton 1988). Here we do not attempt to provide details on the mechanics of these methods, but instead attempt to simply provide an interpretation of the concepts to aid the reader's understanding of why they are used in this book.

The geometric interpretation of the derivative of a function $f(x)$ is the slope of the tangent line to $f(x)$ at the point x, or the instantaneous rate of change in $f(x)$ at x (Fig. A.1). The derivative with respect to x is often denoted as $df(x)/dx$ or $f'(x)$. If the function $f(x)$ is increasing as x increases, then the slope of the tangent line, and hence the derivative, is positive, while if the function is decreasing as x increases, then the derivative will be negative. An important result in the context of this book is that at the point where $f(x)$ is maximized, the derivative of the function will be zero. This result (or the geometric interpretation of it) is used to obtain estimates of the parameter values that maximize the likelihood function (i.e., obtain maximum likelihood estimates).

Integration is simply a technique for finding the area under a curve defined by the function $f(x)$ between an upper limit b and a lower limit a, of x [Fig. A.2(a)]. This is formally notated as $Area = \int_a^b f(x)dx$, where "dx" denotes that the integration is performed "with respect to x" in case the function contains more than one variable. For many simple problems, the integration can be per-

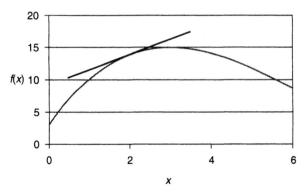

FIGURE A.1 A function $f(x)$ (grey line), and the associated tangent line at $x = 2$ (black line). The slope of the indicated tangent line is the derivative of $f(x)$ at $x = 2$.

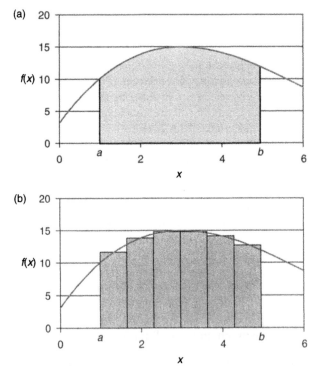

FIGURE A.2 Evaluating the area under the curve $f(x)$ between the points $x = a$ and $x = b$ using integration. Panel (a) indicates the area to be calculated, and panel (b) illustrates how this may be approximated using a number of rectangular areas.

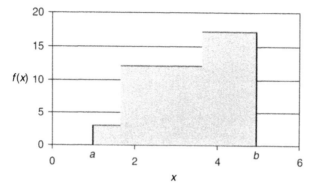

FIGURE A.3 Determining the area under the "curve" by integration for a noncontinuous function between the points $x = a$ and $x = b$.

formed analytically using calculus, but often it must be approximated numerically. Many numerical integration algorithms are based on the view that the area under a curve can be approximated by dividing up the finite interval of x from a to b into many (say, n) small sub-intervals that each form a small rectangular area [Fig. A.2(b)]. The area under the curve between the limits of a and b can thus be approximated by adding together the areas of all the rectangles. Obviously, the approximation improves as the rectangles become narrower by increasing the number of subintervals used (i.e., increasing n). Hence the view, as noted in Chapter 3, that integration can be considered as the sum of a very large number of small terms (as the area of each rectangle will become smaller as n is increased). Also note that while the function illustrated in Fig. A.2 is smooth and continuous between a and b, it does not necessarily have to be so for integration to be used. The function may also be stepwise, as in Fig. A.3, in which case the area under the "curve" can be well approximated by relatively few rectangular areas. Hence integration can be performed on both continuous and discrete random variables.

Finally, one common use of integration in statistics is to calculate the expected value of a random variable or a function of a random variable (see Chapter 3, Section 3.1). This may be done, for example, when for a given data point, the random variable itself is unobserved (i.e., latent). The probability (or the likelihood) of observing that data point is therefore dependent on the unknown value of a random variable, which may take many possible values as defined by the probability distribution for that random variable. The latent random variable can be "removed" from the likelihood by integration, which often amounts to finding the expected value of a function of the random variable (e.g., as used in Chapter 5).

References

Akaike, H. (1973). Information theory and an extension of the maximum likelihood principle. *In* "Second International Symposium Information Theory" (B. N. Petrov and F. Csáaki, eds.), pp. 267–281. Akademiai Kidao, Budapest, Hungary.

Alldredge, M. W. (2004). Avian Point Count Surveys: Estimating Components of the Detection Process. Ph.D. Thesis, North Carolina State University, Raleigh, NC.

Alpizar-Jara, R., Nichols, J. D., Hines, J. E., Sauer, J. R., Pollock, K. H., and Rosenberry, C. S. (2004). The relationship between species detection probability and local extinction probability. *Oecologia* 141, 652–660.

Anderson, D. R., Burnham, K. P., White, G. C., and Otis, D. L. (1983). Density-estimation of small-mammal populations using a trapping web and distance sampling methods. *Ecology* 64, 674–680.

Anderson, R. M., and May, R. M. (1991). "Infectious Diseases of Humans: Dynamics and Control." Oxford Univ. Press, Oxford, UK.

Anderson, R. P. (2003). Real vs. artefactual absences in species distributions: tests for *Oryzomys albigularis* (Rodentia: Muridae) in Venezuela. *J. Biogeography* 30, 591–605.

Andren, H. (1994). Can one use nested subset pattern to reject the random sample hypothesis? Examples from boreal bird communities. *Oikos* 70, 489–491.

Andrewartha, H. G., and Birch, L. C. (1954). "The Distribution and Abundance of Animals." Univ. of Chicago Press, Chicago, IL.

Anton, H. (1988). "Calculus with Analytic Geometry," 3rd Ed. Wiley, New York.

Araujo, M. B., and Williams, P. H. (2000). Selecting areas for species persistence using occurrence data. *Biol. Conserv.* 96, 331–345.

Araujo, M. B., Williams, P. H., and Fuller, R. J. (2002). Dynamics of extinction and the selection of nature reserves. *Proc. Roy. Soc. Lond. Ser. B* 269, 1971–1980.

Arnason, A. N., Schwarz, C. J., and Gerrard, J. M. (1991). Estimating closed population-size and number of marked animals from sighting data. *J. Wildl. Manage.* 55, 716–730.

Arrhenius, O. (1921). Species and area. *J. Ecol.* 9, 95–99.

Azuma, D. L., Baldwin, J. A., and Noon, B. R. (1990). Estimating the occupancy of spotted owl habitat areas by sampling and adjusting for bias. *USDA Gen. Tech. Rep. PSW-124.* Berkeley, CA.

Bailey, L. L., Simons, T. R., and Pollock, K. H. (2004). Estimating site occupancy and species detection probability parameters for terrestrial salamanders. *Ecol. Appl.* 14, 692–702.

Bailey, N. T. J. (1975). "The Mathematical Theory of Infectious Diseases," 2nd Ed. Macmillan, New York.

Barbraud, C., Nichols, J. D., Hines, J. E., and Hafner, H. (2003). Estimating rates of local extinction and colonization in colonial species and an extension to the metapopulation and community levels. *Oikos* 101, 113–126.

Barker, R. J., and Sauer, J. R. (1992). Modeling population change from time series data. *In* "Wildlife 2001: Populations." (N. Cappucino and P. W. Price, eds.), pp. 182–194. Elsevier Applied Sciences, New York.

Bart, J., and Klosiewski, S. P. (1989). Use of presence-absence to measure changes in avian density. *J. Wildl. Manage.* 53, 847–852.

Bock, C. E. (1984). Geographical correlates of rarity vs. abundance in some North American winter landbirds. *Auk* 101, 266–273.

Bock, C. E., and Ricklefs, R. E. (1983). Range size and local abundance of some North American songbirds: a positive correlation. *Amer. Natur.* 122, 295–299.

Bodenheimer, F. S. (1938). "Problems of Animal Ecology." Clarendon Press, Oxford, UK.

Bolger, D. T., Albert, A. C., and Soule, M. E. (1991). Occurrence patterns of bird species in habitat fragments: sampling, extinction, and nested subsets. *Amer. Natur.* 137, 155–166.

Borchers, D. L., Buckland, S. T., and Zucchini, W. (2003). "Estimating Animal Abundance." Springer, New York.

Boulinier, T., Nichols, J. D., Sauer, J. R., Hines, J. E., and Pollock, K. H. (1998a). Estimating species richness: the importance of heterogeneity in species detectability. *Ecology* 79, 1018–1028.

Boulinier, T., Nichols, J. D., Hines, J. E., Sauer, J. R., Flather, C. H., and Pollock, K. H. (1998b). Higher temporal variability of forest breeding bird communities in fragmented landscapes. *Proc. Natl. Acad. Sci. U.S.A.* 95, 7497–7501.

Boulinier, T., Nichols, J. D., Hines, J. E., Sauer, J. R., Flather, C. H., and Pollock, K. H. (2001). Forest fragmentation and bird community dynamics: inference at regional scales. *Ecology* 82, 1159–1169.

Bradford, D. F., Neale, A. C., Nash, M. S., Sada, D. W., and Jaeger, J. R. (2003). Habitat patch occupancy by toads (*Bufo punctatus*) in a naturally fragmented desert landscape. *Ecology* 84, 1012–1023.

Brandle, M., Stadler, J., Klotz, J., and Brandl, R. (2003). Distributional range size of weedy plant species is correlated to germination patterns. *Ecology* 84, 136–144.

Brooks, S. P., Catchpole, E. A., Morgan, B. J. T., and Barry, S. C. (2000). On the Bayesian analysis of ring-recovery data. *Biometrics* 56, 951–956.

Brooks, S. P., Catchpole, E. A., Morgan, B. J. T., and Harris, M. P. (2002). Bayesian methods for analyzing ringing data. *J. Appl. Stat.* 29, 187–206.

Brown, J. H. (1984). On the relationship between abundance and distribution of species. *Amer. Natur.* 124, 255–279.

Brown, J. H. (1995). Macroecology. Univ. of Chicago Press, Chicago, IL.

Brown, J. H. (1999). Macroecology: progress and prospect. *Oikos* 87, 3–14.

Brown, J. H., and Kodric-Brown, A. (1977). Turnover rates in insular biogeography: effect of immigration on extinction. *Ecology* 58, 445–449.

Brown, J. H., and Maurer, B. A. (1987). Evolution of species assemblages: effects of energetic constraints and species dynamics on the diversification of North American avifauna. *Amer. Natur.* 130, 1–17.

Brown, J. H., and Maurer, B. A. (1989). Macorecology: the division of food and space among species on continents. *Science* 243, 1145–1150.

Brown, J. H., Stevens, G. C., and Kaufman, D. M. (1996). The geographic range: size, shape, boundaries, and internal structure. *Ann. Rev. Ecol. Syst.* 27, 597–623.

Brownie, C., and Robson, D. S. (1976). Models allowing for age-dependent survival rates for band-return data. *Biometrics* 32, 305–323.

Brownie, C., Anderson, D. R., Burnham, K. P., and Robson, D. R. (1978). Statistical inference from band recovery data—a handbook. *U.S. Fish and Wildl. Serv. Resour. Publ.* 131.

Brownie, C., Anderson, D. R., Burnham, K. P., and Robson, D. R. (1985). Statistical inference from band recovery data—a handbook, 2nd Ed. *U.S. Fish and Wildl. Serv. Resour. Publ.* 156.

Buckland, S. T., Anderson, D. R., Burnham, K. P., and Laake, J. L. (1993). "Distance Sampling: Estimation of Biological Populations." Chapman and Hall, New York.

Buckland, S. T., Anderson, D. R., Burnham, K. P., Laake, J. L., Borchers, D. L., and Thomas, L. (2001). "Introduction to Distance Sampling." Oxford Univ. Press, Oxford.

Bunge, J., and Fitzpatrick, M. (1993). Estimating the number of species—a review. *J. Am. Stat. Assoc.* 88, 364–373.

Burnham, K. P. (1993). A theory for combined analysis of ring recovery and recapture data. *In* "Marked Individuals in the Study of Bird Populations." (J. D. Lebreton and P. M. North, eds.), pp. 199–214. Birkhauser-Verlag, Basel.

Burnham, K. P., and Anderson, D. R. (1998). "Model Selection and Inference: A Practical Information-theoretic Approach." Springer-Verlag, New York.

Burnham, K. P., and Anderson, D. R. (2002). "Model Selection and Multimodal Inference," 2nd Ed. Springer-Verlag, New York.

Burnham, K. P., and Anderson, D. R. (2004). Multimodal inference: understanding AIC and BIC in model selection. *Socio. Meth. Res.* 33, 261–304.

Burnham, K. P., and Overton, W. S. (1978). Estimation of the size of a closed population when capture probabilities vary among animals. *Biometrika* 65, 625–633.

Burnham, K. P., and Overton, W. S. (1979). Robust estimation of population size when capture probabilities vary among animals. *Ecology* 60, 927–936.

Burnham, K. P., Anderson, D. R., and Laake, J. L. (1980). Estimation of density from line-transect sampling of biological populations. *Wildl. Monogr.* 72.

Burnham, K. P., Anderson, D. R., White, G. C., Brownie, C., and Pollock, K. P. (1987). Design and analysis of methods for fish survival experiments based on release-recapture. *Am. Fish. Soc. Monogr.* 5, 1–437.

Buzas, M. A., Koch, C. F., Culver, S. J., and Sohl, N. F. (1982). On the distribution of species occurrence. *Paleobiology* 8, 143–150.

Cabeza, M., Araujo, M. B., Wilson, R. J., Thomas, C. D., Cowley, M. J. R., and Moilanen, A. (2004). Combining probabilities of occurrence with spatial reserve design. *J. Appl. Ecol.* 41, 252–262.

Cam, E., Nichols, J. D., Sauer, J. R., Hines, J. E., and Flather, C. H. (2000a). Relative species richness and community completeness: avian communities and urbanization in the mid-Atlantic states. *Ecol. Appl.* 10, 1196–1210.

Cam, E., Nichols, J. D., Hines, J. E., and Sauer, J. R. (2000b). Inferences about nested subsets structure when not all species are detected. *Oikos* 91, 428–434.

Cam, E., Nichols, J. D., Hines, J. E., Sauer, J. R., Alpizar-Jara, R., and Flather, C. H. (2002). Disentangling sampling and ecological explanations underlying species-area relationships. *Ecology* 83, 1118–1130.

Carroll, C., Zielinski, W. J., and Noss, R. F. (1999). Using presence-absence data to build and test spatial habitat models for the fisher in Klamath region, U.S.A. *Cons. Biol.* 13, 1344–1359.

Casella, G., and Berger, R. L. (2002). "Statistical Inference." Duxbury/Thompson Learning, CA.

Casula, P., and Nichols, J. D. (2003). Temporal variability of local abundance, sex ratio and activity in the Sardinian chalk hill blue butterfly. *Oecologia* 136, 374–382.

Caswell, H. (1978). Predator-mediated coexistence: a nonequilibrium model. *Amer. Natur.* 112, 127–154.

Caswell, H., and Cohen, J. E. (1991a). Communities in patchy environments: a model of disturbance, competition, and heterogeneity. *In* "Ecological Heterogeneity" (J. Kolasa and S. T. A. Pickett, eds.), pp. 97–122. Springer-Verlag, New York.

Caswell, H., and Cohen, J. E. (1991b). Disturbance, interspecific interaction and diversity in metapopulations. *Biol. J. Linn. Soc.* 42, 193–218.

Caughley, G., Sinclair, R., and Scott-Kemmis, D. (1976). Experiments in aerial survey. *J. Wildl. Manage.* 40, 290–300.

Caughley, G., Grice, D., Barker, R., and Brown, B. (1988). The edge of range. *J. Anim. Ecol.* 57, 771–785.

Ceballos, G., and Ehrlich, P. R. (2002). Mammal population losses and the extinction crisis. *Science* 296, 904–907.

Chao, A., and Lee, S.-M. (1992). Estimating the number of classes via sample coverage. *J. Amer. Stat. Assoc.* 87, 210–217.

Chao, A., Yip, P., and Lin, H.-S. (1996). Estimating the number of species via a martingale estimating function. *Statistica Sinica* 6, 403–418.

Chapman, D. G., and Robson, D. S. (1960). The analysis of a catch curve. *Biometrics* 16, 354–368.

Chivers, D. J. (1991). Guidelines for re-introductions: procedures and problems. *Symp. Zoo. Soc. of London* 62, 89–99.

Chown, S. L., van Rensburg, B. J., Gaston, K. J., Rodrigues, A. S. L., and van Jaarsveld, A. S. (2003). Energy, species richness, and human population size: conservation implications at a national scale. *Ecol. Appl.* 13, 1233–1241.

Clark, C. W., and Rosenzweig, M. L. (1994). Extinction and colonization processes: oarameter estimates from sporadic surveys. *Amer. Natur.* 143, 583–596.

Clinchy, M., Haydon, D. T., and Smith, A. T. (2002). Pattern does not equal process: what does patch occupancy really tell us about metapopulation dynamics? *Amer. Natur.* 159, 351–362.

Cochran, W. G. (1977). "Sampling Techniques." Wiley, New York.

Cole, L. C. (1949). The measurement of interspecific association. *Ecology* 30, 411–424.

Colwell, R. K., and Coddington, J. A. (1994). Estimating terrestrial biodiversity through extrapolation. *Proc. Roy. Soc. Lond. B* 345, 101–118.

Connell, J. H. (1961). The influence of interspecific competition and other factors on the distribution of the barnacle *Chthamalus stellatus*. *Ecology* 42, 710–723.

Connell, J. H. (1980). Diversity and the coevolution of competitors, or the ghost of competition past. *Oikos* 35, 131–138.

Connolly, S. R., and Miller, A. I. (2001a). Joint estimation of sampling and turnover rates from fossil databases: capture-mark-recapture methods revisited. *Paleobiology* 27, 751–767.

Connolly, S. R., and Miller, A. I. (2001b). Global Ordovician faunal transitions in the marine benthos: proximate causes. *Paleobiology* 27, 779–795.

Connolly, S. R., and Miller, A. I. (2002). Global Ordovician faunal transitions in the marine benthos: ultimate causes. *Paleobiology* 28, 26–40.

Connor, E. F., and McCoy, E. D. (1979). The statistics and biology of the species-area relationship. *Amer. Natur.* 113, 791–833.

Connor, E. F., and Simberloff, D. (1978). Species number and compositional similarity of the Galapagos flora and avifauna. *Ecol. Monogr.* 48, 219–248.

Connor, E. F., and Simberloff, D. (1979). The assembly of species communities: chance or competition? *Ecology* 60, 1132–1140.

Connor, E. F., and Simberloff, D. (1984). Neutral models of species co-occurrence patterns. *In* "Ecological Communities: Conceptual Issues and the Evidence" (D. R. Strong, Jr., D. Simberloff, L. G. Abele, and A. B. Thistle, eds.), pp. 316–331. Princeton Univ. Press, Princeton, NJ.

Connor, E. F., and Simberloff, D. (1986). Competition, scientific method, and null models in ecology. *Amer. Scientist* 74, 155–162.

Conroy, M. J., and Nichols, J. D. (1984). Testing for variation in taxonomic extinction probabilities: a suggested methodology and some results. *Paleobiology* 10, 328–337.

Conroy, M. J., and Nichols, J. D. (1996). Designing a study to assess mammalian diversity. *In* "Measuring and Monitoring Biological Diversity: Standard Methods for Mammals" (D. E. Wilson, F. R. Cole, J. D. Nichols, R. Rudran, and M. S. Foster, eds.), pp. 41–49. Smithsonian Institution Press, Washington, D.C.

Cook, R. D., and Jacobson, J. O. (1979). A design for estimating visibility bias in aerial surveys. *Biometrics* 35, 735–742.

Cook, R. R., and Quinn, J. F. (1995). The influence of colonization in nested subsets. *Oecologia* 102, 413–424.

Cook, R. R., and Quinn, J. F. (1998). An evaluation of randomization models for nested subset analysis. *Oecologia* 113, 584–592.

Cornell, H. V. (1993). Unsaturated patterns in species assemblages: the role of regional processes in setting local richness. *In* "Species Diversity in Ecological Communities. Historical and Geographical Perspectives" (R. E. Ricklefs and D. Schluter, eds.), pp. 243–353. University of Chicago Press, Chicago, IL.

Cornell, H. V., and Lawton, J. H. (1992). Species interactions, local and regional processes, and limits to the richness of ecological communities: a theoretical perspective. *J. Anim. Ecol.* 61, 1–12.

Coull, B. A., and Agresti, A. (1999). The use of mixed logit models to reflect heterogeneity in capture-recapture studies. *Biometrics.* 55, 294–301.

Crow, J. F., and Kimura, M. (1970). "An Introduction to Population Genetics Theory." Harper and Row, New York.

Crowell, K. L., and Pimm, S. L. (1976). Competition and niche shifts of mice introduced onto small islands. *Oikos* 27, 251–258.

Curnutt, J. L., Pimm, S. L., and Maurer, B. A. (1996). Population variability of sparrows in space and time. *Oikos* 76, 131–144.

Darwin, C. (1859). "On the Origin of Species by Means of Natural Selection, or the Preservation of Favored Races in the Struggle for Life." John Murray, London.

Deredec, A., and Courchamp, F. (2003). Extinction thresholds in host-parasite dynamics. *Ann. Zool. Fennici* 40, 115–130.

Diamond, J. (1982). Effect of species pool size on species occurrence frequencies: musical chairs on islands. *Proc. Natl. Acad. Sci. U.S.A.* 79, 2420–2424.

Diamond, J. M. (1975a). Assembly of species communities. *In* "Ecology and Evolution of Communities" (M. L. Cody and J. M. Diamond, eds.), pp. 342–444. Harvard Univ. Press, Cambridge, MA.

Diamond, J. M. (1975b). The island dilemma: lessons of modern biogeographic studies for the design of nature reserves. *Biol. Cons.* 7, 129–146.

Diamond, J. M. (1976). Island biogeography and conservation: strategy and limitations. *Science* **193**, 1027–1029.

Diamond, J. M., and May, R. M. (1977). Species turnover rates on islands: dependence on census interval. *Science* **197**, 266–270.

Diamond, J. M., and Gilpin, M. E. (1982). Examination of the "null" model of Connor and Simberloff for species co-occurrences on islands. *Oecologia* **52**, 64–74.

DiCastri, F., Hansen, A. J., and Debussche, M. (1990). "Biological Invasions in Europe and the Mediterranean Basin." Kluwer Academic Publ., Dordrecht, Netherlands.

Dice, L. R. (1945). Measures of the amount of ecologic association between species. *Ecology* **26**, 297–302.

Dice, L. R. (1948). Relationship between frequency index and population density. *Ecology* **29**, 289–391.

Dodd, C. K. (2003). Monitoring amphibians in Great Smoky Mountains National Park. *U.S. Geological Survey Circular* 1258, Tallahassee, FL.

Doherty, P. F., Jr., Sorci, G., Royle, J. A., Hines, J. E., Nichols, J. D., and Boulinier, T. (2003a). Sexual selection affects local extinction and turnover in bird communities. *Proc. Natl. Acad. Sci. U.S.A.* **100**, 5858–5862.

Doherty, P. F., Jr., Boulinier, T., and Nichols, J. D. (2003b). Extinction rates at the center and edge of species' ranges. *Ann. Zool. Fennici* **40**, 145–153.

Dorazio, R. M., and Royle, J. A. (2003). Mixture models for estimating the size of a closed population when capture rates vary among individuals. *Biometrics* **59**, 351–364.

Dorazio, R. M., Royle, J. A., Soderstrom, B., and Glimskar, A. Estimating species richness and accumulation by modeling species occurrence and detectability. *Ecology* (in review).

Dorazio, R. M., and Royle, J. A. (2005). Estimating size and composition of biological communities by modeling the occurrence of species. *J. Am. Stat. Assoc.* (in press).

Dunham, J. B., and Rieman, B. E. (1999). Metapopulation structure of bull trout: influence of physical, biotic, and geometrical landscape characteristics. *Ecol. Appl.* **9**, 642–655.

Dunham, J. B., Rieman, B. E., and Peterson, J. T. (2002). Patch-based models to predict species occurrence: lessons from salmonid fishes in streams. *In* "Predicting Species Occurrences" (J. M. Scott, P. J. Heglund, M. L. Morrison, J. B. Haufler, M. G. Raphael, W. A. Wall, and F. B. Samson, eds.), pp. 327–334. Island Press, Washington, D.C.

Edwards, T. C., Jr., Cutler, D. R., Geiser, L., Alegria, J., and McKenzie, D. (2004). Assessing rarity of species with low detectability: lichens in Pacific Northwest forests. *Ecol. Appl.* **14**, 414–424.

Efford, M. (2004). Density estimation in live-trapping studies. *Oikos* **106**, 598–610.

Efron, B. (1979). Computers and the theory of statistics: thinking the unthinkable. *Soc. Industr. Appl. Math.* **21**, 460–480.

Elliott, P., Wakefield, J. C., Best, N. G., and Briggs, D. J. (2001). "Spatial Epidemiology: Methods and Applications." Oxford Univ. Press, Oxford, UK.

Ellner, S. P., and Fussmann, G. (2003). Effects of successional dynamics on metapopulation persistence. *Ecology* **84**, 882–889.

Elton, C. (1927). "Animal Ecology." Sidgwick and Jackson, London.

Engler, R., Guisan, A., and Rechsteiner, L. (2004). An improved approach for predicting the distribution of rare and endangered species from occurrence and pseudo-absence data. *J. Appl. Ecol.* **41**, 263–274.

Enquist, B. J., Jordan, M. A., and Brown, J. H. (1995). Connections between ecology, biogeography, and paleobiology: relationship between local abundance and geographic distribution in fossil and recent molluscs. *Evol. Ecol.* **9**, 586–604.

Erwin, R. M., Nichols, J. D., Eyler, T. B., Stotts, D. B., and Truitt, B. R. (1998). Modeling colony-site dynamics: a case study of gull-billed terns (*Sterna nilotica*) in coastal Virginia. *Auk* **115**, 970–978.

Farnsworth, G. L., Pollock, K. H., Nichols, J. D., Simons, T. R., Hines, J. E., and Sauer, J. R. (2002). A removal model for estimating detection probabilities from point-count surveys. *Auk* 119, 414–425.

Ferraz, G., Russell, G. J., Stouffer, P. C., Bieregaard, R. O., Jr., Pimm, S. L., and Lovejoy, T. E. (2003). Rates of species loss from Amazonian forest fragments. *Proc. Nat. Acad. Sci.* 100, 14069–14073.

Fertig, W., and Reiners, W. A. (2002). Predicting presence/absence of plant species for range mapping: a case study from Wyoming. *In* "Predicting Species Occurrences" (J. M. Scott, P. J. Heglund, M. L. Morrison, J. B. Haufler, M. G. Raphael, W. A. Wall, and F. B. Samson, eds.), pp. 483–489. Island Press, Washington, D.C.

Field, S. A., Tyre, A. J., and Possingham, H. P. (2005). Optimizing allocation of monitoring effort under economic and observational constraints. *J. Wildl. Manage.* 69, 473–482.

Fienberg, S. E., Johnson, M. S., and Junker, B. W. (1999). Classical multilevel and Bayesian approaches to population size estimation using multiple lists. *J. Royal Stat. Soc. London Ser. A* 163, 383–405.

Fischer, J., Lindenmayer, D. B., and Cowling, A. (2004). The challenge of managing multiple species at multiple scales: reptiles in an Australian grazing landscape. *J. Appl. Ecol.* 41, 32–44.

Fisher, R. A. (1922). On the mathematical foundations of theoretical statistics. *Phil. Trans. Royal Soc. London Ser. A* 222, 309–368.

Fisher, R. A. (1947). "The Design of Experiments," 4th Ed. Hafner, New York.

Fisher, R. A., Corbet, A. S., and Williams, C. B. (1943). The relation between the number of species and the number of individuals in a random sample of an animal population. *J. Anim. Ecol.* 12, 42–58.

Flather, C. H. (1996). Fitting species-accumulation functions and assessing regional land use impacts on avian diversity. *J. Biogeography* 23, 155–168.

Fleishman, E., and Murphy, D. D. (1999). Patterns and processes of nestedness in a Great Basin butterfly community. *Oecologia* 119, 133–139.

Fleishman, E., MacNally, R., Fay, J. P., and Murphy, D. D. (2001). Modeling and predicting species occurrence using broad-scale environmental variables: an example with butterflies of the Great Basin. *Cons. Biol.* 15, 1674–1685.

Forbes, S. A. (1907). On the local distribution of certain Illinois fishes: an essay in statistical ecology. *Bull. Illinois St. Lab. Natur. Hist.* 7, 273–303.

Forman, R. T. T., Galli, A. E., and Leck, C. F. (1976). Forest size and avian diversity in New Jersey woodlots with some land-use applications. *Oecologia* 26, 1–8.

Franklin, A. B., Anderson, D. R., Forsman, E. D., Burnham, K. P., and Wagner, F. W. (1996). Methods for collecting and analyzing demographic data on the northern spotted owl. *Studies in Avian Biol.* 17, 12–20.

Franklin, A. B., Anderson, D. R., Gutuerrez, R. J., and Burnham, K. P. (2000). Climate, habitat quality, and fitness in northern spotted owl populations in northwestern California. *Ecol. Monogr.* 70, 539–590.

Franklin, A. B., Guttierez, R. J., Nichols, J. D., Seamans, M. E., White, G. C., Zimmerman, G. S., Hines, J. E., Munton, T. E., LaHaye, W. S., Blakesley, J. A., Steger, G. N., Noon, B. R., Shaw, D. W. H., Keane, J. J., McDonald, T. L., Britting, S. (2004). Population dynamics of the California spotted owl: a meta analysis. *American Ornithologists' Union Monogr.* 54.

Fretwell, S. D. (1972). "Populations in a Seasonal Environment." Princeton Univ. Press, Princeton, NJ.

Fretwell, S. D., and Lucas, H. R. (1969). On territorial behavior and other factors influencing habitat distribution in birds. I. Theoretical development. *Acta Biotheoret.* 19, 16–36.

Fritts, T. H., and Rodda, G. H. (1998). The role of introduced species in the degradation of island ecosystems: a case study of Guam. *Ann. Rev. Ecol. Syst.* 29, 113–140.

Fujiwara, M., and Caswell, H. (2002). Estimating population projection matrices from multi-stage mark-recapture data. *Ecology* 83, 3257–3265.

Gaston, K. J. (1990). Patterns in the geographical ranges of species. *Biol. Rev.* 65, 105–129.

Gaston, K. J. (1991). How large is a species' geographic range? *Oikos* 61, 434–438.

Gaston, K. J. (1994). "Rarity." Chapman and Hall, London.

Gaston, K. J. (1996). The multiple forms of the interspecific abundance-distribution relationship. *Oikos* 75, 211–220.

Gaston, K. J. (1998). Species-range size distributions: products of speciation, extinction and transformation. *Proc. Roy. Soc. Lond., Ser. B* 353, 219–230.

Gaston, K. J., and Blackburn, T. M. (1996). Global scale macroecology: interactions between population size, geographic range size and body size in the Anseriformes. *J. Anim. Ecol.* 65, 701–714.

Gaston, K. J., and Blackburn, T. M. (1999). A critique for macroecology. *Oikos* 84, 353–368.

Gaston, K. J., Blackburn, T. M., and Lawton, J. H. (1997a). Interspecific abundance-range size relationships: an appraisal of mechanisms. *J. of Anim. Ecol.* 66, 579–601.

Gaston, K. J., Blackburn, T. M., and Gregory, R. D. (1997b). Interspecific abundance-range size relationships: range position and phylogeny. *Ecography* 20, 390–399.

Geissler, P. H., and Fuller, M. R. (1987). Estimation of the proportion of area occupied by an animal species. *Proceedings of the Section on Survey Research Methods of the American Statistical Association* 1986, 533–538.

Gelman, A., Carlin, J. B., Stern, H. S., and Rubin, D. R. (1998). "Bayesian Data Analysis." Chapman and Hall, London.

Gerrard, D. J., and Chiang, H. C. (1970). Density estimation of corn rootworm egg populations based upon frequency of occurrence. *Ecology* 51, 237–245.

Gibson, L. A., Wilson, B. A., Cahill, D. M., and Hill, J. (2004). Spatial prediction of rufous bristlebird habitat in a coastal heathland: a GIS-based approach. *J. Appl. Ecol.* 41, 213–223.

Gilpin, M. E., and Diamond, J. M. (1982). Factors contributing to nonrandomness in species co-occurrences on islands. *Oecologia* 52, 75–84.

Gilpin, M. E., and Diamond, J. M. (1984). Are species co-occurrences on islands non-random, and are null hypotheses useful in community ecology? *In* "Ecological Communities: Conceptual Issues and the Evidence" (D. R. Strong, Jr., D. Simberloff, L. G. Abele, and A. B. Thistle, eds.), pp. 297–315. Princeton Univ. Press, Princeton, NJ.

Glazier, D. S. (1986). Temporal variability of abundance and the distribution of species. *Oikos* 47, 309–314.

Gleason, H. A. (1920). Some applications of the quadrat method. *Bull. Torrey Botanical Club* 47, 21–33.

Gotelli, N. J. (2000). Null model analysis of species co-occurrence patterns. *Ecology* 81, 2606–2621.

Gotelli, N. J., and Graves, G. R. (1996). "Null Models in Ecology." Smithsonian Institution Press, Washington, D.C.

Gotelli, N. J., and McCabe, D. J. (2002). Species co-occurrence: a meta-analysis of J. M. Diamond's assembly rules model. *Ecology* 83, 2091–2096.

Gotelli, N. J., and Simberloff, D. (1987). The distribution and abundance of tallgrass prairie plants: a test of the core-satellite hypothesis. *Amer. Natur.* 130, 18–35.

Gotelli, N. J., Buckley, N. J., and Wiens, J. A. (1997). Co-occurrence of Australian birds: Diamond's assembly rules revisited. *Oikos* 80, 311–324.

Grayson, D. K., and Livingston, S. D. (1993). Missing mammals on Great Basin mountains: holocene extinctions and inadequate knowledge. *Cons. Biol.* 7, 527–532.

Green, R. H. (1979). "Sampling Design and Statistical Methods for Environmental Biologists." Wiley, New York.

Griffith, B., Scott, J. M., Carpenter, J. W., and Reed, C. (1989). Translocation as a species conservation tool: status and strategy. *Science* 245, 477–480.

Gu, W., and Swihart, R. K. (2004). Absent or undetected? Effects of non-detection of species occurrence on wildlife-habitat models. *Biol. Cons.* 116, 195–203.

Haila, Y., Hanski, I. K., and Raivio, S. (1993). Turnover of breeding birds in small forest fragments: the "sampling" colonization hypothesis corroborated. *Ecology* 74, 714–725.

Haldane, J. B. S. (1955). The calculation of mortality rates from ringing data. *Proc. International Congress of Ornithologists* 9, 454–458.

Hall, R. J., and Langtimm, C. A. (2001). The U.S. national amphibian research and monitoring initiative and the role of protected areas. *George Wright Forum* 18, 14–25.

Hames, R. S., Rosenberg, K. V., Lowe, J. D., and Dhondt, A. A. (2001). Site reoccupation in fragmented landscapes: testing predictions of metapopulation theory. *J. Anim. Ecol.* 70, 182–190.

Hames, R. S., Rosenberg, K. V., Lowe, J. D., Barker, S. E., and Dhondt, A. A. (2002). Adverse effects of acid rain on the distribution of the wood thrush *Hylocichla mustelina* in North America. *Proc. Nat. Acad. Sci.* 99, 11235–11240.

Hansen, T. A. (1980). Influence of larval dispersal and geographic distribution on species longevity in neo-gastropods. *Paleobiology* 6, 193–207.

Hanski, I. (1982). Dynamics of regional distribution: the core and satellite species hypthesis. *Oikos* 38, 210–221.

Hanski, I. (1991). Single-species metapopulation dynamics: concepts, models and observations. *Biol. J. Linnean Society* 42, 17–38.

Hanski, I. (1992). Inferences from ecological incidence functions. *Amer. Natur.* 139, 657–662.

Hanski, I. (1994a). A practical model of metapopulation dynamics. *J. Anim. Ecol.* 63, 151–162.

Hanski, I. (1994b). Patch-occupancy dynamics in fragmented landscapes. *Trends Ecol. Evolution* 9, 131–135.

Hanski, I. (1997). Metapopulation dynamics: from concepts and observations to predictive models. *In* "Metapopulation Biology: Ecology, Genetics, and Evolution" (I. A. Hanski and M. E. Gilpin, eds.), pp. 69–91. Academic Press, New York.

Hanski, I. (1998). Metapopulation dynamics. *Nature* 396, 41–49.

Hanski, I. (1999). "Metapopulation Ecology." Oxford Univ. Press, Oxford, UK.

Hanski, I., and Gaggiotti, O. E. (2004). "Ecology, Genetics, and Evolution of Metapopulations." Elsevier Academic Press, Burlington, MA.

Hanski, I., and Gilpin, M. E., eds. (1997). "Metapopulation Biology." Academic Press, San Diego.

Hanski, I., and Ovaskainen, O. (2000). The metapopulation capacity of a fragmented landscape. *Nature* 404, 755–758.

Hanski, I., Kouki, J., and Halkka, A. (1993). Three explanations of the positive relationship between distribution and abundance of species. *In* "Species Diversity in Ecological Communities" (R. E. Ricklefs and D. Schluter, eds.), pp. 108–116. Univ. Chicago Press, Chicago, IL.

Hanski, I., Moilanen, A., and Gyllenberg, M. (1996). Minimum viable metapopulation size. *Amer. Natur.* 147, 527–541.

Hanski, I., Moilanen, A., Pakkala, T., and Kuussaari, M. (1996). The quantitative incidence function model and persistence of an endangered butterfly population. *Cons. Biol.* 10, 578–590.

Hanski, I., Pakkala, T., Kuussaari, M., and Lei, G. (1995). Metapopulation persistence of an endangered butterfly in a fragmented landscape. *Oikos* 72, 21–28.

Harris, L. D. (1984). "The Fragmented Forest." Univ. Chicago Press, Chicago.

Harrison, S., Maron, J., and Huxel, G. (2000). Regional turnover and fluctuation in populations of five plants confined to serpentine seeps. *Cons. Biol.* 14, 769–779.

Havel, J. E., Shurin, J. B., and Jones, J. R. (2002). Estimating dispersal from patterns of spread: spatial and local control of lake invasions. *Ecology* 83, 3306–3318.

Hayek, L.-A. C. (1994). Analysis of amphibian biodiversity data. *In* "Measuring and Monitoring Biological Diversity: Standard Methods for Amphibians." (W. R. Heyer, M. R. Donnelly, R. W. McDiarmid, L.-A. C. Hayek, and M. S. Foster, eds.), pp. 207–273. Smithsonian Institution Press, Washington, D.C.

He, F., and Gaston, K. J. (2000). Estimating species abundance from occurrence. *Amer. Natur.* 156, 553–559.

He, F., and Gaston, K. J. (2003). Occupancy, spatial variance, and the abundance of species. *Amer. Natur.* 162, 366–375.

Hecnar, S. J., and M'Closkey, R. T. (1996). Regional dynamics and the status of amphibians. *Ecology* 77, 2091–2097.

Hedley, S. L., Buckland, S. T., and Borchers, D. L. (1999). Spatial modelling from line transect data. *J. of Cetacean Research Manage.* 1, 255–264.

Hengeveld, R. (1990). "Dynamic Biogeography." Cambridge Univ. Press, Cambridge, UK.

Hestbeck, J. B., Nichols, J. D., and Malecki, R. A. (1991). Estimates of movement and site fidelity using mark resight data of wintering Canada geese. *Ecology* 72, 523–533.

Hewitt, O. H. (1967). A road-count index to breeding populations of red-winged blackbirds. *J. Wildl. Manage.* 31, 39–47.

Hilborn, R., and Mangel, M. (1997). "The Ecological Detective: Confronting Models with Data." Princeton Univ. Press, Princeton, NJ.

Hilborn, R., and Walters, C. J. (1992). "Quantitative Fisheries Stock Assessment: Choice, Dynamics, and Uncertainty." Routledge, Chapman and Hall, New York.

Hirzel, A. H., Hausser, J., Chessel, D., and Perrin, N. (2002). Ecological-niche factor analysis: how to compute habitat-suitability maps without absence data. *Ecology* 83, 2027–2036.

Holling, C. S., ed. (1978). "Adaptive Environmental Assessment and Management." Wiley, Chichester, UK.

Holt, A. R., Gaston, K. J., and He, F. H. (2002). Occupancy-abundance relationships and spatial distribution. *Basic and Applied Ecology* 3, 1–13.

Hosmer, D. W., and Lemeshow, S. (1989). "Applied Logistic Regression." Wiley, New York.

Hurlbert, S. H. (1984). Pseudoreplication and the design of ecological field experiments. *Ecology* 54, 187–211.

Hutchinson, G. E. (1957). Concluding remarks. *Cold Spring Harbor Symposia on Quantitative Biology* 22, 415–427.

Ihaka, R., and Gentleman, R. (1996). R: a language for data analysis and graphics. *J. of Computational and Graphical Stat.* 5, 299–314.

Johnson, N. J., Kotz, S., and Kemp, A. W. (1992). "Univariate Discrete Distributions," 2nd Ed. Wiley, New York.

Johnson, D. H., Nichols, J. D., and Schwarz, M. D. (1992). Population dynamics of breeding waterfowl. *In* "Ecology and Management of Breeding Waterfowl." (B. D. J. Batt, A. D. Afton, M. G. Anderson, C. D. Ankney, D. H. Johnson, J. A. Kadlec, and G. L. Krapu, eds.), pp. 446–485. Univ. Minn. Press, Minneapolis, MN.

Johnson, F. A., Moore, C. T., Kendall, W. L., Dubosky, J. A., Caithamer, D. F., Kelley, J. R., Jr., and Williams, B. K. (1997). Uncertainty and the management of mallard harvests. *J. Wildl. Manage.* 61, 202–216.

Johnson, C. M., Johnson, L. B., Richards, C., and Beasley, V. (2002). Predicting the occurrence of amphibians: an assessment of multiple-scale models. *In* "Predicting Species Occurrences" (J. M. Scott, P. J. Heglund, M. L. Morrison, J. B. Haufler, M. G. Raphael, W. A. Wall, and F. B. Samson, eds.), pp. 157–170. Island Press, Washington, D.C.

Karanth, K. K., Nichols, J. D., Sauer, J. R., and Hines, J. E. Comparative dynamics of avian communities across edges and interiors of North American ecological regions. (in review).

Karanth, K. U., Nichols, J. D., Kumar, N. S., Link, W. A., and Hines, J. E. (2004). Tigers and their prey: predicting carnivore densities from prey abundance. *Proc. Natl. Acad. Sci.* 101, 4854–4858.

Karlson, R. H., and Cornell, H. V. (1998). Scale-dependent variation in local vs. regional effects on coral species richness. *Ecol. Monogr.* 68, 259–274.

Kawanishi, K., and Sunquist, M. E. (2004). Conservation status of tigers in a primary rainforest of peninsular Malaysia. *Biol. Cons.* 120, 329–344.

Keddy, P. A., and Drummond, C. G. (1996). Ecological properties for the evaluation, management, and restoration of temperate deciduous forest ecosystems. *Ecol. Appl.* 6, 748–762.

Kelt, D. A., Taper, M. L., and Mesevre, P. L. (1995). Assesing the impact of competition on community assembly: a case study using small mammals. *Ecology* 76, 1283–1296.

Kendall, W. L. (1999). Robustness of closed capture-recapture methods to violations of the closure assumption. *Ecology* 80, 2517–2525.

Kendall, W. L. (2001). Using models to facilitate complex decisions. *In* "Modeling in Natural Resource Management." (T. M. Shenk and A. B. Franklin, eds.), pp. 147–170. Island Press, Washington, D.C.

Kendall, W. L., and Hines, J. E. (1999). Program RDSURVIV: an estimation tool for capture-recapture data collected under Pollock's robust design. *Bird Study* 46, 32–38.

Kendall, W. L., and Nichols, J. D. (1995). On the use of secondary capture-recapture samples to estimate temporary emigration and breeding proportions. *J. Appl. Stat.* 22, 751–762.

Kendall, W. L., and Nichols, J. D. (2004). On the estimation of dispersal and movement of birds. *Condor* 106, 720–731.

Kendall, W. L., Nichols, J. D., and Hines, J. E. (1997). Estimating temporary emigration using capture-recapture data with Pollock's robust design. *Ecology* 78, 563–578.

Kendall, W. L., Hines, J. E., and Nichols, J. D. (2003). Adjusting multistate capture-recapture models for misclassification bias: manatee breeding proportions. *Ecology* 84, 1058–1066.

Kéry, M. (2004). Extinction rate estimates for plant populations in revisitation studies: importance of detectability. *Cons. Biol.* 18, 570–574.

Kimura, M., and Weiss, G. H. (1964). The stepping stone model of population structure and the decrease of genetic correlations with distance. *Genetics* 49, 561–576.

Klute, D. S., Lovallo, M. J., and Tzilkowski, W. M. (2002). Autologistic regression modeling of American woodcock habitat use with spatially dependent data. *In* "Predicting Species Occurrences" (J. M. Scott, P. J. Heglund, M. L. Morrison, J. B. Haufler, M. G. Raphael, W. A. Wall, and F. B. Samson, eds.), pp. 335–343. Island Press, Washington, D.C.

Kodric-Brown, A., and Brown, J. H. (1993). Incomplete data sets in community ecology and biogeography: a cautionary tale. *Ecol. Appl.* 3, 736–742.

Krebs, C. J. (1972). "Ecology." Harper and Row, New York.

Krebs, C. J. (1991). The experimental paradigm and long-term population studies. *Ibis* 133 suppl. 1, 3–8.

Krebs, C. J. (2001). "Ecology: The Experimental Analysis of Distribution and Abundance: Hands-on Field Package," 5th Ed. Benjamin Cummings, San Francisco, CA.

Krementz, D. G., Barker, R. J., and Nichols, J. D. (1997). Sources of variation in waterfowl survival rates. *Auk* 114, 93–102.

Kunin, W. E. (1998). Extrapolating species abundance across spatial scales. *Science* 281, 1513–1515.

Kunin, W. E., Hartley, S., and Lennon, J. J. (2000). Scaling down: on the challenge of estimating abundance from occurrence patterns. *Amer. Natur.* 156, 560–566.

Lancia, R. A., Nichols, J. D., and Pollock, K. H. (1994). Estimating the number of animals in wildlife populations. *In* "Research and Management Techniques for Wildlife and Habitats" (T. Bookhout, ed.), pp. 215–253. The Wildlife Society, Bethesda, MD.

Lancia, R. A., Kendall, W. L., Pollock, K. H., and Nichols, J. D. (2005). Estimating the number of animals in wildlife populations. In "Research and Management Techniques for Wildlife and Habitats" (C. E. Braun, ed.), pp. 105–153. The Wildlife Society, Bethesda, MD.

Lande, R. (1987). Extinction thresholds in demographic models of territorial populations. Amer. Natur. 130, 624–635.

Lande, R. (1988). Demographic models of the northern spotted owl (Strix occidentalis caurina). Oecologia 75, 601–607.

LaRoe, E. T., Farris, G. S., Puckett, C. E., Doran, P. D., and Mac, M. J. (1995). Our living resources: a report to the nation on the distribution, abundance, and health of U.S. plants, animals, and ecosystems. U.S. Dept. Interior, National Biological Service, Washington, D.C.

Lebreton, J. D., Burnham, K. P., Clobert, J., and Anderson, D. R. (1992). Modeling survival and testing biological hypotheses using marked animals—a unified approach with case-studies. Ecol. Monogr. 62, 67–118.

Lebreton, J. D., and Pradel, R. (2002). Multistate recapture models: Modelling incomplete individual histories. J. Appl. Stat. 29, 353–369.

Leopold, A. (1933). "Game Management." Charles Scribner's Sons, New York.

Levins, R. (1969). Some demographic and genetic consequences of environmental heterogeneity for biological control. Bull. Entomol. Soc. Am. 15, 237–240.

Levins, R. (1970). Extinction. In "Some Mathematical Questions in Biology, Vol. II" (M. Gustenhaver, ed.), pp. 77–107. American Mathematical Society, Providence, RI.

Lichstein, J. W., Simons, T. R., Shriner, S. A., and Franzreb, K. E. (2002). Spatial autocorrelation and autoregressive models in ecology. Ecol. Monogr. 72, 445–463.

Link, W. A. (2003). Nonidentifiability of population size from capture-recapture data with heterogeneous detection probabilities. Biometrics 59, 1123–1130.

Link, W. A. (2005). Individual heterogeneity and identifiability in capture-recapture models. Anim. Biodivers. Conserv. 27, 87–91.

Link, W. A., and Barker, R. J. (1994). Density-estimation using the trapping web design—a geometric analysis. Biometrics 50, 733–745.

Link, W. A., and Nichols, J. D. (1994). On the importance of sampling variance to investigations of temporal variation in animal population size. Oikos 69, 539–544.

Link, W. A., and Sauer, J. R. (1997). New approaches to the analysis of population trends in land birds: comment. Ecology 78, 2632–2634.

Link, W. A., and Sauer, J. R. (2002). A hierarchical analysis of population change with application to cerulean warbler. Ecology 83, 2832–2840.

Link, W. A., Barker, R. J., Sauer, J. R., and Droege, S. (1994). Within-site variability in surveys of wildlife populations. Ecology 74, 1097–1108.

Link, W. A., Cam, E., Nichols, J. D., and Cooch, E. G. (2002). Of BUGS and birds: Markov chain Monte Carlo for hierarchical modeling in wildlife research. J. Wildl. Manage. 66, 277–291.

Lomolino, M. V. (1996). Investigating causality of nestedness of insular communities: selective immigrations or extinctions? J. Biogeography 23, 699–703.

Lomolino, M. V., Brown, J. H., and Davis, R. (1989). Island biogeography of montane forest mammals in the American southwest. Ecology 70, 180–194.

Lynch, J. F., and Whigham, D. F. (1984). Effects of forest fragmentation on breeding bird communities in Maryland, USA. Biol. Conserv. 28, 287–324.

MacArthur, R. H. (1972). "Geographical Ecology." Harper and Row, New York.

MacArthur, R. H., and Wilson, E. O. (1963). An equilibrium theory of insular zoogeography. Evolution 17, 373–387.

MacArthur, R. H., and Wilson, E. O. (1967). "The Theory of Island Biogeography." Princeton Univ. Press, Princeton, NJ.

MacKenzie, D. I. (2006). Modeling the probability of use: the effect of, and dealing with, detecting a species imperfectly. *J. Wildl. Manage.* (in press).

MacKenzie, D. I. (2005a). Was it there? Dealing with imperfect detection for species presence/absence data. *Austral. & New Zealand J. of Stat.* 47, 65–74.

MacKenzie, D. I. (2005b). What are the issues with "presence/absence" data for wildlife managers? *J. Wildl. Manage.* (in press).

MacKenzie, D. I., and Bailey, L. L. (2004). Assessing the fit of site occupancy models. *J. Agri. Biol. Envir. Stat.* 9, 300–318.

MacKenzie, D. I., and Kendall, W. K. (2002). How should detection probability be incorporated into estimates of relative abundance? *Ecology* 83, 3532–3532.

MacKenzie, D. I., and Nichols, J. D. (2004). Occupancy as a surrogate for abundance estimation. *Anim. Biodivers. Conserv.* 27, 461–467.

MacKenzie, D. I., and Royle, J. A. (2005). Designing efficient occupancy studies: General advice and tips on allocation of survey effort. *J. Appl. Ecol.* (in press).

MacKenzie, D. I., Nichols, J. D., Lachman, G. B., Droege, S., Royle, J. A., and Langtimm, C. A. (2002). Estimating site occupancy rates when detection probabilities are less than one. *Ecology* 83, 2248–2255.

MacKenzie, D. I., Nichols, J. D., Hines, J. E., Knutson, M. G., and Franklin, A. D. (2003). Estimating site occupancy, colonization and local extinction when a species is detected imperfectly. *Ecology* 84, 2200–2207.

MacKenzie, D. I., Royle, J. A., Brown, J. A., and Nichols, J. D. (2004a). Occupancy estimation and modeling for rare and elusive populations. *In* "Sampling Rare or Elusive Species: Concepts Designs, and Techniques for Estimating Population Parameters" (W. L. Thompson, ed.), pp. 149–172. Island Press, Washington, D.C.

MacKenzie, D. I., Bailey, L. L., and Nichols, J. D. (2004b). Investigating species co-occurrence patterns when species are detected imperfectly. *J. Anim. Ecol.* 73, 546–555.

MacKenzie, D. I., Nichols, J. D., Sutton, N., Kawanishi, K., and Bailey, L. L. (2005). Improving inference in population studies of rare species that are detected imperfectly. *Ecology* 86, 1101–1113.

Manley, P. N., Zielinski, W. J., Schlesinger, M. D., and Mori, S. R. (2004). Evaluation of a multiple-species approach to monitoring species at the ecoregional scale. *Ecol. Appl.* 14, 296–310.

Manly, B. F. J. (1992). "The Design and Analysis of Research Studies." Cambridge Univ. Press, Cambridge, U.K.

Manly, B. F. J. (1995). A note on the analysis of species co-occurrences. *Ecology* 76, 1109–1115.

Manly, B. F. J. (1997). "Randomization, Bootstrap and Monte Carlo Methods in Biology," 2nd Ed. Chapman and Hall, London.

Manly, B. F. J., McDonald, L., Thomas, D. L., McDondald, T. L., and Erickson, W. P. (2002). "Resource Selection by Animals: Statistical Design and Analysis for Field Studies." Kluwer Academic Publishers, Boston, MA.

Marra, P. P., Griffing, S., Caffrey, C., Kilpatrick, A. M., McLean, R., Brand, C., Saito, E., Dupuis, A. L., Kramer, L., and Novak, R. (2004). West Nile virus and wildlife. *Bioscience* 54, 393–402.

Martinez-Solano, I., Bosch, J., and Garcia-Paris, M. (2003). Demographic trends and community stability in a montane amphibian assemblage. *Cons. Biol.* 17, 238–244.

Maurer, B. A. (1994). "Geographical Population Analysis." Blackwell Scientific, Oxford, UK.

McArdle, B. H. (1990). When are rare species not there? *Oikos* 57, 276–277.

McCullagh, P., and Nelder, J. A. (1989). "Generalized Linear Models." Chapman and Hall, New York.

McCullough, D. R., ed. (1996). "Metapopulations and Wildlife Conservation." Island Press, Washington, D.C.

McNaughton, S. J., and Wolf, L. L. (1970). Dominance and the niche in ecological systems. *Science* 167, 131–139.

Mehlman, D. W. (1997). Change in avian abundance across the geographical range in response to environmental change. *Ecol. Appl.* 7, 614–624.

Merila, J., and Kotze, J. D., eds. (2003). Extinction thresholds. *Ann. Zool. Fennici* 40, 69–245.

Moilanen, A. (1999). Patch occupancy models of metapopulation dynamics: Efficient parameter estimation using implicit statistical inference. *Ecology* 80, 1031–1043.

Moilanen, A. (2002). Implications of empirical data quality to metapopulation model parameter estimation and application. *Oikos* 96, 516–530.

Moilanen, A., and Cabeza, M. (2002). Single-species dynamic site selection. *Ecol. Appl.* 12, 913–926.

Moilanen, A., and Hanski, I. (1995). Habitat destruction and coexistence of competitors in a spatially realistic metapopulation model. *J. Anim. Ecol.* 64, 141–144.

Mooney, H. A., and Drake, J. A. (1986). "Ecology of Biological Invasions in North America and Hawaii." Springer-Verlag, New York.

Moore, N. W., and Hooper, M. D. (1975). On the number of bird species in British woods. *Biol. Cons.* 8, 239–250.

Morris, W. F., and Doak, D. F. (2002). "Quantitative Conservation Biology." Sinauer, Sunderland, MA.

Nee, S. (1994). How populations persist. *Nature* 367, 123–124.

Nee, S., and May, R. M. (1992). Dynamics of metapopulations: habitat destruction and competitive coexistence. *J. Anim. Ecol.* 61, 37–40.

Nichols, J. D. (1991a). Extensive monitoring programs viewed as long-term population studies: the case of North American waterfowl. *Ibis* 133 suppl. 1, 89–98.

Nichols, J. D. (1991b). Science, population ecology, and the management of the American black duck. *J. Wildl. Manage* 55, 790–799.

Nichols, J. D. (1996). Sources of variation in migratory movements of animal populations: statistical inference and a selective review of empirical results for birds. *In* "Population Dynamics in Ecological Space and Time" (O. E. Rhodes, Jr., R. K. Chesser, and M. H. Smith, eds.), pp. 147–197. Univ. of Chicago Press, Chicago, IL.

Nichols, J. D. (2001). Using models in the conduct of science and management of natural resources. *In* "Modeling in Natural Resource Management: Development, Interpretation and Application" (T. M. Shenk and A. B. Franklin, eds.), pp. 11–34. Island Press, Washington, D.C.

Nichols, J. D., and Conroy, M. J. (1996). Estimation of species richness. *In* "Measuring and Monitoring Biological Diversity. Standard Methods for Mammals" (D. E. Wilson, F. R. Cole, J. D. Nichols, R. Rudran, and M. Foster, eds.), pp. 226–234. Smithsonian Institution Press, Washington, D.C.

Nichols, J. D., and Johnson, F. A. (1989). Evaluation and experimentation with duck harvest management strategies. *Trans. N. Amer. Wildl. Nat. Resour. Conf.* 54, 566–593.

Nichols, J. D., and Karanth, K. U. (2002). Statistical concepts: assessing spatial distributions. *In* "Monitoring Tigers and Their Prey: A Manual for Wildlife Managers, Researchers, and Conservationists." (K. U. Karanth and J. D. Nichols, eds.), pp. 29–38. Centre for Wildlife Studies, Bangalore, India.

Nichols, J. D., and Pollock, K. H. (1983). Estimating taxonomic diversity, extinction rates and speciation rates from fossil data using capture-recapture models. *Paleobiology* 9, 150–163.

Nichols, J. D., Johnson, F. A., and Williams, B. K. (1995). Managing North American waterfowl in the face of uncertainty. *Ann. Rev. Ecol. Syst.* 26, 177–199.

Nichols, J. D., Hines, J. E., Lebreton, J.-D., and Pradel, R. (2000). The relative contributions of demographic components to population growth: a direct estimation approach based on reverse-time capture-recapture. *Ecology* 81, 3362–3376.

Nichols, J. D., Kendall, W. L., Hines, J. E., and Spendelow, J. A. (2004). Estimation of sex-specific survival from capture-recapture data when sex is not always known. *Ecology* 85, 3192–3201.

Nichols, J. D., Morris, R. W., Brownie, C., and Pollock, K. H. (1986). Sources of variation in extinction rates, turnover and diversity of marine invertebrate families during the Paleozoic. *Paleobiology* 12, 421–432.

Nichols, J. D., Stokes, S. L., Hines, J. E., and Conroy, M. J. (1982). Additional comments on the assumption of homogeneous survival rates in modern band recovery estimation models. *J. Wildl. Manage.* 46, 953–962.

Nichols, J. D., Boulinier, T., Hines, J. E., Pollock, K. H., and Sauer, J. R. (1998a). Estimating rates of local species extinction, colonization, and turnover in animal communities. *Ecol. Appl.* 8, 1213–1225.

Nichols, J. D., Boulinier, T., Hines, J. E., Pollock, K. H., and Sauer, J. R. (1998b). Inference methods for spatial variation in species richness and community composition when not all species are detected. *Cons. Biol.* 12, 1390–1398.

Nichols, J. D., Hines, J. E., Sauer, J. R., Fallon, F. W., Fallon, J. E., and Heglund, P. J. (2000). A double-observer approach for estimating detection probability and abundance from point counts. *Auk* 117, 393–408.

Noon, B. R., and McKelvey, K. S. (1997). A common framework for conservation planning: linking individual and metapopulation models. *In* "Metapopulations and Wildlife Conservation" (D. R. McCullough, ed.), pp. 138–165. Island Press, Washington, D.C.

Norris, J. L., and Pollock, K. H. (1996). Nonparametric MLE under two closed-capture models with heterogeneity. *Biometrics* 52, 639–649.

Olson, G. A., Anthony, R. G., Forsman, E. D., Ackers, S. H., Loschl, P. J., Reid, J. A., Dugger, K. M., Glenn, E. M., and Ripple, W. J. (2005). Modeling of site occupancy dynamics for northern spotted owls, with emphasis on the effects of barred owls. *J. Wildl. Manage.* (in press).

Otis, D. L., Burnham, K. P., White, G. C., and Anderson, D. R. (1978). Statistical inference from capture data on closed animal populations. *Wildl. Monogr.* 62.

Ovaskainen, O. (2002). Long-term persistence of species and the SLOSS problem. *J. Theor. Biol.* 218, 419–433.

Ovaskainen, O., Sato, K., Bascompte, J., and Hanski, I. (2002). Metapopulation models for extinction threshold in spatially correlated landscapes. *J. Theor. Biol.* 215, 95–108.

Patil, G. P., and Taillie, C. (1979). An overview of diversity. *In* "Ecological Diversity in Theory and Practice" (J. F. Grassle, G. P. Patil, W. Smith, and C. Taillie, eds.), pp. 3–27. Statistical Ecology Vol. 6. International Co-operative Publishing House, Fairland, MD.

Patterson, B. D. (1987). The principle of nested subsets and its implications for biological conservation. *Cons. Biol.* 1, 323–334.

Patterson, B. D. (1990). On the temporal development of nested subset patterns of species composition. *Oikos* 59, 330–342.

Patterson, B. D., and Atmar, W. (1986). Nested subsets and the structure of insular mammalian faunas and archipelagos. *Biol. J. Linn. Soc.* 28, 65–82.

Peltonen, A., and Hanski, I. (1991). Patterns of island occupancy explained by colonization and extinction rates in shrews. *Ecology* 72, 1698–1708.

Peres-Neto, P. R., Olden, J. D., and Jackson, D. A. (2001). Environmentally constrained null models: site suitability as occupancy criterion. *Oikos* 93, 110–120.

Periera, H. M., Daily, G. C., and Roughgarden, J. (2004). A framework for assessing the relative vulnerability of species to land-use change. *Ecol. Appl.* 14, 730–742.

Peterson, C. R., and Dorcas, M. E. (1994). Automated data acquisition. *In* "Measuring and Monitoring Biological Diversity: Standard Methods for Amphibians" (W. R. Heyer, M. A. Donnelly, R. W. McDiarmid, L.-A. Hayek, and M. S. Foster, eds.), pp. 47–57. Smithsonian Institution Press, Washington, D.C.

Petranka, J. W. (1998). "Salamanders of the United States and Canada." Smithsonian Institution Press, Washington D.C.

Pielou, E. C. (1975). "Ecological Diversity." Wiley, New York.

Pielou, E. C. (1977). "Mathematical Ecology," 2nd Ed. Wiley, New York.

Pirsig, R. M. (1974). "Zen and the Art of Motorcycle Maintenance." Bantam Books, New York.

Platt, J. R. (1964). Strong inference. Science 146, 347–353.

Pledger, S. (2000). Unified maximum likelihood estimates for closed capture-recapture models using mixtures. Biometrics 56, 434–442.

Pledger, S., Pollock, K. H., and Norris, J. L. (2003). Open capture-recapture models with heterogeneity: I. Cormack-Jolly-Seber model. Biometrics 59, 786–794.

Pollock, K. H. (1974). The Assumption of Equal Catchability of Animals in Tag-recapture Experiments. Ph.D. thesis, Cornell University, Ithaca, NY.

Pollock, K. H. (1982). A capture-recapture design robust to unequal probability of capture. J. Wildl. Manage. 46, 752–757.

Pollock, K. H., and Kendall, W. L. (1987). Visibility bias in aerial surveys—a review of estimation procedures. J. Wildl. Manage. 51, 502–510.

Pollock, K. H., and Raveling, D. G. (1982). Assumptions of modern band recovery models with emphasis on heterogeneous survival rates. J. Wildl. Manage. 46, 88–98.

Pollock, K. H., Nichols, J. D., Brownie, C., and Hines, J. E. (1990). Statistical inference for capture-recapture experiments. Wildl. Soc. Monogr. 107.

Pollock, K. H., Nichols, J. D., Simons, T. R., and Sauer, J. R. (2002). Large scale wildlife monitoring studies: statistical methods for design and analysis. Environmetrics 13, 1–15.

Pradel, R. (1996). Utilization of capture-mark-recapture for the study of recruitment and population growth rate. Biometrics 52, 703–709.

Preston, F. W. (1948). The commonness and rarity of species. Ecology 29, 254–283.

Pulliam, H. R. (1988). Sources, sinks, and population regulation. Amer. Natur. 132, 652–661.

Rapoport, E. H. (1982). "Areography." Pergamon Press, Oxford, UK.

Repasky, R. R. (1991). Temperature and the northern distributions of wintering birds. Ecology 72, 2274–2285.

Reunanen, P., Nikula, A., Monkkonen, M., Hurme, E., and Nivala, V. (2002). Predicting occupancy for the Siberian flying squirrel in old-growth forest patches. Ecol. Appl. 12, 1188–1198.

Ricklefs, R. E., and Schluter, D., eds. (1993)."Species Diversity in Ecological Communities. Historical and Geographical Perspectives." University of Chicago Press, Chicago, IL.

Robbins, C. S., Bystrak, D., and Geissler, P. H. (1986). The breeding bird survey: its first fifteen years, 1965–1979. U.S. Fish Wildl. Serv. Resour. Publ. 157.

Robbins, C. S., Dawson, D. K., and Dowell, B. A. (1989). Habitat area requirements of breeding forest birds in the middle Atlantic States. Wildl. Monogr. 103.

Robson, D. S., and Youngs, W. D. (1971). Statistical analysis of reported tag-recaptures in the harvest from an exploited population. BU-369-M. Biometrics Unit, Cornell Univ., Ithaca, NY.

Rohde, K. (1999). Latitudinal gradients in species diversity and Rapoport's rule revisited: a review of recent work and what can parasites teach us about the causes of gradients? Ecography 22, 593–613.

Romesburg, H. C. (1981). Wildlife science: Gaining reliable knowledge. J. Wildl. Manage. 45, 293–313.

Root, T. (1988a). Energy constraints on avian distributions and abundance. Ecology 69, 330–339.

Root, T. (1988b). Environmental factors associated with avian distributional boundaries. J. Biogeography 15, 489–505.

Rosenzweig, M. (1995). "Species Diversity in Space and Time." Cambridge Univ. Press, New York.

Rosenzweig, M. L., and Clark, C. W. (1994). Island extinction rates from regular censuses. Cons. Biol. 8, 491–494.

Royle, J. A. (2004). N-mixture models for estimating population size from spatially replicated counts. *Biometrics* 60, 108–115.

Royle, J. A. (2005). Site occupancy models with heterogeneous detection probabilities. *Biometrics.* (in press).

Royle, J. A., and Nichols, J. D. (2003). Estimating abundance from repeated presence absence data or point counts. *Ecology* 84, 777–790.

Royle, J. A., Dawson, D. K., and Bates, S. (2004). Modeling abundance effects in distance sampling. *Ecology* 85, 1591–1597.

Royle, J. A., Dorazio, R. M., and Link, W. A. Analysis of multinomial models with unknown index using data augmentation. *J. of Computational and Graphical Stat.* (in review).

Royle, J. A., Nichols, J. D., and Kéry, M. (2005). Modeling occurrence and abundance of species when detection is imperfect. *Oikos* 110, 353–359.

Samuel, M. D., Garton, E. O., Schlegel, M. W., and Carson, R. G. (1987). Visibility bias during aerial surveys of elk in northcentral Idaho. *J. Wildl. Manage.* 51, 622–630.

Sanathanan, L. (1972). Estimating the size of a multinomial population. The Annuals of Math. *Stats.* 43, 142–152.

Sanders, N. J., Gotelli, N. J., Heller, N. E., and Gordon, D. M. (2003). Community disassembly by an invasive species. *Proc. Natl. Acad. Sci. U.S.A.* 100, 2474–2477.

Schoener, T. (1974). Competition and the form of habitat shift. *Theor. Popul. Biol.* 6, 265–307.

Scott, J. M., Heglund, P. J., Morrison, M. L., Haufler, J. B., Raphael, M. G., Wall, W. A., and Samson, F. B., eds. (2002). "Predicting Species Occurrences." Island Press, Washington, D.C.

Scott, J. M., Davis, F., Csuti, B., Noss, R., Butterfield, B., Groves, C., Anderson, H., Caicco, S., D'Erchia, F., Edwards, T. C., Jr., Ulliman, J., and Wright, R. G. (1993). Gap analysis: a geographic approach to protection of biological diversity. *Wildl. Monogr.* 123.

Seber, G. A. F. (1970). Estimating time-specific survival and reporting rates for adult birds from band returns. *Biometrika* 57, 313–318.

Seber, G. A. F. (1971). Estimating age-specific survival rates for birds from bird-band returns when the reporting rate is constant. *Biometrika* 58, 491–497.

Seber, G. A. F. (1973). "The Estimation of Animal Abundance and Related Parameters." Griffen, London.

Seber, G. A. F. (1982). "The Estimation of Animal Abundance and Related Parameters," 2nd Ed. Macmillan, New York.

Self, S. G., and Liang, K. Y. (1987). Asymptotic properties of maximum likelihood estimators and likelihood ratio tests under non-standard conditions. *J. Am. Stat. Assoc.* 82, 605–610.

Simberloff, D. S. (1969). Experimental zoogeography of islands: a model for insular colonization. *Ecology* 50, 296–314.

Simberloff, D., and Connor, E. F. (1981). Missing species combinations. *Amer. Natur.* 118, 215–239.

Sjogren-Gulve, P., and Ray, C. (1996). Using logistic regression to model metapopulation dynamics: large-scale forestry extirpates the pool frog. *In* "Metapopulations and Wildlife Conservation" (D. R. McCullough, ed.), pp. 111–137. Island Press, Washington, D.C.

Skalski, J. R. (1994). Estimating wildlife populations based on incomplete area surveys. *Wildl. Soc. Bull.* 22, 192–203.

Skalski, J. R., and Robson, D. S. (1992). "Techniques for Wildlife Investigations: Design and Analysis of Capture Data." Academic Press, San Diego, CA.

Skelly, D. K., Werner, E. E., and Cortwright, S. A. (1999). Long-term distributional dynamics of a Michigan amphibian assemblage. *Ecology* 80, 2326–2337.

Smith, A. T., and Gilpin, M. E. (1997). Spatially correlated dynamics in a pika metapopulation. *In* "Metapopulation Biology: Ecology, Genetics, and Evolution" (I. A. Hanski and M. E. Gilpin, eds.), pp. 401–428. Academic Press, New York.

Soberón, J. M., and Llorente, J. B. (1993). The use of species accumulation functions for the prediction of species richness. *Cons. Biol.* 7, 480–488.

Spiegelhalter, D. J., Thomas, A., Best, N. G., and Lunn, D. (2003). "WinBUGS 1.4 User Manual." MRC Biostatistics Unit, Cambridge, U.K.

Stauffer, H. B., Ralph, C. J., and Miller, S. L. (2002). Incorporating detection uncertainty into presence-absence surveys for marbled murrelet. *In* "Predicting Species Occurrences" (J. M. Scott, P. J. Heglund, M. L. Morrison, J. B. Haufler, M. G. Raphael, W. A. Wall, and F. B. Samson, eds.), pp. 357–366. Island Press, Washington, D.C.

Stauffer, H. B., Ralph, C. J., and Miller, S. L. (2004). Ranking habitat for marbled murrelets: a new conservation approach for species with uncertain detection. *Ecol. Appl.* 14, 1374–1383.

Stevens, G. C. (1989). The latitudinal gradient in geographic range: how so many species coexist in the tropics. *Amer. Natur.* 133, 240–256.

Stevens, G. C. (1992). The elevational gradient in altitudinal range: an extension of Rapoport's latitudinal rule to altitude. *Amer. Natur.* 140, 893–911.

Stith, B. M., and Kumar, N. S. (2002). Spatial distributions of tigers and prey: mapping and the use of GIS. *In* "Monitoring Tigers and Their Prey: A Manual for Wildlife Managers, Researchers, and Conservationists" (K. U. Karanth and J. D. Nichols, eds.), pp. 51–59. Centre for Wildlife Studies, Bangalore, India.

Stone, L., and Roberts, A. (1990). The checkerboard score and species distributions. *Oecologia* 85, 74–79.

Stone, L., and Roberts, A. (1992). Competitive exclusion, or species aggregation: an aid in deciding. *Oecologia* 91, 419–424.

Strong, D. R., Jr., Simberloff, D., Abele, L. G., and Thistle, A. B., eds. (1984). "Ecological Communities: Conceptual Issues and the Evidence." Princeton Univ. Press, Princeton, NJ.

Syms, C., and Jones, G. P. (2000). Disturbance, habitat structure, and the dynamics of a coral-reef fish community. *Ecology* 81, 2714–2729.

ter Braak, C. J., and Etienne, R. S. (2003). Improved Bayesian analysis of metapopulation data with an application to a tree frog metapopulation. *Ecology* 84, 231–241.

Thomas, C. D., and Hanski, I. (1997). Butterfly populations. *In* "Metapopulation Biology: Ecology, Genetics, and Evolution" (I. A. Hanski and M. E. Gilpin, eds.), pp. 359–386. Academic Press, New York.

Thompson, R. L., and Gidden, C. S. (1972). Territorial basking counts to estimate alligator populations. *J. Wildl. Manage.* 36, 1081–1088.

Thompson, S. K. (1992). "Sampling." Wiley, New York.

Thompson, S. K. (2002). "Sampling." Wiley, New York.

Thompson, S. K., and Seber, G. A. F. (1996). "Adaptive Sampling." Wiley, New York.

Thompson, W. L., White, G. C., and Gowan, C. (1998). "Monitoring Vertebrate Populations." Academic Press, San Diego, CA.

Thomson, G. M. (1922). "The naturalization of animals and plants in New Zealand." Cambridge Univ. Press, Cambridge, UK.

Tobalske, C. (2002). Effects of spatial scale on the predictive ability of habitat models for the green woodpecker in Switzerland. *In* "Predicting Species Occurrences" (J. M. Scott, P. J. Heglund, M. L. Morrison, J. B. Haufler, M. G. Raphael, W. A. Wall, and F. B. Samson, eds.), pp. 197–204. Island Press, Washington D.C.

Tosh, C. A., Reyers, B., and van Jaarsveld, A. S. (2004). Estimating the abundance of large herbivores in the Kruger National Park using presence-absence data. *Anim. Conserv.* 7, 55–61.

Trenham, P. C., Koenig, W. D., Mossman, M. J., Stark, S. L., and Jagger, L. A. (2003). Regional dynamics of wetland-breeding frogs and toads: turnover and synchrony. *Ecol. Appl.* 13, 1522–1532.

Tyre, A. J., Possingham, H. P., and Lindenmayer, D. B. (2001). Inferring process from pattern: can territory occupancy provide information about life history parameters? *Ecol. Appl.* 11, 1722–1737.

Tyre, A. J., Tenhumberg, B., Field, S. A., Niejalke, D., Parris, K., and Possingham, H. P. (2003). Improving precision and reducing bias in biological surveys: estimating false-negative error rates. *Ecol. Appl.* 13, 1790–1801.

Udvardy, M. D. F. (1969). "Dynamic Zoogeography: With Special Reference to Land Animals." Van Nostrand-Reinhold, New York.

Urquhart, N. S., and Kincaid, T. M. (1999). Designs for detecting trend from repeated surveys of ecological resources. *J. Agri. Biol. and Environ. Stat.* 4, 404–414.

U.S. Fish, and Wildlife Service. (1990). Endangered and threatened wildlife and plants: 16 determinations of threatened status for the northern spotted owl. *Federal Register* 55, 26114–26194.

Van Horne, B. (1983). Density as a misleading indicator of habitat quality. *J. Wildl. Manage.* 47, 893–901.

Verboom, J., Schotman, A., Opdam, P., and Metz, A. J. (1991). European nuthatch metapopulations in a fragmented agricultural landscape. *Oikos* 61, 149–156.

Verheyen, K., Vellend, M., Van Calster, H., Peterken, G., and Hermy, M. (2004). Metapopulation dynamics in changing landscapes: a new spatially realistic model for forest plants. *Ecology* 85, 3302–3312.

Verner, J., Morrison, M. L., and Ralph, C. J., eds. (1986). "Wildlife 2000: Modeling Habitat Relationships of Terrestrial Vertebrates." Univ. of Wisconsin Press, Madison, WI.

Wahlberg, N., Klemetti, T., and Hanski, I. (2002). Dynamic populations in a dynamic landscape: the metapopulation structure of the marsh fritillary butterfly. *Ecography* 25, 224–232.

Wahlberg, N., Moilanen, A., and Hanski, I. (1996). Predicting the occurrence of endangered species in fragmented landscapes. *Science* 273, 1536–1538.

Walther, B. A., Cotgreave, P., Price, R. D., Gregory, R. D., and Clayton, D. H. (1995). Sampling effort and parasite species richness. *Parasitology Today* 11, 306–310.

Walters, C. J. (1986). "Adaptive Management of Renewable Resources." Macmillan, New York.

Warren, M., McGeoch, M. A., and Chown, S. L. (2003). Predicting abundance from occupancy: a test for an aggregated insect assemblage. *J. Anim. Ecol.* 72, 468–477.

Weber, D., Hinterman, U., and Zangger, A. (2004). Scale and trends in species richness: considerations for monitoring biological diversity for political purposes. *Global Ecol. Biogeogr.* 13, 97–104.

Weir, L. A., Royle, J. A. Nanjappa, and Jung, R. E. (2005). Modeling anuran detection and site occupancy on North American Amphibian Monitoring Program (NAAMP) routes in Maryland. *J. Herp.* (in press).

Whitcomb, S. D., Servello, F. A., and O'Connell, Jr., A. F. (1996). Patch occupancy and dispersal of spruce grouse on the edge of its range in Maine. *Can. J. Zool.* 74, 1951–1955.

White, G. C. (1993). Evaluation of radio tagging marking and sighting estimators of population size using Monte Carlo simulations. *In* "Marked Individuals in the Study of Bird Population" (J. D. Lebreton and P. M. North, eds.), pp. 91–103. Birkhauser Verlag, Basel, Switzerland.

White, G. C., and Burnham, K. P. (1999). Program MARK: survival estimation from populations of marked animals. *Bird Study* 46, 120–139.

White, G. C., Burnham, K. P., and Anderson, D. R. (2001). Advanced features of Program MARK. *In* "Wildlife, Land, and People: Priorities for the 21st Century. Proceedings of the Second International Wildlife Management Congress" (R. J. Warren, H. Okarma, and P. R. Sievert, eds.), pp. 368–377. The Wildlife Society, Bethesda, MD.

White, G. C., Anderson, D. R., Burnham, K. P., and Otis, D. L. (1982). "Capture–recapture and removal methods for sampling closed populations." Los Alamos National Laboratory, Los Alamos, NM.

Whittaker, R. H. (1956). Vegetation of the Great Smoky Mountains. *Ecol. Monogr.* 22, 1–44.

Wiens, J. A., Crist, T. O., Day, R. H., Murphy, S. M., and Hayward, G. D. (1996). Effects of the *Exxon Valdez* oil spill on marine bird communities in Prince William Sound, Alaska. *Ecol. Appl.* 6, 828–841.

Wikle, C. K. (2003). Hierarchical Bayesian models for predicting the spread of ecological processes. *Ecology* 84, 1382–1394.

Williams, B. K. (1997). Logic and science in wildlife biology. *J. Wildl. Manage.* 61, 1007–1015.

Williams, B. K., Nichols, J. D., and Conroy, M. J. (2002). "Analysis and Management of Animal Populations." Academic Press, San Diego, CA.

Williams, C. B. (1964). "Patterns in the Balance of Nature." Academic Press, New York.

Williams, M. (1981). " The Duckshooter's Bag." The Wetland Press, Wellington, N.Z.

Williams, P. H., and Araujo, M. B. (2000). Using probabilities of persistence to identify important areas for biodiversity conservation. *Proc. Roy. Soc. Lond. Ser. B* 267, 1959–1966.

Williamson, M. (1996). "Biological Invasions." Chapman and Hall, London.

Willis, J. C. (1922). "Age and Area." Cambridge Univ. Press, Cambridge, UK.

Wilson, E. O., and Willis, E. O. (1975). Applied biogeography. *In* "Ecology and Evolution of Communities" (M. L. Cody and J. M. Diamond, eds.), pp. 522–534. Harvard Univ. Press, Cambridge, MA.

Wintle, B. A., McCarthy, M. A., Parris, K. M., and Burgman, M. A. (2004). Precision and bias of methods for estimating point survey detection probabilities. *Ecol. Appl.* 14, 703–712.

Wright, D. H., and Reeves, J. H. (1992). On the meaning and measurement of nestedness of species assemblages. *Oecologia* 92, 416–428.

Wright, D. H., Patterson, B. D., Mikkelson, G. M., Cutler, A., and Atmar, W. (1998). A comparative analysis of nested subset patterns of species composition. *Oecologia* 113, 1–20.

Wright, S. (1931). Evolution in Mendelian populations. *Genetics* 16, 97–159.

Wright, S. (1951). The genetical structure of populations. *Annals of Eugenics* 15, 323–354.

Yoccoz, N. G., Nichols, J. D., and Boulinier, T. (2001). Monitoring of biological diversity in space and time. *Trends Ecol. Evolution* 16, 446–453.

Yule, G. U. (1926). Why do we sometimes get nonsense-correlations between time-series? A study in sampling and the nature of time-series. *J. R. Stat. Soc.* 89, 1–69.

Zielinski, W. J., and Stauffer, H. B. (1996). Monitoring *Martes* populations in California: survey design and power analysis. *Ecol. Appl.* 6, 1254–1267.

Zonneveld, C., Longcore, T., and Mulder, C. (2003). Optimal schemes to detect the presence of insect species. *Cons. Biol.* 17, 476–487.

Index

Count, incorporating, 280–281
Counts estimates, ratio of, 9–10
Count statistics
  for estimating relative abundance, 10
  treated as indices, 9
Covariate *Browse*, 118–120, 118*t*
Covariates
  absence of site-specific, 270
  effects of, 144–146
  environmental, 15, 223
  incorporating information, 197–198,
    237–238, 252–253
  logit link function, 71–72
  measured, 102
  modeling, 103–104, 278
  non-inclusion of, 241
  season-specific, 262
  site-specific, 40–41, 112, 256, 262
  species-specific, 255
  survey-specific, 162
  that influence detection probability, 134
  time interval, 217
  time of day, 164
  unmodeled continuous, 107
Credible intervals, 68
Cyanocitta cristata, 142, 143*t*–144*t*, 162–163

D

Darwin, Charles, 28
Data
  analysis of capture-recapture, 214
  analyzing ecological, 108–109
  detection heterogeneity, 214
  detection-nondetection, 42
  flexibility in the modeling of, 180
  fossil, 48
  incorporating, 280–281
  mathematical equations
    abundance, 140–141, 148–151
    Azuma-Baldwin-Noon method, 89–90
    colonization and local extinction,
      193–195
    covariate effects, 144–146
    detection probability, 95–98, 120–122,
      135–140, 193–195, 199–200
    estimation, 73–75
    habitat suitability, 274–279
    likelihood of observed data, 186–187
    logit link, 72
    missing observations, 195, 196*t*, 239

  Nichols-Karanth method, 90–91
  occupancy estimation, 124–125,
    166–167, 176–178, 206–208,
    233–237, 257–260, 270–272
  odds ratio, 227–231
  species temperature effect, 255–256
  standard design, 167–170
  survey costs, 170–173
  transition probability matrix, 246
  from multiple seasons, 186–187
  null distribution, 227
  poorly collected, 155
  verbal description
    colonization and local extinction, 193
    detection histories, 186–187, 192–193,
      232–233, 239
    observed data, 92–93
Data, incidence function, 38–39
Data pooling, 107
Data sets, 50
Deinacrida mahoenui, 42–43, 113, 116–122,
  118*tt*
Delta method, 66, 73–74
Density, estimation of, 42
Descriptors, 30
Design efficiency, 173, 174*t*–175*t*
Designs, robust, 180
Detectability, definition of, 9, 85
Detection heterogeneity
  multiple season data, 214
  probability, 134, 137–139, 143, 163
  site occupancy models, 135–142
Detection history, 85, 110*t*
  covariate effects, 144–146
  examples of, 196*f*
  mathematical equations, 186–187
  multiple season, colony, 190–192
  verbal description of, 92–93, 186
Detection-nondetection data, 42
Detection probability, 104, 156
  based on aggregations of species, 48
  constancy of, 88–89, 135–137
  estimation of, 175–176
  factors that influence, 10
  formula for, 9
  heterogeneity in, 107–108
  heterogeneous, 134, 137–139, 143, 163
  variation in, 133
Diamond, J. M., 18, 19, 36, 189–190
Discrete multivariate distributions, 57

Printed and bound by CPI Group (UK) Ltd, Croydon, CR0 4YY

03/10/2024

01040416-0005